VOLUME LXX

PROCEEDINGS OF

THE AMERICAN CATHOLIC

PHILOSOPHICAL ASSOCIATION

PHILOSOPHY OF TECHNOLOGY

Edited by

Thérèse-Anne Druart

Papers for the regular sessions were selected by the program committee:
Gregory R. Beabout (Chair), John D. Jones and Elizabeth Morelli

Issued by the National Office of
the American Catholic Philosophical Association
The Catholic University of America
Washington, D.C. 20064

American Catholic Philosophical Association
Proceedings

Volume LXX   1996

Table of Contents

ACPA REPORTS AND MINUTES

Technology and the Decline of Leisure

by Thomas C. Anderson

Ever since the Industrial Revolution prophets have predicted that increased progress in technology would result in a decrease in human labor and an increase of leisure. Well into the 1970's, authors were forecasting that as more work was turned over to machines, computers, robots, and highly automated factories, the average work week would continue to dwindle reaching twenty or so hours by the end of this century.[1] In support of their prediction, some pointed to the decrease in time on the job that people in the industrialized world have enjoyed since the Industrial Revolution. By the middle of this century, time at work had fallen from a six or seven day week, and from 16, 14 or 12 hour days during the 1800's, to the 8 hour day and 40 hour week,[2] and the trend was expected to continue.

To compare the amount of work, however, in our century to that in the early Industrial Age is to compare us to that period in Western civilization in which working hours were probably at an all time high. A comparison with earlier pre-industrial times shows something quite different. In Medieval Europe, for example, holidays, holy days, took up one-third of the year in England, almost five months of the year in Spain—even for the peasants.[3] Although work was from sunrise to sunset, it was casual, able to be interrupted for a chat with a friend, a long lunch, a visit to the pub or the fishing hole—none of which a modern factory or office worker dare do. The fact is that American workers of the mid-twentieth century with their 40 hour week were just catching up with their medieval counterparts; and American workers at the end of this century have fallen behind their medieval ancestors! Our incredible growth in technology has not resulted in a corresponding increase in leisure.

Recent studies show that time on the job per day and per year for those with full-time employment has been steadily increasing over the last twenty-five years to the point where Americans (not Europeans) now work an average of one month more a year than they did in 1970.[4] (The only people on earth who work more than we do are the Japanese.) That is one reason young parents today spend 40% less time with their

1

children than their own parents spent with them, and that the number of hours donated to volunteer organizations is significantly decreasing, especially among younger adults.[5] Surveys also show that on-the-job pressure and stress have increased dramatically so that heart specialists now cite "time urgency" as a major cause of early heart disease (and Japan is experiencing an epidemic of sudden-death-syndrome from overwork).[6] In this paper I will investigate technology's contribution to our decline of leisure. I will argue that, although it is not the only factor, technology, or, better, the way humans choose to interact with technology, has a great deal to do with the fact that: (i) we have less *free* time, that is, time away from our jobs; (ii) jobs themselves are more hectic; and (iii) we have less genuine leisure even in the scarce free time we do have. I will treat items one and two together, since decreasing free time and the increase of stress on the job are so intertwined.

Let me begin by addressing the part technology plays in both eliminating and creating jobs; first—elimination. Traditionally, manufacturing jobs, especially if unionized, have paid well. As you know, however, manufacturing is an area that has especially felt the brunt of new technology as smart machines and high tech factories and offices have displaced many. Workers are replaced precisely to the extent that human energy and skill can be transferred to a technological device—a transfer we see all around us. The clerk at the checkout counter hardly needs to be able to add, subtract or make change—a scanner reads the prices and feeds a computer which does all that. You take your car to the garage and discover that mechanics don't diagnose the engine's problems, computers do. (Even wills and divorces can now be handled by computer programs.)[7] Not only have workers been displaced by building their skills into machines, the machines themselves may be driven by programmed computers. In an automated brewery, for example, computers control the entire process. They open and close pumps and valves, regulate the flow and mixture of the ingredients, adjust the heat of the liquids to the proper temperature, transfer the mix to fermentation and storage tanks, and eventually to bottles, cans, and kegs. Few humans are involved, and most that are simply mind the machines.

Incidentally, studies show that those jobs which remain after the workplace has been thoroughly transformed by high technology, jobs which simply involve monitoring machines, although physically untaxing, are psychologically enormously stressful. They combine mind deadening boredom, which comes from just watching the endless repetition of the same operations or reading the same dials or computer screens, with the need to pay close attention at every moment to the process or the readings so that any malfunctioning can instantly be stopped.[8] The stress which accompanies such jobs is further heightened because any worker performing such unskilled tasks knows full well that he or she is easily replaceable. In general, any job which technology has rendered almost skilless is readily exportable to regions of this country or to other countries which have an abundant supply of unskilled laborers and few government or union protections, laborers who, therefore, have little choice but to work for meager wages under sweatshop conditions.

Displaced manufacturing workers may move to other jobs, most often in the service industry, for many such jobs have been created in recent years, but these positions are often not unionized, much poorer paying, and may even be only parttime. Thus many have to work two or three jobs, or a lot of overtime, to maintain their standard of living.[9] The end result, of course, is less free time.

Yet, technology has been used since time immemorial to take the place of human labor. Setting wheels under a cart to move heavy objects easily must have meant that fewer humans were needed to push and slide it over rough terrain. A plow pulled by animals or by a tractor decreases the work of human beings, as does machinery driven by wind, water, steam or electric power. Today's computer-controlled machines in factories and computer-run operations in offices are simply the most recent stage of this long continuous history. Accordingly, we often hear even very profitable businesses justify their downsizing, right sizing, or "derecruiting" by the introduction of the latest labor saving technology— an interesting euphemism which really means to decrease or eliminate human labor not to save it. Profits without payrolls have always been most businesses' ideal. Note, by the way, that it is not technology as some autonomous entity that causes the loss of jobs; rather people choose to use technology to eliminate their workers. It makes sense, for machines do not mind working around the clock, nor demand higher wages or benefits, nor threaten to strike. Of course, for a worker to know that he or she may be replaced by an automated machine, or a microchip or computer program, contributes mightily to job insecurity and its concomitant stress and tension. No wonder that many try to preserve their positions by working longer hours and dare not complain when their work load increases after their employer has downsized.[10] No wonder that many, when possible, bring their jobs home with them, something that today's technology renders all too easy.

One's home used to be a haven from the job. Nowadays, not even a vacation trip is safe. Fifteen years ago only obstetricians carried beepers; today engineers and business executives do so that their employer can contact them at any time. Such contact is even easier with cellular phones. We have all seen commuters, who used to have a few minutes of semifree time going to and from work to listen to music or the news or to books on tape, now on their cellular phones with their offices or clients or customers as soon as they get on the road. I suspect that many of you have almost instantaneous access via electronic mail from your homes to individuals and institutions worldwide—which means, of course, that you are also accessible to all of them in your homes. My son- in-law, an engineer, took his lap-top computer with modem and fax with him on a recent vacation to Puerto Rico so that he could continue to have input into the projects he was vacationing from. Corporate executives routinely check their voice mail daily from almost any place on the planet and review documents faxed to them and fax their responses. Clearly, the ever-increasing use of these communication technologies means that many people have decreasingly less time *truly* free from their jobs. Yet in spite of this evidence, many continue to adhere to the ideology of the Industrial Revolution, that increasing technology in the workplace decreases the amount of work to be done by humans.

Some, however, fear that in the long run technological advances will eliminate so many jobs that most people will have no opportunity for work at all. I suspect that such concerns are overstated, and that brings me to the role of technology in creating jobs. There are numerous cases where new technologies have created far more jobs and far more work than they have eliminated.[11] In general, think of the tremendous explosion in computer technology in the last decades; yet this has not resulted in any corresponding dramatic increase in unemployment. One reason the introduction of new technologies often creates more jobs than it eliminates is that those technologies are used not just to do the specific jobs of the workers they replace, but also to do jobs not done before, or not done very much. For example, installing the computer, and before it the typewriter, in the office has dramatically increased the amount of clerical work and so the need for such workers. About twelve years ago our department decided to replace the secretaries' electronic typewriters with computers. At the time we had one and four-fifths secretaries and we feared that once we computerized we would have to let our four-fifths secretary go. What happened was exactly the opposite. We now have two full time secretaries (even though our enrollment and faculty size have decreased), and they are so terribly overworked that we are hoping to hire a third. The reason the new computer technology did not decrease secretarial work, I think, is that the word processor made revising, saving, printing and reprinting, sending and receiving, memos, letters and reports so much quicker and easier than the typewriter, that it encouraged much more written communication and distribution of that communication to many more people. Many of you can recall the days when everything was typed, and we sometimes had to throw away entire documents and start over because a crucial word or phrase was accidentally omitted. And remember the time and pain involved in correcting carbon copies and ditto masters. Because it was so difficult and so time consuming, we just didn't make a lot copies for very many people. Since the computer does everything the typewriter did more quickly and has the capacity to do so much more, things that used to be taken care of in person or by phone, or which simply weren't taken care of at all, because to do so was too cumbersome or took too much time, are now put into written form (that is, processed) and sent to more people. Of course, one doesn't just send, one also receives all this computer-processed information, including responses from the wider audience one has copied, which must in turn be read, analyzed, evaluated and responded to in some way.

Indeed, studies have shown that one major reason for the increased tempo, length, and stress of the work day is that computer technology has made readily available so much more information—by paper, email, internet, fax, over and above the ordinary phone, postal mail, newspaper, etc., that a person cannot possibly sift through it all, let alone digest and respond meaningfully to it, in just an eight hour day or forty hour week.[12] As someone said, trying to cope with all that available information is like trying to drink from a fire hydrant! Furthermore, unlike postal mail which took days or weeks, since the computer transmits information from all corners of the world almost instantly, people tend to expect a prompt reply. In business, a quick response to information made instantly available, and almost as quickly obsolete, may be absolutely necessary to beat the competition in a world wide market. No wonder business executives never dare be without their beepers or cellular phones or very far from their laptops with modems and faxes. (By the way, psychologists now speak of *pseudo* attention deficit disorder which some people develop because of the tremendous number of stimuli continually bombarding them, stimuli which demand a prompt response whatever their real importance.)[13] In any case, computerizing the office has certainly increased the work to be done and the need for secretarial staff, which illustrates my general point that increased technology often results in an overall increase of work because it opens up the possibilities of doing, not just the jobs replaced, but tasks not done before or not done very much. A final example. Sensing devices feeding computers, like the scanners used in most stores, provide up-to-the-second data about a business's sales, inventory, production, purchases, available credit, etc.—information that used to be painstakingly gathered by numerous workers and presented only in monthly or quarterly reports. No doubt many of those human tabulators have been displaced by the new information technology, yet the easy availability of so much data has also caused an explosion of positions at various levels in marketing, purchasing, inventory control, finance, production, etc.[14] You may have noted that similar growth has also occurred in academia. The increase in information technology is partly responsible for the growth of administrators and their staffs, sometimes even while student and faculty numbers have remained stable or decreased. Whether or not this has improved education, it illustrates again, that increased technology in these areas has not meant an overall decrease in work. Yet, some businesses still believe that introducing more technology will decrease human labor, or at least they justify reducing their workforce by introducing new technology, even when the consequences are clearly that the fewer remaining employees, now using that new technology to perform more tasks than before, are forced to work for longer hours and at a more hectic pace.

Another reason workers who remain after downsizing are often under great pressure to perform their tasks more quickly is because their work has apparently been made so easy, for after all, it is said, the hard part (and sometimes the interesting part) has been turned over to the computer or to computer-driven machines. Thus, the factory job that involves just tending machines is sped up simply by speeding up the machines. In highly automated offices, the only skill left for some secretaries is entry of data into word processors, what we used to call typing. Since their job is so simplified, data entry clerks are expected to significantly increase the volume of their work---the limit is simply how fast they can make their fingers move. A secretary used to average 30,000 keystrokes an hour on a typewriter; the average data entry clerk is expected to execute 80,000 strokes an hour.[15] Similarly, since so much information is now delivered to customers by recordings, live telephone operators are expected to handle significantly more live customers per hour than before---and to spend less than thirty seconds per call. Post Office workers are supposed to enter one zip code a second.[16] For the technocratic mentality this is efficiency, the more that can be crammed into a given unit of clock time the better. This mentality, however, totally disregards the human organism's natural time, its physical and psychological limits and how much uniformity, repetition, and speed it can stand without breaking down. No wonder there is a stress epidemic in this country and continual complaints from employees in the high tech office about eyestrain, nervous fatigue, back pain, headaches, and a plague of repetitive motion disorders.[17]

Another major technological contributor to this epidemic of stress is the prevalence of sensing devices used by management to monitor worker performance. Many word processing systems include a means of counting the number of key strokes a typist makes. The grocery store scanner not only adds up prices but records the speed at which the clerk dispatches goods. The response time of telephone operators is monitored by voice activated sensors and printed out for their supervisors.[18] *The Wall Street Journal* reported that some factory workers have their movements electronically timed and compared with a set standard to determine their pay.[19] Estimates by the Office of Technology Assessment are that twenty-six million workers are presently monitored by sensing devices and the number is growing.[20]

I have been speaking about technology being used to create and eliminate jobs. Before I leave this point, I want to briefly address the oft heard claim that the jobs created by new technologies involve more skill than those rendered obsolete---and the concomitant demand that to keep up with advances in technology workers will just have to be willing to undergo repeated retraining and be prepared for frequent job changes. (Our last commencement speaker, the chairman of Sears, told the graduates that they should expect to have six or more *major* career changes in their lives.)

In the first place, I am suspicious about the assertion that new technology creates jobs demanding high skills, for it is often based on the assumption that those who *use* very sophisticated technologies have to be themselves technologically sophisticated. On the contrary, the task of many computer engineers is to make the human operation of computer-driven machines so easy, or in contemporary jargon so "user friendly," that such jobs become practically idiot friendly, requiring only minimal intelligence and dexterity.

Be that as it may, to the extent that ever changing technology does demand continually learning different skills and does involve frequent job changes, and it does, it is obvious that one result will be a decrease in both time off the job and time available for leisure. One might also wonder if it is even possible for all workers to repeatedly "reinvent" themselves. This sounds like a version of Lake Woebegone, the belief that everyone is above average and so has the ability and intelligence to continually attain new expertise. Many of the skills necessary, however, for using information technologies are of a more abstract, intellectual nature than the concrete, intuitive, sensual know-how used in traditional hands-on production. Such skills are hermeneutic in character; that is, they involve interpreting symbols on a wide array of dials or computer screens and through them understanding the complex physical processes and machines they monitor, and then manipulating these symbols to indirectly control the operations of those machines and processes.[21] In order to translate symbols into actual physical events occurring in the real world, one must have a general understanding of the whole operation being monitored and its many interconnected steps, and this is not easy. The workers at Three Mile Island, for example, were not stupid nor inattentive. Rather, they were not able to interpret correctly certain readings they had, nor to infer what specific information they needed to look for, within the vast amount of data available to them.[22] Most people have never been in a control room of a highly automated factory, but almost everyone has peeked into the cockpit of a jumbo jet—and wondered how the pilots could interpret and keep track of all the data on all of those instruments. They answer is, they can't! They would be overwhelmed by all that information if they tried. They can only concentrate on a small number of variables at a time and to know just what to focus on and what to ignore, that is, to know what data or information is relevant at specific times, they have to have a general understanding of the entire plane and its various interrelated systems. The same thing is true about the operators in a factory control room and in a nerve center of a large office.

Can traditional blue collar workers, displaced brewery workers, for instance, used to using all their bodily senses to perceive and direct the brewing process, attain the abstract understanding and hermeneutic skills necessary to do the new jobs created by information technology? Thirty years of teaching philosophy to some students who seem to have insurmountable difficulties grasping abstract concepts does not give me a lot of optimism. This is one reason some, including the present Secretary of Labor, fear we are moving to a two-tier society in which there will be many challenging well-paid jobs for those who can attain the necessary hermeneutic capabilities, but only very menial low paying ones for those who can not.[23]

Moreover, how are these demands for constant change compatible with our human needs for stability, security and roots? Must these goods be sacrificed to so-called technological progress—which, of course, is so often portrayed as inevitable, as if we have no alternative except to climb on board or be left out. Surely, we already have ample evidence of the toll that lack of stability and of security takes on the family.

Besides, where did we get the idea that humans are or should be so flexible in the first place? My hunch is that we are modeling people on the technology that is most glamorous today. In early industrial times, society tended to model people on machines. Thus, humans were expected to willingly become their servants, and, like them, work long hours without complaint under the rule of the clock; the ideal business organization was itself one that operated "like a well-oiled machine." Today our society is enthralled by the computer, and one difference between it and earlier machines is its flexibility. Unlike those machines which perform a few functions over and over to produce a concrete *finished* product, computers can manipulate information in innumerable ways, none ever "finished," and the machines they drive are also quite flexible. You might think of the difference between the typed, or better the old printed, text which possessed a certain fixity and definiteness making it very difficult and annoying to redo or change once it was in print. But an entire text on a word processor can be thoroughly modified so easily that it seems arbitrary to consider any version of it definitive or "finished."[24] In the heyday of the assembly line, human beings were expected to gladly perform mindless repetitive tasks like the machines they tended; so today humans are expected to be as perfectly flexible and programmable as the computer and the machines it drives.

While all of these things I have mentioned contribute to more work and job stress, and result in less free time, I suspect I have not yet touched on some of the most important reasons why increased technology has meant a decline of leisure.

Over the long haul, due in no small part to technological advances, this country has seen a continual growth in productivity—generally between 1 and 2% a year since World War II. Increased productivity means that more is produced per unit of work; therefore, cost per unit decreases and profits increase. Some of the profit, of course, goes to investors, stock holders, an awful lot nowadays goes to top executives, some is reinvested, and some finally trickles down to the workers. But how does the worker get it, as a decrease in the amount of work or as an increase in pay? In principle it could be either, or some combination. If profit is translated into free time, then, in theory, future costs and productivity will remain the same. If profit is translated into more pay with no decrease in work, then productivity will continue to rise.[25]

Since the war, increase in productivity and profit has almost always been translated entirely into increased income, not into free time. Workers rarely have any choice about it. Of course, increased income does mean increased buying power, and that is why some complain that our technological advances have served primarily to promote increased consumption. It is true that we Americans are the most consumer-oriented society in history, spending three to four times as many hours a year shopping as do Europeans. Most homes are retail outlets with cable shopping channels, mail order catalogues, toll free numbers, and computer hookups. We can even shop in airplanes! The average American consumes twice as much now as forty years ago.[26] We all have technological devices and gadgets in our homes and offices that we had no idea we needed twenty years ago. People used to justify extra work by saying they did it to purchase time saving appliances, especially for their homes: dish washers, automatic washers and dryers, microwaves, etc. We don't hear that justification as much any more because it has turned out that people worked more to purchase things that saved housework, so they could save time in order to—in fact, work more, to be able to buy things that save more housework, giving them more time—so they can work more, to buy more labor saving devices, ad infinitum. Of course, because we must work longer hours to earn the income to purchase all the high-tech things we desire for the home and also for our recreation (sailboats, ski paraphernalia, huge in-home movie screens, high definition TV, jacuzzies, the latest tennis and jogging equipment), we have even less time to enjoy any of them! (People have taped programs they seldom have time to watch, CD's rarely listened to, high tech exercise equipment, swimming pools and boats hardly used.) Having purchased so many devices, we find we must cram even more activities into our diminishing free time in order to use and maintain them all. No wonder our lives are so hectic even when we are not at our jobs.

Speaking of maintenance, experience bears out what studies show, that a large portion of time away from our jobs is spent not in leisure activities but in necessary maintenance of our houses, yards, cars, TVs, computers, etc.[27] (Either we perform the maintenance ourselves, or we work more to earn the money to hire others to do it for us.) It has been said that many American families have three jobs—the job of the husband, the job of the wife, and the job of maintaining their home, car and other possessions. If one defines leisure as doing and enjoying things for their own sake, not for any exterior purpose—not because we *have* to in order to achieve some end, in other words—then time devoted to this kind of personal work is not genuine leisure time. By *personal* work I mean what we *have to* do to maintain our possessions, and also to maintain ourselves. By the latter I am thinking of such things as jogging, aerobics, exercising on Nordic track and so forth. Personal work also includes the things we *have to* do for others—such as, chauffeuring our children to school, lessons, doctors, games as well as the things we *must* do to maintain our jobs, such as commute, take courses, recharge

our batteries in order to work better. To repeat, if leisure means activity
done for its own sake, none of these activities, however important and
necessary, is one of true leisure.[28] I will return to this point.

Overarching everything is the requirement that we quickly become
dissatisfied with whatever we have and are, a dissatisfaction continuously stimulated by what Eric Fromm called "the boredom producing
industry," namely the media, the best funded educational system in this
country.[29] Our worth comes from what we have, it constantly reminds
us, yet what we have is always inadequate. No wonder we have difficulty
accepting the Christian message that we are loved simply for who we
are. Yet our standard of living requires that we discard the not quite
new long before it is deficient, and replace it with the new, which we
must soon become bored with and replace in its turn—endlessly. (The
area of computer hard and software seems particularly prone to this
hype.) Of course, we have to work more to afford the unending consumption of the new, which decreases our opportunities for leisure. But
enough of this; critiques of American society's materialism and consum-
erism abound.

Besides, I think that for many people, especially professionals, there
is a much different reason for their workaholism, for either they have
sufficient money and consumer goods, or they just aren't all that interested in that sort of thing. (I think, in fact, many feel uneasy about a life
centered on nothing higher than consumption.) What moves such
people more than the desire to consume, I believe, is the need to be busy,
or more accurately, to be *productive*. They genuinely feel the obligation
to contribute something to society, and work seems to do so, even when
its goal is to produce still more consumer goods of dubious value. Our
society almost seems to value all work whatever its goals, and to place
little importance on taking the time to enjoy things good in themselves,
things such as an evening at home conversing with family or friends,
preparing and leisurely eating a meal, strolling in the country, playing
family games, contemplating the beauties of art or nature, learning a
musical instrument, time spent enjoying our children as well as partici-
pating in liturgical celebrations.[30] These leisure activities become increasingly difficult to fit into one's busy schedule because they are
considered to be less important than "getting things done." After all, as
we say, we have so much to "do."[31] (Furthermore, after a hectic nine or
ten hour work day, entertaining guests at home or practicing on the
piano requires too much energy.) "Getting things done," being productive, is so important, I would suggest, because our society takes as its
standard the kind of productive activity associated with the machine
and with technology generally. Just as any technological device has
value only in its use, in what it does and produces—a machine that
stands idle is worthless—so, I think, many see their own value in terms
of *their use*, their output, their work. In fact, even in their free time they
feel they must be busy doing something productive—such as keeping
their body fit, improving their work skills, servicing their lawns, chauf-

feuring their children. I think the current national debate over welfare reform glaringly illustrates this valuation of work above all else. In my home state a single mother on welfare is now required to get a job as soon as her infant is 12 weeks old! Able-bodied men and women are considered downright immoral if they are not always engaged in some *useful* occupation, defined as one that produces goods and services, and so contributes to the Gross Domestic *Product*. To repeat, since, like technological devices, work clearly has an external product, it is highly valued. Leisure, since it is useless, having no purpose outside itself, is often degraded, being identified with idleness or laziness or, horror of horrors, with wasting time! No wonder that many are simply too busy to read a good book or enjoy a concert, to take the time to smell the roses, or to develop long-lasting intimate relationships, even with their children.[32] Maybe the denigration of leisure partly accounts, too, for the low esteem in which philosophy is held, for if Aristotle is correct that philosophy begins in wonder, that means, I think, that it requires leisure, since wonder is the fruit of leisure.[33]

This brings me to a final suggestion about why our advances in technology have led to decreased leisure. As the heirs of Francis Bacon, we moderns believe that the purpose of knowledge is control, and, of course, modern technology furnishes us unparalleled mastery over both nature and human beings. Imbued by the "spirit of technology" (a phrase of Marcel's), our ultimate goal, then, seems to be to so thoroughly control both nature and our human nature that by *our own efforts* we will create the good life for ourselves. Some science-fiction utopias capture this well, portraying genetically engineered humans living in a totally man-made world. Within that spirit, it is not surprising that leisure is dismissed, for it invites us to "let being be." In leisure we humbly and gratefully accept reality in all its intractable unruliness. We acknowledge, in awe and wonder, the ultimate mysteriousness and fragility of life, of creativity and beauty, of friendship, love and truth. Josef Pieper has pointed out that traditionally, even in pre-Christian societies, leisure was joined to religious ceremonies. He concludes that the "soul" of true leisure is the joyful celebration of the wonders of creation and of their Divine source.[34] To value leisure is to appreciate the unexpected and unmanageable, the unmerited, the *gift*; it is to acknowledge that grace, not work, not even good works [sic], is the source of our salvation.

Yet, for someone today whose real wages have declined and who is in danger of being replaced by a high tech machine, leisure is a luxury he or she literally cannot afford. As I have noted, he or she has little option except to work overtime or at a second job. That is why it is crucial, if we believe that leisure is essential to the human spirit, to address the growing disparity of incomes in this country. As you know, the United States has seen dramatic increases in the inequality of wealth in the last two decades, an inequality that is now greater than any time since the Depression.[35] Since the 1970's our wealth has become increasingly concentrated in the top 1/2% of the population, and the richest 5% of Americans now have more wealth, much more, than the other 95%.[36] Concern over this growing disproportion prompted the American Catholic Bishops just last fall to repeat their call, from their 1986 pastoral on the economy, for a more just distribution of this country's resources.[37]

It is also essential that employees be given the option of taking the fruits of their increased productivity as free time or as some combination of higher wages and free time. In some recent surveys a majority of full-time workers have said they would be willing to trade a portion of higher salaries for more free time—if they had the choice.[38] Just as we now have federally mandated maternity leave, perhaps we should consider federally mandated vacation time, as many European countries do—a minimum of one month per year. The last major decline in the average work week, to 40 hours, occurred in this country in the 30's, and it was done not primarily to give people more leisure but to give more people jobs.[39] Isn't it strange that we rarely hear support for that today, even though we have increasing numbers of un-and-under employed who desire more work, at the same time that large numbers of others work mandatory overtime and fifty or more hour weeks. A recent AF of L study claimed that three million new jobs would be created if businesses just returned to 1982 levels of work time.[40]

Yet, for any of this to happen, I believe there has to be genuine empowerment of workers. Historically, with a few enlightened exceptions, businesses have fought every reduction in the work day (even from 14 to 12 and from 12 to 10 hours); they have fought every increase in vacation time and genuine sharing of power. Yet each time work was reduced, the long-term result has been increased productivity.[41] Likewise increased power-sharing in the workplace has usually resulted in no decline in productivity and often in dramatic increases of it.[42] Whether empowered employees making good wages would in fact change business structures to enable them to translate increased productivity into leisure time, or would become workaholics like so many well-paid professionals and executives, remains to be seem. At least they should have the power to exercise their options. Still, as I have suggested, more than economic changes are necessary. A paradigm shift in the way we look at ourselves, at work and leisure, and at our relationship to technology is needed.

Let me conclude by suggesting one small step towards such a shift. Those of you my age can remember when the Church, and most Catholics and other Christians, took seriously the prohibition of work, including most commercial transactions and housework, on Sunday, the sabbath.[43] For Catholic Christians that prohibition was not joined to a puritanical belief that pleasure or play was improper on holy days, quite the contrary. Holy days, and that included Sundays, were days where people, freed from their service to the machine, could *enjoy* leisure. They were properly times for pageants and festivals, games and celebrations, for creativity and re-creation. They were days on which humans, like the Creator, could delight in the goodness and beauty of a world charged with the grandeur of God. We hear a great deal today about the importance of the work ethic and the dignity of work. What about the dignity of leisure? Perhaps we could begin to restore its prestige and to develop a "leisure ethic" by again taking seriously the admonition to keep holy, with grateful remembrance and joyful celebration, the Lord's day. That would, at least once a week, dethrone technology and reestablish the human being's proper dignity as steward, not master, of the universe.

Marquette University
Milwaukee, Wisconsin

* * * * *

Endnotes

1. See, for example, the opinion of Andre Gorz cited in *The Philosophy of Leisure*, eds. T. Winnifirth and C. Barrett (New York: St. Martin's Press, 1989), 34-36. Also, many authors in *Technology, Human Values, and Leisure*, eds. M. Kaplan and P. Bosserman (Nashville: Abingdon Press, 1971), express the view that the work week will decrease significantly by the end of this century due to technology.

2. J. Schor, *The Overworked American* (New York: Basic Books, 1991), chapter 3; S. de Grazia, *Of Time, Work and Leisure* (New York: Vintage Books, 1994), chapter 2.

3. *The Overworked American*, 43-56; *Of Time, Work, and Leisure*, 63-82, 89-90; L. Mumford, *The Myth of the Machine*, vol. I (New York: Harcourt, Brace, Javanovich, 1967), 271; W. Rybczynski, *Waiting for the Weekend* (New York: Viking, 1991), 215.

4. *The Overworked American*, chapter 2; *Waiting for the Weekend*, 216-17.

5. S. A. Hewlett, *When the Bough Breaks; The Cost of Neglecting Our Children* (New York: Basic Books, 1991); S. Roberts "Alone in the Vast Wasteland," *The New York Times*, 15 Jan. 1996, 3(E).

6. "Running Out of Time," a Public Broadcasting Corporation movie, produced by Oregon Public Broadcasting in partnership with KCTS, Seattle, 1994; S. Linder, *The Harried Leisure Class* (New York: Columbia Univ. Press, 1970), 22-25; excerpts from "Inside American" by Louis Harris in the *Milwaukee Sentinel*, 7 Sept. 1987, 8.

7. L. Uchitelle and N.R. Kleinfield, "On the Battlefields of Business, Millions of Casualties," *The New York Times*, 3 March 1996, 14-17.

8. S. Zuboff, *In the Age of the Smart Machine* (New York: Basic Books, 1988), 120; R. Howard, *Brave New Workplace* (New York: Penguin Books, 1985), 69.

9. "Compensation drop is largest in 8 years," *Milwaukee Journal*, 24 June, 1995. The shrinking of the middle class has been addressed by many. See for

example: J. Beatty, "Who Speaks for the Middle Class?" *Atlantic Monthly*, May, 1994; M. Lind, "To Have and Have Not," *Harper's Magazine*, June, 1995; "What People Earn," *Parade Magazine*, 18 June 1995; L. Thurow, "Companies Merge; Families Break Up," *The New York Times*, 3 Sept. 1995, 11(E).

10.   "Breaking Point," *Newsweek*, 6 March 1995, 56-62; J. Bardwick, *Danger in the Comfort Zone* (New York: Amacom, 1996).

11.   David Noble gives examples in his *Forces of Production: A Social History of Industrial Automation* (New York: Knopf, 1984). See also, *In the Age*, 115 & ff.

12.   According to the Office of Technology Assessment, businesses must handle 400 billion documents a year, and this increases 72 billion each year—*In the Age*, 417. See also L. Winner, *The Whale and the Reactor* (Chicago: Univ. of Chicago Press, 1994), chapter 6, "Mythinformation."

13.   E. Hallowell and J. Ratey, *Driven to Distraction* (New York: Simon and Schuster, 1994), 191-94.

14.   R. Volti, *Society and Technological Change* (New York: St. Martin's Press, 1988), 126-28; *In the Age*, introduction, chapter 5, and appendix.

15.   J. Rifkin, *Time Wars* (New York: Henry Holt, 1987), 117-20; *In the Age*, 133-36; *Brave New Workplace*, chapters 2 and 3 have many examples.

16.   *Society and Technological Change*, 126; "Running Out of Time."

17.   A study by The National Institute of Occupational Safety and Health found that clerical workers who used video display terminals full-time exhibited the highest levels of stress ever reported; even higher than air traffic controllers: *Brave New Workplace*, chapter 3; *In the Age*, 141.

18.   *Brave New Workplace*, 63.

19.   *Society and Technological Change*, 127; *Brave New Workplace*, 29 ff.; D. Noble, "Social choice in machine design; the case of automatically controlled machine tools," in *The Social Shaping of Technology*, eds. D. MacKenzie and J. Wajcman (Philadelphia: Open Univ. Press, 1985), 118-20.

20.   "Running Out of Time." A 1984 U S Department of Labor survey estimated that nearly two-thirds of those who work at video-display terminals were monitored by their employers: B. Garson, *The Electronic Sweatshop* (New York: Penguin, 1988).

21.   *In the Age* has a good explanation of the hermeneutic skills needed to use contemporary information technology, chapters 2, 3 and appendix. See also L. Hirschhorn, *Beyond Mechanization* (Cambridge: MIT Press, 1990).

22.   M. Martin and R. Schinzinger, *Ethics in Engineering*, 2nd ed. (New York: McGraw Hill, 1989), 146-49; *Beyond Mechanization*, chapters 8 and 9.

23.   C. Greer, "We Can Save Jobs," *Parade Magazine*, 21 May 1995, 4-5: R. Kuttner, *The Economic Illusion* (Boston: Houghton Mifflin, 1984), 168-83; *Society and Technological Change*, 128-29.

24.   *Time Wars*, 155-58.

25.   *The Overworked American*, 146-50.

26.   *The Overworked American*, 107-12.

27.   *The Harried Leisure Class*, chapter 4.

28.   *The Harried Leisure Class*, chapters 1 and 4; *Of Time, Work and Leisure*, chapter 10; *Waiting for the Weekend*, 218 ff; *The Philosophy of Leisure*,

chapter 1. In a very interesting article, E. Telfer argues that even useful activities can be leisure if they can be practiced and valued for their own sakes. See "Leisure" in *Moral Philosophy and Contemporary Problems*, ed. J.D. Evans (Cambridge: Cambridge Univ. Press, 1987).

29. E. Fromm, *The Revolution of Hope: Toward a Humanized Technology* (New York: Bantam Books, 1968), 40.

30. I must acknowledge here the influence of A. Borgmann's notion of focal things set forth in his *Technology and the Character of Contemporary Life* (Chicago: Univ. of Chicago Press, 1987). See also *The Harried Leisure Class*, 83-93.

31. F. Buckley, "The Everyday Struggle for the Leisurely Attitude," *Humanitas*, 8 Nov. 1972, 312-317; *Of Time, Work and Leisure*, 307-10, 404-09.

32. *The Harried Leisure Class*, 24-27 and chapter 8; J. Pieper, *Leisure the Basis of Culture*, trans. A. Dru (New York: Pantheon, 1952), 40-47.

33. For Aristotle's view of leisure see: *Of Time, Work and Leisure*, 11-21; "Leisure," 157-60; J. Owens, "Aristotle on Leisure," *Canadian Journal of Philosophy*, XI (Dec., 1981).

34. *Leisure the Basis of Culture*, 42, 52-57, 71-76; "The Everyday Struggle," 312-18; *Of Time, Work and Leisure*, 421, 435.

35. *The Overworked American*, 113-14, 150-51; L. Thurow, "Why Their World Might Crumble," *The New York Times Magazine*, 19 Nov. 1995, 78-79; K. Bradsher, "Inequalities in Income are Reported Widening," *The New York Times*, 29 Oct. 1995, 2(E). See also texts cited in note 9 above.

36. L. Michel and J. Bernstein, *The State of Working America, 1992-1993* (Armonk, NY: M E Sharpe Inc., 1993), 42-53 and Figure 5C; H. Sklar, "Real Wages Drop," *Milwaukee Journal Sentinel*, 3 Sept. 1995, 2(J).

37. U.S. Bishops' Pastoral Letter, *Economic Justice for All*, paragraphs 181-183, published by the *Catholic Herald*, Archdiocese of Milwaukee, Oct. 7, 1985.

38. "Running Out of Time;" *The Overworked American*, 148, 164.

39. *Waiting for the Weekend*, 143; *The Overworked American*, 72-77.

40. "Running Out of Time."

41. *The Overworked American*, 72-78, 152-57.

42. Institute for Social Research, *Productivity and Worker Participation* (Ann Arbor: Univ. of Michigan, 1977) cited in M. Carnoy and D. Shearer, *Economic Democracy* (Armonk, NY: ME Sharpe, Inc., 1980), 409 n. 19; B. Stokes, *Helping Ourselves* (New York: Norton, 1981), 35-36; D. Jenkins, *Job Power* (New York: Penguin Books, 1974), chapter 12.

43. The Old Testament forbids work, not just servile or manual work, on the sabbath: *Exodus*, Ch. 20, vv. 8-11 and *Leviticus*, 5, vv. 12-15.

Jacques Ellul on the Technical System and the Challenge of Christian Hope

by Vincent C. Punzo

In his 1975 defense of the West against the post-modernist critique, Jacques Ellul writes, "All the works and creations, all the intellectual, economic, and technical advances of the West have been the result of this tension and conflict, this head-on collision between man who wants to be himself and God who also wants man to be himself. The difficulty is that 'himself' does not mean the same thing in both cases; in fact, the one meaning contradicts the other."[1] My purpose in this paper is to sketch out an illustration of Ellul's view of this conflict and tension by considering how he brings Christian hope to bear on contemporary humanity's alienation within a social order ruled by the demands of technique. I decided on a paper of this scope because I wanted to make a presentation that would bring together the two fundamental sides of Ellul's legacy, namely, his exploration and critique of the place of technique in human life and his contribution to the Christian intellectual tradition. The topic is thus meant to serve as an introduction to the spirit of Ellul's thought. There is no pretense to a full, in-depth treatment of either his critique of the technical system or his account of how hope is to be lived in today's world. Rather, I shall try to develop an exposition of the underlying spirit and thrust of his understanding and critique of the technical system and a similar exposition of what he thinks Christian hope has to offer a humanity trapped in this system.

The paper will proceed in three sections. The first will provide an account of the bibliographical setting within which Ellul develops his treatment of technique, along with his definition of technique. The second section will develop the fundamental outlines of his view of the workings of the technical system and of their impact on persons. The third section will deal with his understanding of the nature of Christian hope and of the role it has to play as humanity seeks to confront the challenge of the technical system.

I

The centerpiece of Ellul's writings on technique is a trilogy consisting of *La technique ou l'enjeu du siècle*, 1954 (*The Technological Society*, 1964), *Le système technicien*, 1977 (*The Technological System*, 1980), and *Bluff technologique*, 1986 (*The Technological Bluff*, 1990). A comparison of the French and English titles calls for a word regarding translation. Ellul notes that the word 'technologie' refers properly to "a discourse on *technique*," including "a discourse on different *techniques* (English, technologies) ....then an attempt at discursing on *technique* (English, technology) in general."[2] Having made the point, Ellul gave the translator permission to follow American usage and translate the French *technique* as 'technology.'[3] I think once the reader has been informed regarding the difference, it is easy enough to distinguish by context when the English 'technology' is being used to refer to technique and when it is being used to mean discourse on technique. Thus, it seems to me clear that when one speaks of 'the technological society,' one is referring to *la société technicienne*.

Ellul describes the focus of *The Technological Society* as follows:

> We shall be looking at technique in its sociological aspect; that is, we shall consider the effect of technique on social relationships, political structures, economic phenomena. Technique is not an isolated fact in society...but is related to every factor in the life of modern man; it affects social facts as well as all others. Thus technique itself is a sociological phenomenon, and it is in this light that we shall study it.[4]

The Technological System is essentially an up-dating and re-working of the position taken in *The Technological Society*, with the significant exception that the foundational concept of the earlier book is what Ellul refers to as the 'technical phenomenon,' whereas it is the notion of the technical system (système technicien) that is central to the later work.[5] We shall return to the notion of the technical phenomenon later in this section as we move to Ellul's definition of technique.

The Technological Bluff takes a somewhat different turn from the other two. It sets out to unmask the gigantic bluff being played on modern humanity by the public discourse regarding the role of technique in its life. Ellul is not offering any sort of conspiracy theory, but is saying that the media, politicians, advertisers, intellectuals, and technicians when they leave their work and come out to speak are engaging in a discourse that envelops us all in a bluff by talking exclusively about the successes of technique, leaving out any meaningful considerations relating to costs, risks, or human utility, by advancing technique as the only possible solution for all the world's problems, and, finally, by presenting technique as the only chance that societies have for progress and development.[6] The outcome of this bluff is that humanity seems to have lost any sense of conflict between itself and the demands of modern technique. It has produced an "outflanking or encirclement," of the conflict that is the great innovation of our time: "People have stopped looking for direct means to resolve conflicts [engendered by the technical system]. They have stopped trying to adapt politics or economics to technique by force. At the same time, they have stopped trying to produce mutants, that is, people perfectly consistent (flawless) with a technical universe. They have stopped trying to crash head on with obstacles and refusals. They have stopped trying to rectify technical malfunction by direct action. There has been a transformation the results of which we are not yet able to measure."[7] Ellul refers to this as the "true technical innovation," because it provides the support of the social body and of individuals in that body which enables the technical system to develop.[8] The task which Ellul sets for himself in *The Technological Bluff* is to unmask the bluff by bringing out the conflicts and dangers involved in the technical system. The trilogy is essentially a warning of the fundamental and profound threat that the technical system constitutes for modern humanity now and in the future.[9]

We can now turn toward what Ellul means by the technical phenomenon as a way of moving toward the definition of technique that is at work in the trilogy. He uses the phrase, 'the technical phenomenon,' to capture what it is about the place of technique in the West since the eighteenth century that distinguishes it from its place pre-eighteenth century.[10] He distinguishes the technical phenomenon from "technical operation," with the latter phrase being used to refer to the fact that technique has always been a part of the human condition. "The technical operation includes every operation carried out in accordance with a certain method in order to attain a particular end."[11] Primitive humanity found ways to fish, to hunt, ways to make fires, and ways to skin a bear. A change began to occur with respect to such spontaneous, pragmatic techniques as Western humanity moved through the industrial revolution. The phrase "technical phenomenon" is meant to capture the outcome of that change.

With the industrial revolution, human beings began to put reason to work to scrutinize more critically the great diversity of techniques, taking "what was previously tentative, unconscious and spontaneous," and bringing it "into the realm of clear voluntary, and reasonable concepts."[12] As Ellul puts it, "Two factors enter into the extensive field of technical operation, consciousness and judgment. This double intervention produces what I call the technical phenomenon."[13] His account of how this intervention worked to produce the technical phenomenon is none too clear, but I think we can get at his point by drawing the following ratio. People's thinking and measuring are to logic and the mathematical sciences as the technical operation is to the technical phenomenon. People thought and measured long before they made thinking and measuring objects of reflective inquiry. The growth of logic and the mathematical sciences has had an impact on the activities of thinking and measuring. What Ellul is claiming is that when reason became reflective with regard to traditional and pragmatic technologies (technical operation), it opened the possibility of releasing ways and means from the limits of customary practice. Humans could now focus on the creation of new tools, new ways of operating. In short, the technical phenomenon was born which "can be described as the quest for the one best means in every field....This 'one best means' is in fact the technical means. It is the aggregate of these means that produce technical civilization."[14] Humans are now in a position to deal with means simply as means, to establish a universe of discourse in which means are evaluated simply from the perspective of means so as to be able to come up with the technical means, the one best means.

It is this notion of technical phenomenon that is incorporated into his definition of technique: "The term technique,....does not mean machines, technology, or this or that procedure, for attaining an end. In our technological society, *technique* is the *totality of methods* rationally arrived at and having absolute efficiency (for a given stage of development) in every field of human activity. Its characteristics are new; the technique of the present has no common measure with that of the past."[15] The qualification with which Ellul introduces this definition, "In our technological society," and his emphasis on the point that the characteristics of technique in the present have no common measure with that of the past are significant for situating his treatment of technique in its appropriate universe of discourse. He distinguishes three levels on which social issues can be discussed.

There is, first, the level of current events. Here we are dealing with society simply in terms of the ever-changing events and personalities that greet us every morning when we turn to our daily newspapers. There is, in addition, a third level to which we may refer as the philosophical level which deals with social realities on such an abstract level that its concepts may apply indifferently to all societies existing at many different times. We are dealing with the unchanging essences of things on this level. Between these there is what may be termed an historical level, which Ellul describes as follows: "Underneath the events and above the fundamental constants, there are structures, the movements and temporary regularities which go to make up the actual history, and which produce an epoch or a regime with its characteristic features."[16] It is on this historical level that Ellul conducts his critique of the technical system. He is not dealing with the timeless essence of technique. Thus, he is willing to acknowledge that on a purely hypothetical level, technique could be conceived as being a pure principle of human liberation.[17] He is not dealing in hypotheticals or in purely logical possibilities. He is dealing with technique as he finds it at a particular time in history, from the eighteenth century with the rise of the technical phenomenon to more recent times when, thanks to the

computer, the technical phenomenon became systematized in the technical system.[18] Thus, it is not appropriate to view Ellul as a critic of
technique *per se*. His criticism is directed to the impact of the technical
system on humanity.

Ellul's thumb-nail etymology of technique will help tie this treatment
of his definition together. Originally, people spoke of technique as a way
of doing something, a process, or an interplay of processes. With the
coming of the industrial revolution, technique became linked with
different energy sources, coal, petroleum, electricity. We are now in a
period in which technique is a way of organizing people and undertak-
ings, a way of processing information, a way of decision-making. As
technique came to include such ways and means, it began to be apparent
that technique had significant impact on human beings and their
societies, thus opening the way to speak about the technological society
and the technological system. Finally, as one considers the development
of the meaning of technique through these various stages, it is clear that
there was always implicit in these various meanings the notion of
efficiency. Thus, in *The Technological System*, Ellul gives us a reword-
ing of the definition supplied in the first book of the trilogy, namely, "the
ensemble of the absolutely most efficient means at any given moment."[19]

In closing this section, it is worth noting that for Ellul a link remains
between technique and machines even after we have denied a simple
identification between the two. The machine is paradigmatic for technique: "The machine is deeply symptomatic; it represents the ideal
toward which technique strives. The machine is solely, exclusively,
technique; it is pure technique, one might say. For wherever a technical
factor exists, it results almost inevitably in mechanization: technique
transforms everything it touches into a machine."[20] This quotation
constitutes an excellent foreshadowing of the following section.

II

Ellul's conception of the place of technique in the modern world
rejects the view that humanity's social situation is to be understood as
involving an amalgam of those dimensions of social life grasped on the
"current events" level of discourse and those grasped on the philosophical level. On the one hand, there is the philosophical level: "Society (the
same old society) is thought of consisting, still, of classes (with similar
class relationships) and obeying the same old dialectics.... In other
words, there is a permanent reality undergoing surface modifications,
the reality of man for some, the reality of society for others."[21] On the
other hand, there is the current events perspective: those realities that
distinguish modern society from the past, for example, express highways, computers, video games, space travel. Modern society is understood as a bringing together of these two dimensions without any
significant unification or bonding. Hence, the ensemble of techniques
we experience every day of our lives is understood to "Change one or two

aspects of society," but as simply adding to it. Modern society is understood as being "merely the traditional society *plus* technologies."[22] The technologies which so distinguish today's world from the past unite with human society as water and oil.

In opposition to this view, Ellul maintains that the technical system has become the environment of modern humanity.[23] It has replaced nature, which was the environment of primitive humanity, and society, which has been humanity's environment for approximately five thousand years.[24] As our environment, the system plays a fourfold function in human life today: "it enables us to live, it sets us in danger, it is immediate to us, and it mediates all else."[25] Modern humanity could not live if deprived of its electrical, gas, and water utilities, or of its transportation and communication facilities, or of the structures of the state with its police, fire, and welfare organizations. At the same time, this technical system is full of dangers for us: busy highways, various forms of pollution, terrorist attacks, accidents involving nuclear power plants. It is the third characteristic of this environment that is especially crucial to Ellul's critique. As immediate, the system is already "here," present to us when we come to consciousness. We are instantly in a universe of machines, products, media, organizations.[26] "The social or individual consciousness today is formed directly by the presence of technology, by man's immersion in that environment without the mediation of thought, for which technology would only be an object, without the mediation of culture."[27] Granted the immediacy of the system, it follows that our experience of nature and of society, our two previous environments, is mediated through the system.

If we are to understand the impact of the system in its immediacy in the shaping of human consciousness, we must try to get a grasp of its fundamental character. It would appear that the system as an environment consists of an endless array of new objects and gadgets to fill human life. Acknowledging that it is an environment of objects, Ellul calls attention to the transient and evanescent standing of these objects. There is an endless parade of objects coming into and passing out of existence. Hence, it is not objects and gadgets that constitute the ultimate character of this environment: "The invasion by objects is accompanied by scorn for these very objects.... Things are made to be destroyed, bought to be discarded, multiplied in order to be eliminated."[28] In the last analysis, the environment shaping human consciousness is not one of objects, but rather one of means. "What characterizes this society is not the object, but the means."[29] Although ultimately rejecting structuralism, Ellul finds that a structural analysis fits the technical system. Structuralism shows that there is neither subject nor object in a world given over to means. The "to be" of subjects is to be obedient to means, while objects are merely the result of means. "We thus reach the decisive conclusion that our universe is not a universe of objects ... but a universe of means and a technological system."[30] Objects are significant in such a universe only as factors in the movement of means, that is, as manifesting means and as material for the further exercise of means.

The remainder of this section will concentrate on the place of human subjects in the technical system by sketching out Ellul's account of two key factors that characterize the system, namely, automatism and autonomy. The section will conclude with his description of what he refers to as the "great design," which represents the vision of human happiness and fulfillment underlying the technological bluff.

The characteristic of automatism focuses on the determining role played by the demands of technique on human choice and decision. One may say formally that human beings are the agents of choice, choosing one technique over another. Human beings in a technical system, however, can no more be said to be agents of such choices than they can be said to have a choice to make when deciding whether four is quantitatively more than three. Faced with questions of technical progress, humans caught up in the system are more appropriately understood to be apparatuses for registering the results to be obtained by different techniques. The choices are made for them by the requirements of the technical milieu in which they find themselves. Once one line of action is seen to entail maximum efficiency, the decision has been made.[31] In a system ruled by means and efficiency, the choice is either efficiency or lose, with efficiency being understood as either that which increases human dependence on new technologies such as nuclear power plants, or that which makes new technologies the basis for introducing more order, control, and system into human life. Ellul's position regarding the automatism of technique does not deny that one will find individuals refusing to accept new technical procedures when these enter a domain in which things were done in accord with tradition and from certain moral or empirical perspectives. His point is that such an individual will find herself vanquished.[32]

The automatism of the technical system also calls for a social order perfectly malleable to the demands of technique, requiring that political, economic, and educational structures be constantly open to meet these demands. Thus, it calls for an economic order in which workers must be prepared to change occupations any number of times over the course of their working lives. Educational structures must be available to change their courses of study to meet the changing demands of the technical system. A social order thus fully compliant to the demands of the technical system is ultimately a social order whose underlying principle of unity and cohesion is no longer a matter of moral or traditional commitments. It has become a mechanical ordering of institutions and social affairs.[33] In closing this sketch of the automatism of technique, it is appropriate to note that Ellul holds that his characterization of the technical system is no more guilty of anthropomorphism than are liberal economists when they speak of the laws of the market.[34]

The autonomy of technique brings out what is implied in the automatism of technique. The technical system is an automatism determining human choice because as autonomous, it is not a purely neutral reality. The autonomy of technique means that technique "has its own weight, which goes in technology's direction."35 The limited example of our use of that supposedly neutral means, the automobile, may serve to illustrate this characteristic. If we are to use automobiles, we need techniques to find oil and other raw materials required to make an automobile and to use it; highways must be built, requiring the manufacture and development of materials needed for the task; service stations must be developed; governments must set up licensing procedures, must take steps to insure safety, legislative and judicial determinations must be made regarding punishments to be meted out to drivers who injure or kill others. In short, the automobile as a means has its own weight, sending us out in search of a vast array of other means. Multiply this example to bring into play all the other supposedly neutral means available to us, and one may begin to appreciate the definition of technique which Ellul advanced in our opening section, "the ensemble of the absolutely most efficient means at any given moment."

Insofar as it is autonomous, "Technology in itself does away with limits."36 The technical system as autonomous is a system of means in search of other means. The only sort of "limit" intrinsic to such a system is the provisional limitation of that which is beyond today's technical capabilities, but which at the same time looks toward a future of new means which render such a limitation obsolete. The technical universe is thus a boundless universe. Boundaries, limits have no essential or ultimate standing within such a universe. It, therefore, stands beyond good or evil, being by its very nature incapable of transgression since the notion of limits is essentially alien to it.37 To try to regulate this universe of means on the basis of some sort of moral standard is to seek to make it submit to heteronomously imposed demands.38

Granted that human life would be impossible without the recognition of some sort of moral limit, there are two ways in which the technical system can find a place for moral considerations. First, moral problems can be translated into technical problems.39 For example, the problem of teen-age pregnancies is a matter to be resolved by sex-education programs which may mention moral issues regarding sex out of marriage and which may have a word to say about sexual abstinence. We all know, however, that dependence on moral appeals and on the exercise of self-control does not hold out much promise of being effective. Hence, the ultimate basis for dealing with out-of-wedlock births is education in the use of condoms. It is a short step from this sort of education to the conclusion that the only significant "moral" consideration involved in consensual sexual relations is whether or not a condom has been properly used. A moral problem has become a problem in making use of the technical instruments at our disposal.

A second way in which to domesticate moral problems within a technical system is to equate morality with the normal, always leaving open the possibility to re-define the normal in terms of whatever stage of technical development we may have reached.[40] This way may not be substantially different from what I have described as the first way, but it deserves separate mention as serving to clarify Ellul's position. Ellul provides a fine example of the equation of the moral with the normal in quoting from a film review in *Le Monde* which praised the film's depiction of an abortion: "'By presenting the images of an interrupted pregnancy as a normal phenomenon, normal because it is clearly explained, approached without fear, in full liberty of individual choice and under medical supervision, this film removes the drama, the guilt from abortion.'"[41] There was no need for guilt or fear, no place for drama, because the technology of the day (medical supervision) has incorporated interrupted pregnancies within the system as normal occurrences. Granted that medical supervision assures the woman that there are no significant problems or dangers involved in the procedure, granted that the procedure is seen as a normal practice within the system, granted the explanation given was in the jargon of an "interrupted pregnancy," how free was the woman's choice in the last analysis? Such a "choice" appears to be a perfect example of the automatism of technique: a choice determined by how the system presents the abortion "option" to the woman.[42]

As autonomous, technique not only does away with any morality other than one based on its own demands, but it also plays a desacralizing role, creating a vacuum in human life which is filled ultimately by itself becoming the sacred.[43] There is intrinsic to human existence a sense of mystery, "a great deep above which lie his reason and his consciousness."[44] Technique as autonomous recognizes no such mystery or secret. There is no sacred to be respected. Acknowledging no rule or norm outside of itself to which it must submit, it lays claim to its intrinsic right to turn everything into means. "Technique worships nothing, respects nothing. It has a single role: to strip off externals, to bring everything to light, and by rational use to transform everything into means."[45] Nevertheless, just as we have seen that humanity cannot live without a morality, so also it cannot live without a sacred, without something to respect and to give meaning to life. Hence, the descralizing power itself takes on the character of the sacred.[46]

Humanity finds its standing and meaning in the world through its empowerment in the technical system. It is the system that assures it its future. To be cut off from this empowerment is to feel "poor, alone, naked, deprived."[47] Having itself become the "great deep," in which human life lives, and moves, and has its being, it is the unquestioned foundation with which all the problems, difficulties, and challenges of life are to be met. As Ellul puts it, "Technology is not one of the terms of the dialectics: it is the universe within which dialectics operates."[48] Ellul sees the great design of the technological bluff as being directed to sealing human life within this universe.

It is humanity's failure to conform to the "manifest excellence of technical progress," that is seen to be the fundamental problem in this universe holding to the sacredness of technique. Hence, humanity must be brought to fit itself to the demands of technique, if it is to find its happiness and well-being.[49] The great design envisions such happiness as calling for a conflict-free existence. There are to be no conflicts within individuals, among neighboring communities, between corporations and their workers, within political life.[50] The achievement of this task depends on the following four factors. First, matters of social policy and of collective life must be put in the hands of experts. Secondly, the media must assume responsibility for determining the opinions people are to hold, making determinations regarding what they need to know and what they need not know, what is significant and what is not significant. Thirdly, individuals are to limit their concerns to their work not necessarily because of an intrinsic fulfillment or meaning in the work, but in order to earn wages so as to enjoy the blessings of being a consumer, which is the fourth key factor upon which the completion of the great design depends.[51]

The society of the great design is to be populated by two types of personalities who are ideally suited to this conflict-free social order and who are among us today: the fascinated and the diverted. The fascinated are those who relate to technique with so passionate an interest and are so exclusively fixated on it that they find it impossible to turn away. Ellul maintains that it is the most educated part of our population that fits this personality type: journalists, intellectuals, technicians, scientists. They are caught up in all the talk about progress and technology fulfilling humanity's greatest dreams. The diverted are those who lose themselves in the many attractions of the technical system, having neither the time nor the energy to reflect on themselves and the human condition or to pursue humanly lofty ideals.[52] Ellul explores five such attractions: (1) games, (2) sports, (3) the automobile, (4) mechanistic arts, and (5) what he terms "ultimate idiocies," under which he lists Disneyland and the "Idol," meaning by this such celebrities as Michael Jackson.[53]

Finally, Ellul provides a picture of the sort of society envisioned in the great design by describing it in terms of three panels that make up "a magnificent triptych."[54] The central panel depicts those who have been trained from youth to be scientists and technicians. Their mission is to promote the technical system. The left panel depicts the fascinated; the right, the diverted. Closing the side panels on the central, "we have the representation of a perfectly balanced, happy, and fulfilled humanity, never protesting, knowing no trouble, calmed by hypnotics, *mens sana in corpore sana*, kept healthy by jogging and other kinds of exercise."[55]

III

The humanity depicted in the triptych is a humanity for whom the technical system is destiny or fate. It is the humanity that is "meant to be," in a world determined by the automatism and autonomy of technique. This section will sketch out Ellul's account of the role of Christian hope in providing an opening whereby humanity can break through this fate to live in that freedom promised by Christian revelation. Ellul sees the task of the Christian in this technical age as being not "to reject technology, but...to cause hope to be born again."[56] Such a re-birth is needed not to condemn technique as such, but to challenge the "fabric of the technological society," specifically, "the dominance of technology."[57] Put in terms of the distinction drawn in the first section of this paper, a re-birth of hope is needed not to condemn technique understood as technical operation, but rather to challenge technique in its modern reality, the technical phenomenon or the technical system. The challenge is to the sacred character taken on by technique as automatism and autonomous. Subjecting the system to the criticism of revelation, Christians must become iconoclastic toward it, "must destroy the deified religious character of technology."[58]

So long as humanity accepts technique as sacred, it "has no intellectual, moral, or spiritual reference point for judging and criticizing technology."[59] It is Christian hope that overcomes the bondage of technique as destiny or fate. Christian hope is not to be confused with human hope or optimism understood as a feeling that whatever crisis may be confronting us today will be overcome either because we have resources readily available or because there are factors at work in the crisis itself that will bring it to a desirable conclusion. Where there is such human hope there is no room for Christian hope. It is when we find ourselves confronted by necessities which we have no human hope of overcoming that Christian hope comes into play. It is not a hope grounded in a belief in humanity.[61]

Christian hope is rather the link joining two poles of a Christian's existence: "One is the divine pole, the pole of the Kingdom of God. If hope does not rivet us to that power we are not bearers of anything. We represent nothing but ourselves....The other is the pole of socio-political action. If hope does not rivet us to that action, then the Kingdom of God within us becomes a meaningless, sterile contemplation, a repetitive automatism, and we abort the Kingdom."[62] This second pole means that Christian hope is not a way of escaping the problems of today by looking toward a future to come. It is rather a living and active hope seeking to bring the promise of the Kingdom to bear on the situation that confronts us in our own time.[63]

It is because it is rooted in the divine pole that Christian hope is able to introduce a breach, a heteronomy in a closed age. For as rooted in that pole, Christian hope finds its life and sustenance in the life of the transcendent Lord, Creator of heaven and earth, who sent His only begotten Son to free us from the wages of sin and death. Jesus' victory over death, "the most decisive form of fate," means that, though hemmed in by all sorts of conditions and necessities that control and direct its life, humanity need not bow before them as "ineluctable necessities" or as destiny and fate.[64] A hope grounded in the Kingdom of the risen Christ tells us: "We can escape destiny. We have scope for life."[65] Christian hope recognizes and acknowledges those necessities and determinations that the technical system imposes on human life, but it rejects the technological bluff which accepts these necessities as destiny and which seeks to shape humanity in ways that will enable it to find whatever semblance of "happiness" or "freedom" this destiny will permit.

As situated in the Kingdom, hope relativizes and desacralizes the system, while at the same time acknowledging technique as a significant contributor to human life. There is for Christian hope only one Absolute, the Lord of Revelation who created heaven and earth. It, therefore, rejects all sacreds that would insert themselves as Absolutes giving ultimate meaning and value to human life.[66] "It is evident that if our hope is in Christ it cannot be in other things."[67] Hope nonetheless takes desacralized technique seriously because the Christ in whom hope rests is the Christ of the Incarnation whose presence in the world means that hope must give due significance to the things of this world, including human achievements. A life of hope is directed to challenging the automatism and autonomy of technique in order that technique will take its proper role in human life as a legitimate instrument for human health, comfort, and ease. Hope empowers the fascinated and diverted to turn from technique and its magic to confront the meaning and mystery of their lives. Technicians will continue to make their contributions to mankind, disabused of the notion that the ultimate fulfillment of humanity rests on their technical expertise.[68]

Finally, insofar as it seeks to live according to the promise of the Kingdom, "Hope arouses us to an ethos which is not the same as that which would have existed if it were not there."[69] It is an ethos "plainly and realistically impossible," to those for whom there is no Kingdom.[70] As we have seen in our previous section, the ethos of the technical system seeks to reduce all moral questions to questions of the effective use of means. Power effectively used is power morally used. In contrast to this approach, the ethos of hope calls for "a special examination of means. They must be justified not just by their efficiency or end but by themselves and in themselves."[71] The ethos of hope is opposed to that of the system because hope exercises human power and makes use of the resources of the earth, "in order that they might manifest the glory of God, which is their true, if not their only purpose."[72] It is an ethos which at one and the same time recognizes that the created world in which we live witnesses the glory of the Lord and that our task is to have our technological creations serve as witnesses.[73]

"Hope," Ellul tells us, "is believing in the beatitudes in spite of appearances."[74] To live the beatitudes is to follow Christ "who chose the way of nonpower, nonforce, nondomination," and thereby glorify God, to witness the love of God.[75] The West's technological venture has put humanity on an opposite course, searching for mastery in all areas of life, leading it to treat the order of nature with harsh possessiveness and to seek to control and dominate other human beings.[76] From the perspective of hope seeking to glorify God by witnessing the love of God, this attitude of power and control which is the driving spirit of the technical system constitutes a negation of hope.[77] The substitution of the will to power for love changes the significance of everything. God's creation ceases to be seen as a free gift to be treasured and treated with respect and caution, becoming instead an alien restraint to be dominated and exploited for whatever purpose may cross our minds.[78] To treat ourselves and our world as glorifying God is to act in keeping with that Kingdom founded in a freedom which finds its fulfillment in love and sharing, and not in dominating power and possessiveness.

In conclusion, we are now in a position to contrast the tension-free world of the great design with a world dedicated by hope to the glory of God. The latter is a world that returns the West to the roots of its greatness by embodying that creative tension between "man who wants to be himself and God who also wants him to be himself." The following provides a fine description of a world alive with hope, seeking to glorify God:

> The West was chosen to bear witness to the gift of self amid the lust for possession, to self-humbling amid the quest of power, to the Spirit amid a world of rigid structure, to freedom amid a civilization shot through with rationalism. It is precisely here that we have the great dramatic conflict of the West. The West has never been able to reach its logical end because it was pierced to the heart by a gospel that was its utter opposite and constantly undermined its grandiose projects. Christianity, on the other hand, has never been able to be fully itself because it has been tangled in a network of systems that have constantly been endeavoring to assimilate it.[79]

Because this world is not the fullness of the Kingdom that awaits us in eternity, it is not ultimately satisfying. But satisfying or not, it is a world graced by Christian hope that keeps open the route of true human freedom and creativity both as a witness to the true Kingdom and as a warning against the false kingdom of a technical system centered in power and possessiveness, and in a rationalism drunk with efficiency.

St. Louis University
St.Louis, Missouri

* * * * *

Endnotes

1. Jacques Ellul, *The Betrayal of the West* (New York: Seabury Press,

1978), 77. Hereafter cited as *BW*.

2. Jacques Ellul, *The Technological System* (New York: Continuum, 1980), 33. Hereafter cited as *TS*.

3. *TS*, 33.

4. Jacques Ellul, *The Technological Society* (New York: Vintage Books, 1964), xxv-xxvi. Hereafter cited as *T. Society*.

5. *T. Society*, xxx. *TS*, 1.

6. Jacques Ellul, *The Technological Bluff* (Grand Rapids, Michigan: William B. Eerdmans), xvi. Hereafter cited as *TB*.

7. *TB*, 18.

8. *TB*, 18.

9. See Jacques Ellul, *Living Faith: Belief and Doubt in a Perilous World* (San Francisco: Harper and Row, 1983), 201.

10. *TS*, 79.

11. *TS*, 19.

12. *T. Society*, 20.

13. *T. Society*, 20.

14. *T. Society*, 21.

15. *T. Society*, xxv.

16. Jacques Ellul, *Hope in Time of Abandonment* (New York: Seabury Press, 1977), 280-81. Hereafter cited as *Hope*.

17. *BW*, 137. Also *TB*, 116.

18. *TS*, 98-99, 101-02.

19. *TS*, 26, 24-25.

20. *T. Society*, 4.

21. *TS*, 88.

22. *TS*, 88.

23. *TS*, 38.

24. Jacques Ellul, *What I Believe* (Grand Rapids, Michigan: William B. Eerdmans, 1989), 132, 104-32. Hereafter cited as *WB*.

25. *WB*, 133.

26. *TS*, 311.

27. *TS*, 38.

28. *TS*, 42-43.

29. *TS*, 43.

30. *TS*, 43.

31. *TS*, 239.

32. *TS*, 250-51.

33. *TS*, 244.

34. *TS*, 240.

35. *TS*, 154.

36. *TS*, 154.

37. *TS*, 154.

38. *T. Society*, 134.

39. *TS*, 146.

40. Jacques Ellul, *To Will and To Do*, (Boston: Pilgrim Press, 1969),

192-93.

41. *TS*, 319.
42. *ST*, 322.
43. *T. Society*, 141.
44. *T. Society*, 141.
45. *T. Society*, 142.
46. *T. Society*, 143. See also Jacques Ellul, *The New Demons* (New York: The Seabury Press, 1975), 158.
47. *The New Demons*, 74. *T. Society*, 145.
48. *TS*, 318.
49. Jacques Ellul, *Autopsy of Revolution*, (New York: Alfred A. Knopf, 1971), 245.
50. *TB*, 405.
51. *TB*, 405-06.
52. *TB*, 323, 358.
53. *TB*, 358-83.
54. *TB*, 405.
55. *TB*, 405.
56. *Hope*, 232.
57. *Hope*, 232. *Autopsy of Revolution*, 275. Also *BW*, 138-39.
58. Jacques Ellul, *Perspectives on Our Age* (New York: The Seabury Press, 1981), 108. Hereafter cited as *POA*.
59. *TS*, 318.
60. *Hope*, 248.
61. *Hope*, 275. *POA*, 109.
62. *Hope*, 251.
63. *POA*, 107.
64. Jacques Ellul, *The Ethics of Freedom* (Grand Rapids, Michigan: William B. Eerdmans, 1976), 14. Hereafter cited as *EF*.
65. *EF*, 18.
66. *Hope*, 242.
67. *EF*, 18.
68. *Hope*, 237.
69. *Hope*, 201.
70. *Hope*, 201.
71. *EF*, 405.
72. *Hope*, 233.
73. *Hope*, 236.
74. *EF*, 18.
75. *BW*, 131.
76. *BW*, 35.
77. *EF*, 18.
78. *To Will and To Do*, 60-61. Also *Hope*, 237.
79. *BW*, 76.

Technology and the Crisis of Contemporary Culture

by Albert Borgmann

Social theorists tend to find society in a state of crisis more often than in a period of flowering. In recent years, however, findings of crisis and peril have been especially numerous. More significant still, what has been particularly worrisome to the critics is the health of the middle class. Given the social diversity and economic volatility of the United States, the middle class is not just the middle segment in the hierarchy of status and wealth, but the steadying and anchoring force of society. The middle class has traditionally provided the power if not always the leadership to keep the excesses of the upper class in check and the miseries of the lower class within bounds. Now this centering force seems itself in danger of attenuation and dislocation.

Already in 1989, Barbara Ehrenreich published her passionate and provocative study of the perils besetting the middle class and striking it with the *Fear of Falling*.[1] Meanwhile the genteel *New Yorker* has taken pity on the middle class, asking plaintively "Who Killed the Middle Class?" on October 16, 1995 and lamenting that there "Ain't No Middle Class" on December 11, 1995.[2]

Nevertheless, these assessments and plaints are strangely inconclusive. They miss both the appearance and the heart of contemporary culture. To all appearances, American middle class culture has, at least since the seventies, been tranquil and steady, free of social unrest and material deprivations. In the olden days a crisis was marked by marauding bands, widespread starvation, or lethal epidemics. Nothing even approaching such calamities has lately been seen in our society.

As for the heart of contemporary culture, the social maps that the critics give us have a blank space where one should expect to find the living center of society. The moral and economic critics fear for the stability and prosperity of the social structure but appear to take it for granted that as long as firmness and affluence are secured, the good life will take care of itself and naturally fill out the framework.

That consumption is at the center of the supposedly good life and casts a shadow of doubt on the actual goodness of our lives is noted occasionally. Barbara Ehrenreich has chronicled the uneasiness that rises occasionally when we contemplate our devotion to consumption, and she has well described the mutations that consumption undergoes without ever losing its infectious force.[3] The elusiveness of the cultural center of contemporary society is illustrated by Charles McGrath's attempt to find enlightenment about our cultural state of affairs in a "brand-new genre," one he calls "the prime-time novel" — the weekly network dramatic series.[4] Fifty-four point four million Americans watch them every night, an impressive number when you consider that if one percent of them bought and read a copy of the old style genre novels once a month, we would have a literary revolution on our hands. What McGrath values in these programs is that "they frequently attain a kind of truthfulness, or social seriousness, that movies, in particular, seem to be shying away from these days."[5] What he found

> were stories involving abortion rights and affirmative action; a murder, very similar to a famous Westchester case, in which a young man, suffering an alcoholic blackout, killed two people he mistakenly took to be his parents, and the apprehension and conviction of a Katherine Ann Power-like fugitive.[6]

Is this what typically happens to you or your parents, your neighbors or your colleagues? Does not the real significance of television lie not so much in *what* people watch but in the fact *that* they watch it so religiously? McGrath does note that the "failure of TV drama to take itself into account is one of the great oddities of the medium."[7] On reflection this does not seem odd at all, for who would want to stare down an infinite regress? A realistic portrayal of television would come to showing a family watching television. And if what *they* are watching is a slice of typical family life it would be another family watching TV— television watching families watching television watching families ad infinitum.

The suspicion that there is nothing but an infinite abyss at the heart of American culture has haunted more than a few students of the subject and has made them recoil from the heart of the matter. Journals such as *Material Culture* and *Technology and Culture* cling to safely circumscribed historical phenomena. Norman J.G. Pounds' *Hearth and Home: A History of Material Culture* stops short of the 20th century.[8] Robert Crunden's *A Brief History of American Culture* sticks to politics and the high, yet inconsequential, culture of literature and the fine arts.[9]

The crisis of contemporary culture is as elusive as the heart of the culture that is ailing. The evidence of trouble that we are presented with is nothing but the symptom of a hidden, if chronic, malady. One begins to suspect that contemporary culture cannot heal because its injury is so concealed.

To act on this suspicion is the beginning of philosophy. Bringing the malaise of contemporary culture into focus is a philosophical enterprise, and the notion of *culture* is a good vehicle to get the enterprise underway. Culture is something more and other than politics, economics, or aesthetics. Helpfully it has a descriptive and a normative sense. In the former, as E.B. Tylor put it in 1871, it means "that complex whole which includes knowledge, belief, art, morals, law, custom, and any other capabilities and habits acquired by man as a member of society."[10] At least as old is the notion of culture as an ideal of human flourishing and perfection.

To lift the veil of indistinctness from contemporary culture in the descriptive sense, let us look at it against the background of normative conceptions of culture, the culture of the word and the culture of the table. To begin with the culture of the word, I mean by it practices of conversation and reading. Though such practices have normative status today, they are by no means elitist. They were fully realized in the middle class of the first half of this century. Consider the testimony of Norman Maclean, author of the celebrated *A River Runs Through It*. Here is what he says of his youth in Missoula, Montana:

> After breakfast and again after what was called supper, my father read to us from the Bible or from some religious poet such as Wordsworth; then we knelt by our chairs while my father prayed. My father read beautifully. He avoided the homiletic sing-song most ministers fall into when they look inside the Bible or edge up to poetry, but my father overread poetry a little so that none of us, including him, could miss the music.[11]

Young Norman had to do a lot of his own reading as well as writing.[12] The culture of the word, however, extended beyond writing and literature into his daily life. Here is more testimony:

> But when I was young---certainly no older than 17 or 18---I was telling Montana stories myself. There was a small bunch of us of the same age who would sit in the evening on the steps of the First National Bank in Missoula, then owned by the Jacobs family, and tell Montana stories. We were all young, but we all worked in the woods in the summer, and I don't think you can be a Montana storyteller unless you have worked in the woods or on ranches.[13]

As Maclean's life teaches us, words beget words, the culture of reading and narrating issues in literature.

Obviously the culture of the word has ceased to recreate itself. "I need hardly tell you," Maclean notes sorrowfully, "that families no longer read to each other. I am sure it leaves a sound-gap in family life."[14] Culture, however, abhors a vacuum and has filled the gap with television, or has not rather television insinuated itself into family life and choked off reading? How could that happen?

One way of setting off reading against television is to borrow from information theory and ask, "How many bits of information do you need for a minute of reading as opposed to a minute of television?" There is no hard and fast way to determine this, but 7,000 bits per minute of reading is a defensible answer. On a CD, a minute of sound requires about 100 million bits. Video, as we know, is a hog when it comes to bits; so, depending on how compressed, it is a multiple of the 100 million sound bits per minute. In an obvious way, then, video is infinitely richer than a text, and the people who turn from books to television turn from what is austere to what is rich as who would not?

Television is not only richer than reading but, in a different way, richer than storytelling too. If today one feels the need to be entertained by the presentation of a captivating event, one does not have to wait till it is evening and one's friends show up at the First National Bank, nor is one at the mercy of a particular story teller and the hard and cold steps of the bank. Instead you can at any time summon the dramatic rendering of whatever story and enjoy it in the comfort of your couch. Thus technology not only enriches our lives but also furnishes it with a kind of freedom that differs importantly from the political liberty that normally comes to mind when we talk of freedom. The promise of technology is one of material and social liberty, the promise of disburdenment from the pains and limits of things and the claims and foibles of humans.

Thus the full promise of technology has always been one of a special liberty and prosperity. The promise inaugurated the modern era and has to this day animated our society's most coordinated and strenuous efforts. It comes to the fore in advertisements, the public proclamations of our furtive aspirations. Most every advertisement stages versions of two formulas, "Now you no longer have to ..." --- the promise of liberation---and "Now you can ..." --- the promise of enrichment.

When over 54 million people in this country sit down in the evening to take in one of the "prime-time novels," who is it that is laying such a wealth of spectacles at their feet? Obviously not the remaining 200 million Americans the way the troupe of players did for Hamlet and the royal court at Elsinore. Instead contemporary folk are entertained by a complex machinery that touches their private sphere through TV sets and VCRs and extends from there to transmission devices, senders, studios, and scriptwriters and extends laterally to utility grids, manufacturing facilities, and R & D labs.

Technology has delivered liberty and prosperity not through magic as fairy tales have it, nor through servitude as feudal lords enjoyed it, but through increasingly sophisticated and powerful machineries. As their development has progressed, the goods that these machineries are procuring have become available more instantly, ubiquitously, easily, and safely---more commodiously in an older sense of the word. Such goods may well be called commodities, not just because they are commercially available but, more important, because of their comfort and convenience as the word commodity, again in a less common sense, suggests.

Once it is clear how the culture of the word has been replaced by the commodities of entertainment through an elaborate machinery, one can see a like pattern of displacement throughout the culture. The culture of the table, the careful preparation and the daily or festive celebration of meals, has been invaded by the commodious flexibility and variety of foods that are bought ready-made, stored safely and easily, and prepared in an instant.[15] Underlying this commodity are the vast machineries of agriculture, food industry, and supermarkets, and the ubiquitous machineries of cars, refrigerators, and microwaves. Wherever you turn, whether to matters of transportation, health, the arts, or, increasingly, education, you see the same conjunction of commodification and mechanization having transformed and still transforming our culture.

The pattern of this transformation is well instantiated by any of the technological devices that surround us in daily life. The commodity of a TV set is the smoothly moving, brilliantly colored picture and its contents. The machinery is known to exist somewhere behind the screen, beyond the competence and intelligibility of ordinary people. It can change significantly, say from analog to digital transmission or from cathode ray tubes to active matrix screens, unbeknownst to the consumer who merely registers an improvement in the size, reliability, resolution, and space requirements of the commodity.

To have a fixed vocable for this cluster of technological phenomena— the cultural displacements, the commodification and mechanization and their embedding in contemporary culture—we may speak of the pattern of the technological device or the *device paradigm* for short.[16] We can now ask, "What is the cumulative cultural effect of the device paradigm and who is responsible for that effect?" In important part the effect has been most beneficial. There are burdens of hunger, sickness, and confinement that the device paradigm has lifted from the shoulders of the advanced industrial countries, and no responsible person would want those afflictions to return.

This constructive layer in the development of modern technology has always overlain a more profound feeling of loss and betrayal; but during its most constructive phase, roughly from 1850-1950, technology was mostly seen as a beneficial force. Increasingly, however, the layer of doubt and sorrow has grown, and the benefits of technology have become thinner.

Who is responsible for this development? To speak of the device paradigm or of technology as doing this and preventing that is to imply a premature and tendentious answer to this question. It is to subscribe to technological determinism, the view that technology is a force in its own right, irresistibly forcing our hand as we shape our culture. Though many a critic of technology has lapsed into the determinist position on occasion, it is normally and rightly repudiated. In the prevailing view, technology is an ensemble of neutral structures and procedures that can be used for well or ill—the instrumentalist view of technology.[17]

Instrumentalism no doubt applies to many situations and quandaries in the technological culture. It nevertheless deflects attention from the crucially different ways cultural decisions are made in our society. Our daily decisions are for the most part channeled and banked by fundamental decisions that are no longer at issue. Using or not using the interstate highway system is not a matter of choice anymore for most of us, and neither are the moral consequences of long commutes, the neglect of family, neighborhood, and inner city. When we finally come home, late and exhausted, greeted by a well-stocked refrigerator, a preternaturally efficient microwave, and diverting television, there is little choice when we fail to cook a good meal and summon the family to the dinner table.

What the instrumentalist fails to see is that we live in a world that is patterned after the device paradigm, a life where we pay our dues to the machinery of the device through labor and where in our leisure time we surrender to the diversions of commodities. We value work because it is, more so than citizenship or education, the crucial certificate of membership in this society. We value leisure because it is still ringing with the echoes of liberty, prosperity, and self-realization. At a deeper level, however, we sense that a life, divided between labor that is not fulfilling and leisure that is not ennobling, is not worth living. We are all implicated in this way of life, and our implication in the device paradigm is a difficult and complex relation. Contrary to what the technological determinists would have us believe, we are not simply and entirely under the sway of technology. And, pace the instrumentalists, we are not normally elevated above technology as wakeful and rational choosers.

Our implication in the device paradigm of technology, I believe, is at the heart of the crisis of contemporary culture. This implication, often shading over into complicity, accounts for the sporadic sorrow we feel at the loss of cherished things and practices that were part of an older culture, now being swept aside by the spreading of technology. It also accounts for the nervousness we feel at the prospect of being pushed to the margins of the pattern of technology when our skills become obsolete or our job is eliminated. Outside of the device paradigm there is increasingly little else. Moreover, our implication in technology explains the invidiousness that has invaded our culture. If the blessings of technology fail to provide genuine happiness, one may still hope to be relatively happy, securing a disproportionate claim on the fruits of technology and contemplating with some sort of satisfaction the unhappiness of the less fortunate. As long as our implication in technology is not explicated, considered, and transformed, the heart of contemporary culture will be ailing—not mortally, for better or worse, but chronically and sullenly.

Is there an alternative to the pattern of technology? As Thomas Kuhn has taught us, people will not let go of a ruling paradigm unless or until a viable counterparadigm is on the horizon. To see that there are in fact counterforces to the device paradigm abroad in contemporary culture, let us return to the distinction between reading and watching television. People continue to read good books. Norman Maclean's *A River Runs Through It* first became a bestseller through one reader's recommendation of it to another.[18] Many readers have professed, moreover, that the book or its lead story has made a more profound and enduring impression on them than the film of the same name. How can it be that something so austere as a text can be more powerful than a film when a film provides information that is richer by several orders of magnitude than the corresponding text?

If we assume that a large effect requires a large if distributed cause, where does the powerful effect of a printed story come from? The answer is that literacy on the part of the reader generates the wealth of information a viewer receives without charge. Literacy is a many-storied skill, rising from word recognition via parsing to comprehension. To read comprehendingly is to follow the author's instruction in the construction of an imaginary world. The author gives us the blueprint, but we must supply the materials and situate the structure. The materials are our experiences as well as our aspirations, in Maclean's case our knowledge of streams and forests, of family, of helplessness and consolation. The location of the structure is somewhere in the life of our imagination, that realm of pregnant possibility that surrounds and informs our actual life. Thus to read is to gather our past and illuminate our present. It is a focal activity that collects our world as a convex lens does and radiates back into our world as does a concave mirror.

A great film undoubtedly can do the same thing, and a cheap novel no doubt will fail to do it. The brain of every mainstream philosopher has been programmed to react to whatever general claim by racing through its instances in search of devastating counterexamples. While this is a fine exercise for graduate students, it prevents the mature philosopher from comprehending the subtle and crucial tendencies of contemporary culture. One such trend is that commodities by their very structure tend to lull and dull our senses and talents. Another trend is that certain things and practices provoke and engage our physical and moral gifts. The things I have in mind are good books, musical instruments, athletic equipment, works of art, and treasures of nature. The practices I am thinking of are those of dining, running, fishing, gardening, playing instruments, and reciting poetry. On closer inspection some such thing and a practice are always correlated. I call these focal things and practices in the small and communal celebrations in the large.[19]

Once the notion of focal concerns and communal celebrations has been clarified, it becomes evident that the pattern of the technological device is the ruling but certainly not the sole cultural force today. There is a literature of focal things and practices and of communal celebrations that testifies to the extent and vigor of an alternative reality.[20] It will be clear also that in most cases consumption of commodities and engagement in focal practices are found in one and the same life. Nevertheless, they cannot substitute for one another, and if one expands, the other must shrink.

The history of modern culture is obviously one of the expansion of the device paradigm and the fracturing and scattering of focal practices and communal celebrations. Focal things today are found in a diaspora. They are for the most part officially ignored and diffidently defended. On occasion concern for them surfaces unexpectedly. In the *New York Times* of December 12, 1995 the architect Hugh Hardy is fulsomely praised, as well he might be, for his restoration of architectural treasures. All this is done in the genteel discourse of high culture consumption when suddenly the language gains in gravity, and we read:

> Mr. Hardy said he believes that it is the vitality of its public spaces that keeps a city healthy enough to counter the mounting popularity of simulated experience—theme parks as well as enclosed shopping malls—over the real thing.
>
> "People used to understand that gathering in public was good; that's what democracy meant," said Mr. Hardy ...[21]

The distinction between "simulated experience" and "the real thing" is one that seems easily understood and is often made. But unlike the distinctions between liberalism vs. conservatism, pro choice vs. pro life, environmentalism vs. industrialism, the distribution vs. the aggregation of wealth, this one has no standing in the public forum, and this is what one should expect given the imbalance between the prominence of commodities and the dispersion of focal things.

It would be churlish to belittle the centering and animating power that can flow from the culture of the word and the culture of the table, but the orienting force of regularly reading poetry and sitting down to dinner trails off all too quickly in the thickets of moral confusions and cultural ambiguities. For a Catholic, however, it is but a short step from the culture of the word to the Word of God and from the culture of the table to the Breaking of the Bread. The history of salvation that is set out in scripture and centered in the Eucharist certainly provides for the scope and coherence that the diaspora of focal things and communal celebrations is lacking.

In a practical sense religion enables some Americans to do the right thing. Ever since Anselm prevailed over Peter Damian, however, the Church has agreed that doing the right thing benefits from thinking through the natural and cultural conditions that provide the setting for faithful action. Considering those conditions today, one cannot help but notice how thin the compass of religion has become. Why is its formative power so slight in our culture?

A culture, informed by the device paradigm, is deeply inhospitable to grace and sacrament. The productive side of technology is an enterprise of conquering and controlling reality. The notions of human incompleteness and deficiency that signify a primal condition for the advent of grace are mere grist for technological mills. On the side of consumption, the paradigmatic object is the instantaneously and easily available commodity. The notion that some *opus*, properly *operatum*, could exert power over us rather than us controlling it, is foreign to the culture of consumption.

The ontological status of a sacrament is a problem I am not in a position to shed much light on. I merely want to note how problematic it has become under the rule of technology. The device paradigm, I am afraid, has not just taken cultural territory from the sacraments and left them untouched in their narrower sphere. By transforming the very notion of what is real it has likely clouded the very reality of the sacraments. In the recent *Catechism of the Catholic Church*, little is left of the tangible heft and gravity of the *opus*, and the emphasis has shifted to the *operatio* that can be conveniently abstracted from its material setting.[22] This theological attenuation is unhappily mirrored in contemporary Catholic liturgy where prayerful attitudes are safely elevated above the standards of excellence by which we used to measure tangible and audible things such as architecture and music.[23]

How then do we go about the task of thinking through the relation of scripture and the sacraments to the culture of technology? To begin with the sacraments, the goal cannot be somehow to improve or adjust them so that they better meet the requirements of contemporary culture. The task is rather to make room for them and to do so in the circumspect way that Heidegger half a century ago sketched this way:

> Only from the truth of Being can the essence of the holy be thought. Only from the essence of the holy is the essence of divinity to be thought. Only in light of the essence of divinity can it be thought or said what the word "God" is to signify.[24]

The truth of Being, we might say, is that reality today is ruled by the device paradigm and therefore inhospitable to the holy. In a society where the sense for the sacred has largely atrophied, sacraments are unlikely to have a focal place. Thus as Catholics we must be concerned to strengthen reverence and piety wherever we find it, the natural piety of environmentalists, the aesthetic reverence of the arts community, the sense of wonder in the sciences. Proceeding from the holy to the divine, we should support the intimations of divinity that are diffidently celebrated in our society. I mean those occasions where people gather to be entertained and end up being inspired by the grace of an athletic event or an artistic performance that supervenes on everyone as an unexpected and undeserved gift.

As for scripture, Catholics cannot be unconcerned about the decay of the culture of the word and the thoughtless dismissal it is suffering at the hands of cyberspace enthusiasts. If generally to read is to gather one's past and to illuminate the present, this is eminently true of reading the Bible. Though sacred scriptures do not have the crucial function for Catholics that they have in other denominations and religions, for us too holy texts are a bond that unites the generations of believers into the people of God. That bond is likely to wither if not die in a culture that neglects or derides thoughtful reading and listening.

It is most important to recognize that making contemporary culture more hospitable to sacrament and scripture is not primarily a labor of sensitizing, of making people more receptive in a passive sense. What needs to be recovered is the skill of celebration. The union of discipline and grace that marks celebration has been divided by the device paradigm. Discipline has been assimilated to the machinery side and has turned into the strain and exertion of labor. Grace has been absorbed by the commodity side and degenerated into the gratification of consumption. Celebration has, as a consequence, lost much of its discipline and is too often thought to be a matter of gratification rather than grace. This development has enfeebled Catholic liturgy. There is no grace without discipline. Self-indulgence now pervades much of the musical and tangible setting of the Eucharist. Thus as we move from the culture of technology via the secular culture of the sacred and the divine to the precincts of the sacraments, one thing we need to acquire and bring along is a sense of discipline and excellence when it comes to celebration.

Unless we join Heidegger in his coy reserve about the presence of God, we must move not only inward from the culture at large toward the center of religion but also outward from the focal point of the Breaking of the Bread to its cultural context. This is no doubt the more difficult move. At the level of partners and parents it comes to affirming on Monday what has been professed on Sunday. If the sacrament of the Eucharist is not reenacted in the sacramental of the dinner table, the Breaking of the Bread has a precarious place in contemporary culture. It is "intrinsically evil," we are told, to use contraception.[25] I have never heard that said of the failure regularly to sit down to dinner, a morally much more consequential calamity. Men, as a rule, are the sinners here. Their contributions to the preparation and scheduling of dinner tend to be deplorable or worse.[26]

The culture of the word and its reflection of the Word of God require still more attention. A family needs to eat every evening and so is daily reminded of the dinner table if not regularly gathered around it, but there is no similarly persistent reminder of reading.

To move, as a matter of public policy, from the center of faith to the culture at large, we need to lower the wall between church and state. Such a suggestion, to be sure, runs squarely into the liberal fear of intolerance and the protestant disdain of secular culture. These, again, are complex issues that I will only engage to conclude my remarks. We should certainly try to oblige liberal apprehensions by renouncing the zealous and imperial postures that religious people sometimes strike when they address the culture at large. Like the Italian Bishops, we should allow the liturgy to be seen "as a cultural event."[27] Mindful of the religious diversity in the United States, we should invite our Jewish and Muslim friends and all religious people of good will to do likewise.

Leaving the wall between church and state intact is to reduce the force of religion in everyday life to an often ineffective overlay of moral exhortation. Meanwhile the material culture on the secular side of the barrier is left to the commandments and blessings of the device paradigm. This predicament is made worse by the in-your-face Christianity of some critics of contemporary culture, where preaching the Gospel comes mainly to scolding the middle class—a practice Catholics have adopted too many times.[28]

What the middle class is calling for is a promise of daily freedom and well-being that breaks with the device paradigm and holds out a sacramental life invigorated by a continuity of sacraments and sacramentals, of worship, focal things and practices, and of communal celebrations. Let us be bringers of good news.

University of Montana
Missoula, Montana

* * * * *

Notes

1. Barbara Ehrenreich, *Fear of Falling: The Inner Life of the Middle Class* (New York: Pantheon, 1989).
2. John Cassidy, "Who Killed the Middle Class?" *New Yorker* (October 16, 1995): 113-24. Susan Sheehan, "Ain't No Middle Class," ibid., (December 11, 1995): 82-93.
3. Ehrenreich, 29-38 and 223-31.
4. Charles McGrath, "The Triumph of the Prime-Time Novel," *New York Times Magazine*, 22 October, 1995, 52.
5. McGrath, 55.
6. Loc. cit.
7. McGrath, 86.
8. Norman J.G. Pounds, *Hearth and Home: A History of Material Culture* (Bloomington, IN: Indiana University Press, 1989).
9. Robert M. Crunden, *A Brief History of American Culture* (New York: Paragon, 1994).
10. Edward B. Tylor, *Primitive Culture*, vol. 1 of 2 vols., 7th ed. (New York: Brentano, 1924 [first published in 1871]) 1.
11. *Norman Maclean*, eds. Ron McFarland and Hugh Nichols (Lewiston,

ID: Confluence Press, 1988), 84.

12. Ibid., 25-26, 82-83.

13. Ibid., 70. See also 26 and 90-91.

14. Ibid., 26.

15. Marjorie L. DeVault, *Feeding the Family* (Chicago: University of Chicago Press, 1991).

16. See my *Technology and the Character of Contemporary Life* (Chicago: University of Chicago Press, 1984).

17. A view, by the way, shared by the *Catechism of the Catholic Church* (Mahwah, NJ, 1994), 552 (sections 2293-94).

18. Norman Maclean, *A River Runs Through It* (Chicago: University of Chicago Press, 1976).

19. For communal celebrations, see my *Crossing the Postmodern Divide* (Chicago: University of Chicago Press, 1992), 126-47.

20. *Technology and the Character*, 201.

21. Julie V. Iovine, "Tenacity in the Service of Public Culture," *New York Times*, 12 December, 1995, 4(B).

22. What little is left can be found in the sections titled "Signs and Symbols" (sections 1145-52, pp. 296-98). The stress is entirely on the referent or tenor. Consider, to the contrary, the emphasis that is given to the sign or vehicle in *Universalis Catechismus Romanus* (Augsburg: Matthaeus Rieger, 1761), 176-86.

23. On music see Thomas Day, *Why Catholics Can't Sing* (New York: Crossroad, 1993).

24. Martin Heidegger, "Letter on Humanism" (first published in 1947), in *Basic Writings*, ed. David Farrell Krell (New York: Harper, 1977), 230.

25. See *Catechism*, section 2370 (570).

26. See DeVault.

27. "Italian Bishops Insist Mass Is a 'Cultural Event'," *National Catholic Reporter*, 16 July, 1993, 10.

28. For example, in *Economic Justice for All: A Pastoral Letter on Catholic Social Teaching and the U.S. Economy* (Washington, DC: National Conference of Catholic Bishops, 1987).

The Philosophical Challenge of Technology

by Carl Mitcham

A thousand or two thousand years ago the philosophical challenge was to think nature---and ourselves in the presence of nature. Today the great and the first philosophical challenge is to think technology. A second is like unto it: to think ourselves in the presence of technology.

I

Thinking Technology as Difference: The French anthropologist Claude Levi-Straus, following Aristotle, was surely correct. We think in terms of difference, from myth to science. It is in and through structural distinctions that we encounter the realities that surround us and that we are. With what structures of difference, then, are we able to think technology?

Historically, not to say logically, the first difference is that between nature and artifice. *Techne* is initially thought as that which is not *phusis*. Nature is that which is independent of human activity, and is always and everywhere the same. Technology is the work of human beings, and as such is not always and everywhere the same.

Paradoxically, technology is thereby thought as natural to human beings. *Techne* is thereby associated with *nomos*, as a second *phusis* for man. It is by nature that human beings move about, graze, and change in what were once the pastures, the provinces, of *nomos*.

Another way of speaking about technology as that which is not always and everywhere the same is to say that it has a history. As Mircea Eliade and Eric Voegelin, among others, have argued, history is from the Jews---and those renegade Jews known as Christians.[1] According-ing to Lynn White, Jr., and before him Max Weber, so is technology.[2] Whether this technology which springs from the root of Jesse is also our salvation has become a secular question.

The precise parameters of this secular salvation history are subject to debate, but not the historicity of technology itself. The difference between those things made by animals such as bird nests and beaver dams, things which are sometimes suggested to be continuous with

45

human beds and houses, is surely that the latter have and help define a history, whereas the former do not. In the history of animals there is nothing like a stone age, a bronze age, or an iron age—temporal pastures and provinces demarcated by distinctive technologies. With animals the technologies come and go with the species, even when partially learned.

Once it appeared that nature, like the animals, did not have a history, or had a history that was essentially cyclical. Linear or progressive history was thought peculiar to some aspect of the human realm—spiritual or political—and even that history was entangled with cyclical reflections of nature. The memory of salvation history is repeatedly re-enacted in a liturgy of the years, weeks, and hours. The changing dynasties of ancient empires aligned their actions with the cycles of seasons and stars. Yet today it appears that linear history is not limited to the human realm, that nature too has its history. Nature has a different history, in that its patterns reach out over much more extensive reaches of space and longer stretches of time, in that its periodization is structurally different from the periodization of human history. It nevertheless has a history. The stars are more—or perhaps less— than a cyclical or moving image of eternity.

Nevertheless, for thousands of years the historicity of technology, despite its progressively insinuated presence from at least the 5th millennium BCE in certain increasingly urbanized pastures of *nomos*, was not a prominent feature of human affairs. Technology was no more than stones marking out the periphery, hardly noticed. Among those who dwelled in the leading cities of those provinces, what appeared much more historically prominent than technology was the political order. In contrast to the political and its dominating history were both technology and nature and their marginal history or non-history, respectively. The discovery of technology as of prominent and founding importance to history anticipates by no more than a century or so the discovery of history in nature. Indeed, from the point of view of natural history the two discoveries could be described as coeval.

These two virtually contemporaneous discoveries occurred in conjunction with a further development that served to undermine the traditional nature/technology distinction: that is, the radical, post-medieval bloom of artifice throughout the European *nomos*. The enclosure of the English commons was not simply an economic transaction; it involved ripping into the earth with the moldboard plow, constructing factories of fire in cities that breached their own walls, and turning the water and air into industrially processed chemicals. In the historiological debates about the first causes of modernity surely it is neither the scientific, nor the economic, nor the political revolutions that most disturbed the traditions of these provinces. Instead it was the steam-powered locomotive of the Industrial Revolution that tore through the countryside, never stopping, driving and depriving the senses of those who watched as well as those who rode.

Imitating the phenomenologists, today, right now, let us simply look at the "things" themselves. Our conference takes place in the Crowne Plaza Redondo Beach and Marina Hotel, in a hermetically framed room named the "Pacific." Not even one window pierces its plastic-coated walls. If there is a moral law within me, it is surely not matched by the strobe-like lights bouncing off the HVAC-rustled movement of the synthetic glass chandelier above me.

Outside this hotel is the room's namesake, the Pacific Ocean. In the words of Robinson Jeffers, who died 250 miles north building Tor House on the rock-strewn coast,

... this dome, this half-globe, this bulging
Eyeball of water, arched over to Asia,
Australia and white Antarctica: [these] are the eyelids that
   never close; this is the staring unsleeping
Eye of the earth; and what it watches is not ...[3]

our oil-consuming internal combustion engines, coal-fired electric power generation plants and transmission lines, the slow lethal dance of controlled nuclear ruptures and would-be rapture—our rising tides of waste, our television sets, the distributed networks of telephones, faxes, and computers, fixed and portable, cable and cellular. That is there, while we are here, inside the cocoon of what the undulating oceanic eye does not see—and from our vantage point neither can we see it. In the more prosaic if somewhat unfashionable difference of Jacques Ellul, the natural milieu has been pushed aside, overlaid, by technique, the technical milieu.[4] Where is there any longer difference to think within this postmodern theater of engineered materials and electronic programs in which we propose to read from hyper-edited, folded-in and cut-up papers on technology? Are we serious?

Consider again: Following close upon the explosion of technology in size and extent came new penetrations in depth and power. For Aristotle, as for Plato, technology is distinguished from nature—or, more carefully, technological pseudo-entities are distinguished from natural entities—by their ontologies. Artifice is less real than nature. In Platonic terms, artifice constitutes no more than the shadowed furniture of our theatrical cave, a copy of a copy or a participation in a participation. For Aristotle, even in the sunlight, artifice yields a weaker unity of form and matter than that found in nature. "If a bed were to sprout," says the Stagarite, "what would come up is not a bed but an oak."[5]

Can the same still be said for those synthesized organic compounds that spring forth from our petrochemical plants or the transuranic (also called TRU) elements emerging from our nuclear reactors? If polyvinyl chloride were to sprout—which it certainly seems to do throughout our condos, office complexes, and entertainment centers—would it not be more PVC that would come up? When that technological cross between a goat and sheep called the "geep" gives birth, is it not another geep that baas. Although animal artifice may not have its own history, are we not in the process of giving it one, as beer cans and milk cartons become incorporated into beaver dams and pesticides alter the web patterns of spiders?

In the face of this manifest breakdown in the traditions of difference for thinking technology there has arisen a counter-difference—not between nature and history but within history itself: the distinction between ancient and modern technologies. This historical difference has been formulated in at least four ways.

First, according to Lewis Mumford, there is a historical difference among technological objects.[6] Premodern tools (instruments under immediate human power and guidance) have been replaced by modern machines (with nonhuman power sources but continuing human guidance) and automatons (nonhuman power and no immediate human guidance). For the French mechanologist Gilbert Simondon, too, there is a progressive integration of technical elements that leads toward the modern, internally integrated and self-subsistent artifact that, unlike traditional tools, marginalizes the significance of human use.[7]

Second, in terms of knowledge, there is the historical distinction between craft and scientific engineering. The former rests on the intuitions of embodied skill, the latter on the explicitly articulated technological sciences and their epistemologically distinct knowledge bases. Going beyond the epistemology of the technological sciences, Martin Heidegger, in his alethiology, argues that modern *Technik* is a kind of truth or disclosure of the world as *Bestand*, resource, as opposed to the premodern disclosure through *poiesis* of the world as things.[8] The traditional world with its differentiation of thing and place is replaced by one of no-thing and non-place. Pastures become first mines, then factories, and finally shopping malls and airports. Then, too, there is the difference between natural and artificial intelligence—which can be rendered as no more than one between before and after silicon design, or before and after information theory.

According to Michel Foucault, however, truth and intelligence are not simply an epistemic disclosure or information processing but a discourse site constituted as the result of organized social disciplines in which knowledge and information are the new masks of power.[9] An emphasis on historical discontinuity in technology as activity, or as a complex of activities, is typical of social scientific differentiations in the historical trajectory of technology. The classical thesis is that what defines modernity is "technicization"—that is, the disengagement of making and using from socio-cultural norms and the independent elaboration of techniques which then come to exercise a dominant influence over social institutions. Within this perspective Ellul's analysis of the central role played by the rational pursuit of (a no doubt illusory) efficiency in the economy, the state, education, and medicine along with Herbert Marcuse's analysis of the societal dominance of one-dimensional instrumentality[10] square off against Jürgen Habermas, Wiebe Bijker, and Andrew Feenberg on the extent of the social construction of technology.[11]

Fourth, and finally, there are differences between premodern and modern technology in terms of volition. For both Mumford and Heidegger, again, modern technology is defined not only by a distinctive artifice and truth but also by a will to power absent from the rational or thinking appetite that animates traditional, premodern technical phenomena. In spiritual terms this was formulated in the early 20th century as an issue of pious versus impious technology, an acceptance or rejection of divinely given bounds to human making and using. In the late 20th century the environmental movement has represented this contrast as one of anthropological self-assertion versus a variously named "deep ecology," "biophilia," and "ecocentric ethics"—the latter of which all seem more characteristic of archaic and indigenous peoples than of those of us who find our lives on the computer screen.

It is not clear, however, that such (what may be called) historicist attempts to think technology in terms of historical difference as object, as knowledge, as activity, and as volition are fully adequate substitutes for failure to think technology as different from nature. Consider no more than the following difficulty: Just how different are the historical differences? Are these historical differences cumulative or exclusive?

On a cumulativist interpretation, premodern tools, intuitive craft skills, traditional cultural orders, and piety are simply supplemented with modern computers, engineering science, technicization, and power. Yet in each case the premodern is not simply being complemented with new objects, knowledge, activities, and needs but technologically infected and transformed as well. Tools are as good an example as any: The traditional hammer is not only progressively assisted by the nail gun and glue stick, but even when picked up is discovered to be an artifact transformed by high tensile steel and composite fiber handle with shock-absorbing synthetic polymer grip.

On a radically exclusivist interpretation, however, there seems no way to understand that difference over against which the different is to be thought. The discontinuities in history becomes so pronounced that, cut loose from all moorings, we are cast adrift in the postmodern present—absent any touchstone or sense of the fitting, such as that nature with which technology was ultimately called upon to harmonize within the nature/technology difference. In its radical historicity, difference disappears behind its own screen of difference, leaving modern technology triumphantly alone in the world—and unthought. To depend upon thinking radical difference in history is to commit the fallacy of misplaced difference.

II

Difference in Philosophy of Technology: Independent of the alleged adequacies or inadequacies of such articulations of the four modes of the manifestation of historical difference in technology there is, in the philosophy of technology, often an implicit criticism of the unqualified modern affirmation of the near difference of the present over against the far difference of the past. The defense of modernity, however, need not be any simple affirmation of present difference, but can seek self-justification by relativizing its own uniqueness through some principle of continuity.[12] If one moment of modernity was to argue for a replacement of premodern history by a new and more power-filled experience of the world, a second moment has been to assert that this new experience is no more than the public seizure of a right that lay hidden as possibility and longing within the premodern. Did not Icarus desire to fly? Have not medical ministrations always been in great demand? Thus even when it disclaims any anti-technological bias or pessimism— as it does in both Heidegger and Ellul—the attempt to think technology as historical difference constitutes a spontaneous but fundamental challenge to the continuity axiom of modernity and, *mutatis mutandis*, to modern technology. Insofar as it fails to establish (or over-establishes) its differences, philosophy of technology presents its profile of impotence in the presence of a continuing and multifaceted trajectory of technological power.

Even if technology itself cannot be thought as difference either in contrast to nature or within history, yet remains as an expanding presence that would still be questioned, this very dialectic of success and failure suggests that our thinking of technology can nevertheless be differentiated as thinking. More deeply, perhaps, than the four-dimensional historical differentiation is the bifurcation between an attempt to think technology as difference and another thinking of technology as non-difference. The former has been termed "humanities philosophy of technology," the latter "engineering philosophy of technology." We are thus confronted with a difference not of or in technology but in the philosophy of technology: between humanities and engineering philosophies of technology.

Thinking through Technology was an extended attempt to think this philosophical difference.[13] Briefly rehearsing the distinction: Engineering philosophy of technology takes technological thought and action as the model for all human thought and action, and attempts to explain or reconfigure all apparently non-technological thought and action in technological terms. All technological amplifications, extensions, enhancements, and modifications of the human senses, for instance, are seen as perfectly legitimate if not actual perfections.[14] Humanities philosophy of technology, by contrast, argues that technological thought and action are only one aspect or dimension of human thought and action, and seeks to delimit or restrict the technological within a more expansive or inclusive framework. Technological modifications and extensions of the human sensorium are to be critically assessed if not outright rejected.[15] Engineers are, it is maintained, unreflective agents of a deformed and deforming technological eros, thinking only in terms of technicism and leading only toward technocracy and techno(per)ception. In doing only what they know, they know not what they do, as uninformed followers of science or, better, misinformed disciples of scientism.

As a self-conscious activity, however, the philosophy of technology so named actually emerged barely a hundred years ago among engineers trying to reflect upon and give more general meaning to their work. These engineer-philosophers—from Ernst Kapp (1808-1896) to Friedrich Dessauer (1881-1963), and their successors—have largely succeeded in two tasks: first, they have stimulated engineers to reflect more generally on their work. In so doing, philosophers have been forced to see technology as something distinct from science, and as deserving of its own epistemological, metaphysical, ethical, and political analyses. Second, they have countered the grand humanities criticisms of technology put forth by Mumford, Heidegger, Ellul, and their successors. In opposition to broad humanities criticisms, the spirit of engineering philosophy of technology manifests itself in numerous piecemeal criticisms of applied philosophy, especially applied ethics.

The power of engineering philosophy of technology is manifest not only in the dynamism of the technological world but also in the vigorous life that currently exists in such interrelated specializations as biomedical ethics, environmental ethics, engineering ethics, computer ethics, and more. This two-fold achievement is not to be belittled. Nevertheless, the power exercised by this engineering philosophy and applied ethics exists only on the margins or at the periphery of the technological enterprise. From the perspective of the grand critiques of Mumford, Heidegger, and Ellul such philosophy of technology exhibits a historical impotence. It is a sideshow that belies the originative thrust of modern philosophy.

Modern technology grew out of a powerful and historically consequential critique of the received philosophical tradition. During a two- or three-century period, from the Renaissance through the Enlightenment, Francis Bacon (1561-1626), Galileo Galilei (1564-1642), René Descartes (1596-1650) and their followers initiated a comprehensive criticism of traditional philosophy, a criticism that became the basis for a world-historical transformation and the rise of the modern technological way of being in the world.

Beginning in the 18th century, throughout the 19th century, and continuing into our own 20th century, counter philosophers from Jean-Jacques Rousseau (1712-1778) through Karl Marx (1818-1883) and Frederich Nietzsche (1844-1900) to phenomenologists and existentialists have subjected the modern tradition to its own withering critique. This critique has, however, had no more than a marginal—not to say a reactionary—influence on modernity. Indeed, the dynamism of modernity is manifest precisely in its continuing ability to absorb its critics, even to the point of declaring itself transformed into a postmodernity.

In one of the ironies of history, modernity repeatedly declares victory for its critics, sends them home thinking it has been vanquished, only to continue to expand its domain.

What is one to conclude from this failure, this impotence of the humanities philosophy of technology? This is an question which has yet to be addressed in any comprehensive manner by the humanities tradition in the philosophy of technology. Instead, like Christians faced with the historical delay of the *parousia*, members of the humanities philosophy of technology community have, searching for signs of the times in everything from economic collapse and worker revolt to Third World revolution and environmental limits, only too repeatedly replied "not yet—but soon." The exercise of sign-reading likewise seems regularly to conflate the issues of possibility and probity. Is it that the technological project ultimately cannot triumph, or that it should not?

In reading much cultural criticism of technology, for instance, one cannot help but be reminded of a TV evangelist, such as the strong-bodied Rev. Jack Van Impe and his well-wrought wife, Rexella. Each week, in a program syndicated on various Christian cable networks, Rexella breathlessly reads the newspapers, and then Jack explains and comments. From the soap opera of the British monarchy to Hubble telescope discoveries, the Rev. Van Impe sees sign upon sign of the imminence of the Second Coming—although his bets are always hedged by Rexella's cautionary note that we can never know the day nor the hour. In salvation history it is both impossible for evil to triumph, and it should not triumph, although we can never know for sure when this conjunction of modality and deontology will finally manifest itself.

By contrast, engineering philosophy of technology, not troubled by the impotence of its analyses, finds itself already at a synergistic intersection of possibility and duty, and has been able to interpret the broadscale failure of humanities philosophy of technology as a confirmation of its own tradition. Even stronger, according to Julian Simon, one of the firmest advocates of the engineering school, humanities criticism of technology actually serves a dialectical purpose to advance the achievements of technology. Humanities philosophy of technology is not so much wrong as simply lacking in self-understanding. To adapt one of Simon's statements of his position:

> The process goes like this: More ... and increased [technology] cause[s] problems in the short run [which humanities philosophers or critics of technology readily point out]. These problems present opportunity, and prompt the search for solutions [by engineers and engineering philosophers of technology]. In a free society, solutions are eventually found, though many people fail along the way at cost to themselves. In the long run[, however,] the new developments [in technol-

ogy] leave us better off than if the problems had not arisen.[16]

To continue with Simon's engineering philosophy of technology vision:

> We now have in our hands ... the technology to feed, clothe,
> and supply energy to any ever-growing population for the
> next 7 billion years. Most amazing is that most of this specific
> body of knowledge developed within the past hundred years
> or so.... Indeed, the last necessary additions to this body of
> knowledge—nuclear fission and space travel—occurred dec-
> ades ago. Even if no new knowledge were ever invented ...
> we would be able to go on increasing forever, improving our
> standard of living and our control over our environment. The
> discovery of genetic manipulation certainly enhances our
> powers greatly, but even without it we could have continued
> our progress forever.[17]

The affinity between such an engineering positivism and postmodernist
celebration of the electro-media culture from movies to television and
the new cult of the World Wide Web should be obvious—and it extends
beyond the internal references and ironies that must be an expected
(and thus, despite its postures, non-ironic) aspect of an expanding and
intensified artifice now present in a rich, playful panoply of information,
images, simulacra, and virtualities.

III

Ourselves as Difference in the Presence of Technology: In a world that
human beings have made, what is our difference? Attempts to restate
the difference of man in the presence of the un-made in terms that still
have force in the presence of the made strike an inevitably hollow note,
exhibiting the very impotence that would be disproved.[18]

What then is our difference in the presence of this difference between
power and impotence in the philosophy of technology? Can one seriously
say both that the applied ethics of technology is no mean achievement—
and that it is ultimately a sideshow? Can one work with and desire to
contribute to the engineering philosophical tradition, yet at the same
time entertain fundamental doubts about and criticisms of the modern
technological project? Can one live this dialectical reality in a way that
is anything other than a bad faith attempt to straddle the fence? Indeed,
is there any real fence left to straddle? Is not the difference of engineering versus humanities philosophy of technology not itself becoming
subject to doubt, as that which once held power in the presence of
impotence loses control over its own project?[19]

Consider again the moment of impotence, as illustrated by a recent
attempt to re-claim power for the humanities philosophy of technology.
Lorenzo Simpson's *Technology, Time, and the Conversations of Moder-*

nity pivots on a distinction between meaning and value. An action is meaningful if it fits into or repeats some life pattern. An action has value when it is an efficient means to some end value. To adapt one of Simpson's examples: to help a neighbor paint a house can actualize "a background cultural commitment to communalism."[20] The *meaning* of helping the neighbor is in "enacting, sustaining, extending, reinforcing and reaffirming communal notions of solidarity."[21] By contrast, the end or use *value* of this action is simply getting the house painted.

Technology is an attempt to develop means for the ever more effective realization of use values. To extend the example, imagine the development of a house painting technology that just drops a large paint bomb from a helicopter. Maybe it's a smart cluster-bomb that divides up the house surface and spray paints it in seconds with thousands of tiny wire-guided spray can bullets. Although the end-result value of this 60-second wonder-process by paint-bomb engineers and high-tech helicopter pilot would be the same as a week-long paint-and-brush body sweat cooperation among neighbors, its meaning would be quite different. If the two technological alternatives of paint-bomb versus paint-brush cooperation are evaluated simply on the basis of the values they achieve, there is no difference. That is, there is no meaningful difference (or difference in meaning), because meaning is excluded. Assuming rough dollar-price equivalent, the paint-bomb miracle-process will be chosen because it's quicker, takes less time.

Indeed, the two approaches to house painting—the meaning perspective and the means-end or use values perspective—have different implications for the experience of time. When the focus is on the meaning of the house painting, the fact that the painting takes a week is not necessarily a negative factor, because during the whole time the participants are actualizing the meaning of cooperation. The length of time may even be a positive factor in deepening communal bonds among the neighbors. From the means-end use values perspective, however, this week is just a waste of time. "Let's get this frigging project over with by paint bombing the mother house!"

Simpson's thesis is that "[T]he growing hegemony of the temporality of making (*techne*), at the expense of temporalities of doing (*praxis*), stands as a threat to the continued presence of meaningful differences in our lives and to there being meaning in a life as a whole."[22] This is the classic humanities philosophy of technology critique of the modern technological project. Simpson argues that technological action and its coordinate experience of time constitute only one aspect or dimension of human action and human temporality, and seeks to delimit or restrict the *tempus technologicus* within a more expansive *tempus humanus*.

When Simpson, however, argues that modern technology is devoid of meaning, he overlooks the delight people take in engaging with technology for its own sake, never mind the end value that may result. This is most vividly presented in the attractiveness of the Internet and its graphic interface, the World Wide Web. People readily spend hours and hours hang-gliding their client computer screens from server to server—or catching the wave of a search engine that surfs the electronic noosphere—hours that lead to no increase in marginal utility, and which in fact in many instances lower it.[23] Value is not enhanced by spending time on the World Wide Web but meaning is. This is the wave of the future.

Gary Wolf, channeling the late Marshall McLuhan through the Internet in the January 1996 issue of *Wired* magazine, asks "Do you still believe that the medium is the message?" McLuhan replies,

> The real message [or "meaning"] of media today is *ubiquity*. It is no longer something we do, but something we are part of.[24]

Does anyone believe that Jack and Rexella would ever choose to exchange cable network syndication for some little brick church in the suburbs—even if by doing so they could actually reach more eyes and ears or the collection basket were larger?

Technology has, as it were, gone beyond itself and become its own meaning, a meta-technology—that in which we participate, even as it participates in us. To think this meta-technology as something at our service, as even engineering philosophy of technology would try to do, is as much an illusion as the delimitation dream of humanities philosophy of technology. We are nothing outside our technologies. With the world transformed into digitalized information, this information represented in images, and the images animated and multiplied into self-referential simulacra and perceptually displayed, interactive virtualities, one is surrounded and drawn into meta-technology. Once people of the book, we have become people of the screen, who issue cyborg manifestos and live in virtual communities protected by an electronic frontier foundation.

As significant as Simpson's effort to revive humanities philosophy of technology is Andrew Feenberg's attempt to point toward what might be called a philosophy of meta-technology. In *Alternative Modernity*, Feenberg provides case studies of the differential manifestation of technical rationality in culture in the United States, in France, and in Japan. His argument is that "Technical competence, like linguistic competence, is realized only in concrete forms."[25] For Feenberg, postmodernism and constructivism have shown that technoscience is not a pure independent formative factor but what he calls "a dependent variable, intertwined with other social forces."[26]

Such an observation need not obliterate the distinction between pretechnological and technological societies, but it relativizes that distinction and points up the need for us to seize the opportunities for difference within the necessary illusions of (negative and positive) autonomy that have been presented as the autonomy of technology in both humanities and engineering philosophies. In Feenberg's words, although "no modern society can forego basic technical discoveries ... significant innovations are possible with respect to what has been the main line of development up to now."[27] Or again: "[M]odern technology is neither a savior nor an in inflexible iron cage; rather, it is a new sort of cultural framework, fraught with problems but subject to transformation from within."[28] Whether or not and to what extent we may be adrift in this new post-technological technology, remains as the challenge to think ourselves as difference—a difference which may well entail selective rejections or modifications, not just celebrations, of ourselves as cyborgs, systems, or artifacts.

Indeed, along with thinking, we must also be acting. In the midst of this particular meeting, it is therefore appropriate to conclude by recalling that religious communities have made such choices in the past and will surely be called upon to make such choices again in the future. The philosophical challenge to think technology and ourselves in its presence can hardly be separated from the practical challenge to live in the post-technological world; but as St. Augustine once argued, Platonism can only be practiced with the assistance of true religion.[29] Insofar as true religious practice may be, in the presence of meta-technology, as problematic as our thinking, it too becomes part of the philosophical challenge of technology—which makes it all the more appropriate that technology should have been taken as a theme of the annual meeting of the American Catholic Philosophical Association. We can thus take as our task not only at this meeting but across the decades to come to continue to reflect on the absence of difference that surrounds us, in order to insinuate and assert ourselves as difference in its midst.

Pennsylvania State University
  University Park, Pennslyvania

* * * * *

Endnotes

1.   Mircea Eliade, *Le Mythe de l'éternel retour: Archétypes et répétition* (Paris: Gallimard, 1949). Eric Voegelin, *Order and History*, vol. 1: *Israel and Revolution* (Baton Rouge, LA: Louisiana State University Press, 1956).

2.   Lynn White, Jr., *Medieval Religion and Technology: Collected Essays* (Berkeley, CA: University of California Press, 1978). Max Weber, *Die protestanische Ethik und der Geist des Kapitalismus*, Archiv für Sozialwissenschaft und Sozialpolitik 20-21 (1904-1905).

3.   Robinson Jeffers, "The Eye," *The Double Axe and Other Poems* (New York: Random House, 1948), 126.

4.   Jacques Ellul, *La Technique, ou l'enjeu du siècle* (Paris: A. Colin, 1954).

5.   Aristotle, *Physics* II, 1 (193b10).

6.   Lewis Mumford, *Technics and Civilization* (New York: Harcourt Brace,

1934).

7.   Gilbert Simondon, *Du Mode d'existence des objets techniques* (Paris: Aubier, 1958).

8.   Martin Heidegger, "Die Frage nach der Technik," *Vorträge und Aufsätze* (Pfullingen: Neske, 1954), 5-36.

9.   Michel Foucault, *Les Mots et les choses* (Paris: Gallimard, 1966); and *L'Archéologie du savoir* (Paris: Gallimard, 1969).

10.   Herbert Marcuse, *One-Dimensional Man: Studies in the Ideology of Advanced Industrial Society* (Boston: Beacon, 1964).

11.   Jürgen Habermas, *Technik und Wissenschaft als "Ideologie"* (Frankfurt am Main: Suhrkamp, 1968). Wiebe E. Bijker, Thomas P. Hughes, and Trevor Pinch, eds., *The Social Construction of Technological Systems: New Directions in the Sociology and History of Technology* (Cambridge, MA: MIT Press, 1987). Andrew Feenberg, *Critical Theory of Technology* (New York: Oxford University Press, 1991).

12.   See, for example, Hans Blumberberg, *Die Legitimität der Neuzeit (erweiterte und überarbeitete neuausgabe)* (Frankfurt am Main: Suhrkamp, 1966).

13.   Carl Mitcham, *Thinking through Technology: The Path between Engineering and Philosophy* (Chicago: University of Chicago Press, 1994).

14.   Cf., for example, the phenomenology of technics in Don Ihde, *Technology and the Lifeworld: From Garden to Earth* (Bloomington, IN: Indiana University Press, 1990).

15.   See, for example, Ernesto Mayz Vallenilla, *Fundamentos de la metatécnica* (Caracas, Venezuela: Monte Avila, 1990).

16.   Norman Myers and Julian L. Simon, *Scarcity or Abundance? A Debate on the Environment* (New York: W.W. Norton, 1994), 65.

17.   Myers and Simon, *Scarcity or Abundance?*, 65.

18.   See, for example, Mortimer Adler, *The Difference of Man the Difference It Makes* (New York: Holt, Rinehart and Winston, 1967).

19.   See, e.g., Kevin Kelly, *Out of Control: The Rise of Neo-Biological Civilization* (Reading, MA: Addison Wesley, 1994).

20.   Lorenzo Simpson, *Technology, Time, and the Conversations of Modernity* (New York: Routledge, 1995), 45.

21.   Simpson, *Technology*, 188, note 6.

22.   Simpson, *Technology*, 63.

23.   For proof of the inefficiencies, see Douglas H. Harris, ed., National Research Council, Commission on Behavioral and Social Sciences and Education, Committee on Human Factors, Panel on Organizational Linkages, *Organizational Linkages: Understanding the Productivity Paradox* (Washington, DC: National Academy Press, 1994).

24.   Gary Wolf, "Channeling McLuhan," *Wired* 4.01 (January 1996), 128-31 and 186-87.

25.   Andrew Feenberg, *Alternative Modernity: The Technical Turn in Philosophy and Social Theory* (Berkeley, CA: University of California Press, 1995),

222.

    26.   Feenberg, *Alternative Modernity*, 221.

    27.   Feenberg, *Alternative Modernity*, 231.

    28.   Feenberg, *Alternative Modernity*, 2.

    29.   Augustine, *De vera religione*, i, 1 - vi, 11.

Philosophy ... Artifacts ... Friendship—and the History of the Gaze

by Ivan Illich

I

I speak as a xenocryst.: "Deus in adjutorium meum intende." Majid Rahnema, to you I say, "bismillahi rahmani rahim." Let me thus launch out on this triple extravagance. I am seventy, and this is the first time I address an assembly of philosophers. Second, twenty-five years ago, I promised Pope Paul VI to abstain from talking to groups of priests or nuns. This is the first time since then that I face a Catholic association. Third, I speak after my teacher. Carl Mitcham has been my main guide in the field of technosophy, shepherding me for years before we met. Since I wrote *Tools for Conviviality*, his periodic and opinionated bibliographic forays have mapped my route into the philosophy of tools. What I have to say now grew out of the seven years I have been privileged to philosophize with him, Lee Hoinacki, and a growing circle of friends at Penn State University.

Further, I am a hedge-straddler, a *Zaunreiter*, which is an old name for witch. With one foot I stand on my home ground in the tradition of Catholic philosophy in which more than two dozen generations have been fed on pagan thinkers served up in a prayerful stew. My other foot, the one dangling on the outside, is heavy with mud clots and scented by exotic herbs through which I have tramped.

I am here to argue for an approach I did not find on your agenda; I want to plead for recognition of the philosophy of technology as an essential element for *askesis*, and specifically for Christian *askesis*, in 1996. By *askesis* I mean the acquisition of habits that foster contemplation. For the believer, contemplation means the conversion to God's human face. Such conversion means Exodus, which is no longer just the aversion from the fleshpots of Egypt, nor even new power tools that increase my range. On the other shore of today's Nile lies a still unexplored anthropogenetic desert that we are called to enter. Understanding the characteristic features of new artifacts has become the necessary preface to this step: to dare chaste friendship, intransitive

59

dying, and a contemplative life in a technogenic world.

I plead for an epistemology of artifacts as the antecedent to virtues that can flower into what Hugh of St. Victor calls gifts. I know that with this plea for a philosophical propedeutics I may appear to fit into your program like a xenocryst—for the crystallographer a mineral foreign to the rock in which it is embedded. But the occasion is special: You have made the philosophy of artifice into your anniversary cake. Hence, I can ask you to be patient with me.

Things are what "matter." Be this thing bread or keyboard, condom or car, things are forever at the center of belief-shaping rituals, and things are inevitably determinants of each moment of our incarnation. This has always been so. However, during the twentieth century, so-called development has increasingly turned the world into a man-made thing. Progress and growth have meant more things—more things which, as artifacts, are made to matter more, and to matter in unprecedented ways. Experientially, even though only tangentially, a fifth post-Thomist transcendental quality has been added to Being: *ens aunque arte factum*. For seeking the one thing that matters in the Gospel sense, namely, the *itinerarium nostrae vitae in Deum*, the flood of consequences following on artifacts has so far not been a central concern of passionate philosophical inquiry. Believe me, Professor Anderson, your invitation to this seventieth anniversary of "Catholic philosophy" makes this date a red-letter day in my own seventy years, because I can address an audience in front of whom it is possible to place the foundation of Christian *askesis* in the philosophy of *techné* ... and hope to be understood.

Philosophia ... ancilla temperantiae: Circumstances have made this new type of propaedeutic fundamental for the *intellectus quaerens fidem*. Objects were once relatively unproblematic. What they were, how they affected our appetites and distracted our minds was obvious to Elias on Mt. Carmel, to the Gregories, to Benedict and Ignatius, and to other masters of Christian prayer. This is no longer so. Economic/technological development has had at least two effects:

1. It has shifted the ontic balance from cosmic entities and objects contingent on the Creator to artifacts on which our existence has become dependent.

2. Further, development has shifted the epistemic balance from objects that can be synaesthetically grasped toward objects whose shadows appear, usually with a halo of context-sensitive help that makes them subtly irresistible, so that we become addicted to them.

The person today who feels called to a life of prayer and charity cannot eschew an intellectual grounding in the critique of perceptions, because beyond things, our perceptions are to a large extent techno-genic. Both the thing perceived and the mode of perception it calls forth are the result of artifacts that are meant by their engineers to shape the users. The novice to the sacred liturgy and to mental prayer has a historically new task. He is largely removed from those things—water, sunlight, soil, and weather—that were made to speak of God's presence. In comparison with the saints whom he tries to emulate, his search for God's presence is of a new kind.

Please do not take me for a technophobe. I argue for detachment from artifacts, because only by abstention from their use can I perceive the seductiveness of their whispers. Unlike the saintly models of yesterday, the one who begins walking now under the eyes of God must not just divest himself of bad habits that have become second nature; he must not only correct proclivities toward gold or flesh or vanity that have been ingrained in his *hexis*, obscuring his sight or crippling his glance. Today's convert must recognize how his senses are continuously shaped by the artifacts he uses. They are charged by design with intentional symbolic loads, something previously unknown.

The things today with decisively new consequences are systems, and these are so built that they co-opt and integrate their user's hands, ears, and eyes. The object has lost its distality by becoming systemic. No one can easily break the bonds forged by years of television absorption and curricular education that have turned eyes and ears into system com-ponents.

This was not so when, a long lifetime ago, this very association was founded. Then, back in 1926, Jacques Ellul's technological bluff that "grips" human perception could scarcely be imagined. Virtual spaces that cannot be entered were around, but they were oddities. The very concept of context-sensitive help was unknown, information theory and systems analysis had not yet been conceived. Those monsters with whom we now rub shoulders—I think of diagnosed lives that must be saved or immune systems that must be protected—were only therato-genic phantasies then.

The *ob-jectum* was routinely perceived as something real, external, separate—a *res* or an *aliquid*—and, at least analogically, as something that, in my *conditio humana*, my traditional reality, had a history. Architects drafted on paper or modeled in clay, not on a screen. True, in the time of Ford's Model A, when Thérèse of Lisieux was canonized, and I was born, the instrumental artifact moved toward its apogee; it was becoming increasingly dominant in the sensual environment. But tech-nology was still conceived as a tool for the achievement of a *telos*, a final cause set by its user, not as milieu. Technology had not yet redefined *homo* from tool-user to co-evolved product of engineering. The nature of the object was not a quandary; it was something more or less what it had been for generations. This is no longer so. The old rules for the discernment of good from evil spirits must be complemented by new rules for the distinction of things from zombies, and objects from pictures. Temperance, what the Cappadocians call *nepsis*, must now guard the heart, not only from real things like sweet skin and weighty bullion, but also guide one to the sound recognition of the allurements of mere images and so-called needs.

Philosophia ... ancilla caritatis.: The rational distinction among things is equally basic for the relationship to persons. The faithful have been deprived of their millenary embodiment in traditional ways of life, each generating a second nature. These ways may have been vastly different from one another, but each was rooted, sustained, and perpetuated through its respective material culture. Each ethos, which means gait or way of life, shaped all human actions to a certain taste. These exercises of common sense, decency, fairness, and styles in the art of cooking, suffering, and swiving provided the seedbed for a set of virtues culminating in the principal one, love. With amazing speed, the hardware and software of the 1980s bulldozed the material milieu that had been generated by human action, and replaced it with a mostly technogenic, increasingly virtual, standard environment.

Paradoxically, the Church began to define her mission as inculturation in the very decade when all that was left of local folkways had been castrated, becoming raw material for a bureaucratically staged facsimile folklore. The critical grasp of the characteristics that distinguish ethnic artifacts from those that are system-engineered is arguably one of the demands of contemporary ecclesiology.

In my own pilgrimage, I engage philosophy as *ancilla*: on the one hand, to resist—how should I call it?—algorithmic reductionism and, on the other, to dispel the illusion that power or organization can ever enhance the practice of charity. This double conceptual shield against loving misplaced *concreta* and belief in benevolent management inevitably implies the rejection of those genetic axioms from which the topology of technological thinking arises. This topology is well protected, if not hidden, by a self-image meant to give comfort to life beyond virtue and the good. The aim to make life always better has crippled the search for the appropriate, proportionate, harmonious or simply good life— claims easily written off as simplistic or irresponsible. Only sober, unsentimental, vernacular rhetoric can possibly demonstrate the incompatibility of mathematical modeling or systems management with the quest for faith and love. The typical artifacts of our decade are at once more intimately and deviously connected to the understanding of revealed truth than hearth or arms or mill, the *res agricola, res bellica,* and *scientia mechanica* of earlier times.

I analyzed schooling as the secularization of a uniquely Catholic ritual because I wanted to grasp the mystery of the *corruptio optimi.* I went into the history of hospitality and care to oppose the Church-initiated sterilization of charity through its institutionalization as service. I wrote on the degeneration of water into H_2O as an instance of the disintegration of bodies and the dissolution of sacramental matter. I got

myself into deep trouble with a pamphlet, *Gender,* on the social history
of duality and its corrosion by sexuality. I wrote that piece, driven by
love for Our Lady who gave birth to that Brother through whom my
fraternity with ... well, a guy like Mitcham, is subsumed in the mystery
of the Trinity. In writing these books, I found the same mysterious
pattern repeated again and again: A gift of grace was transformed into
a modern horror: over and over, the *corruptio optimi quae est pessima.*
Further, I saw that my reliance on the *ancilla* opened scores of unexpected perspectives on the symbolic, ritual, magic, and aesthetic properties that the artifact has acquired by its recent transcendence of
instrumentality.

As you see, I engage philosophy as *ancilla,* not just to avoid blunders
on the path to the good life, but to avoid perverting the Gospel. I engage
philosophy in the late twentieth century—which we may increasingly
imagine, with Ellul, as one all-encompassing artifact—in order to live
in such a way that I go beyond loving my neighbor "just" as myself and
accept the vocation to love him as God enfleshed has done and wants to
do through me.

> To love your neighbor as yourself—that was the rule God laid
> down before the Incarnation; he knew what a powerful motive
> self-love was, and he could find no higher standard by which
> to measure the love of one's neighbor. But this wasn't the "new
> commandment" Jesus gave to his apostles, his own command-
> ment, as he calls it ... I am not just to love my neighbors as
> myself; I am to love them as Jesus loves them, and will love
> them till the end of time ...
> (Thérèse of Lisieux, *The Story of A Soul*)

This is the question that put me to search for those characteristics
in contemporary artifacts that need to be faced fearlessly as issues of
charity. In face of this one assembly today in Los Angeles, I can say these
things without fear of being misunderstood. In this company they are
trivial.

They were not trivial, you can be sure, on those tightropes on which
I must do my balancing act as a teacher. When speaking in Philadelphia
or Bremen, I feel called to shroud my ultimate motive in apophasy. I do
not want to be taken for a proselytizer, a fundamentalist—or worse, a
Catholic theologian; I do not have that mission. Therefore, I do not relate
the unprecedented characteristics of the modern artifact to the new
commandment recorded by St. John, but to the *philia* traditionally
understood as the flowering of *politeia.*

Carl and Lee and I have never found any difficulty presenting a
philosophical inquiry into the nature of the modern artifact as a prereq-
uisite for a dignified, affectionate, passionate life in the 1990s. Our
students show an amazing interest in the practice of *philia,* the more
so, the more clearly they understand the sadness of having lost all

moorings. They follow with surprising attention our doubts about the possibility of ethics in the absence of shared forms of hospitality, and after the loss of respect for the art of suffering. Intuitively, they are ready to grasp our hypothesis that it is *le milieu technique* that conditions, reinterprets, and possibly thwarts the acquisition of ethical habits for a *hexis* shaped by the repetition of good actions. I never cease to be surprised by the readiness of serious students to accept my claim that the philosophical grasp of the nature of technology has become a fundamental condition for ethics in a milieu symbolized by Windows 95.

The ethical awakening in historical research: I have spoken so far from the experience of a medievalist who, year after year, interprets twelfth-century texts on monastic community and friendship, an activity that inevitably leads my students into aporetic bafflement; they experience their own impotence to sympathize intellectually, sensually and bodily with the notions of chastity, humility, prudence, fortitude, and the other virtues we have to translate for them. Simultaneously, many of them are shocked by the amoral sterility of their hearts, livers, and loins when they attempt to address another person as Thou—to borrow an expression from Emmanuel Levinas. In other words, I try to teach philosophy as a discipline of *intellectus quaerens amicum*.

In my seminars, I have seen many a student look up from the exegesis of a passage by Aelred of Rivaulx, Héloïse, or Hugh of St. Victor and search for a correspondence in his or her own twenty-two year-old heart, and recognize what the notions related to process, field, feedback, loop, and context sensitivity have done to their grasp. At such moments of disciplined alienation, it is then possible to foster the insight that it is almost impossible for an inhabitant of "the system" to desire an I-Thou relationship like that cultivated in Talmudic or monastic communities. Following such an awakening and finding themselves at a loss to recapture this past experience, a thirst is incited.

The desire for a self comes into existence through the respectful love of an Other. This longing, characteristic of some in 1996, is utterly different from the spirit of commitment with which the generation of 1968 awoke to a thirst for justice. That generation wanted to atone for privilege by making the Other an object of development, an object of economic, pedagogical, or ideological transformation.

Only ten years ago, when you spoke of the need for a philosophy of things, the conversation veered toward the immanent power of modern objects to polarize society, to destroy the environment, or to control other people. But this is no more. The students I meet have a new readiness to listen to what objects *say*, rather than *do*.

The authority of "science says … " has dwindled. With nausea, many people have become capable of recognizing that participation in systems castrates and sterilizes the heart, enervates ethical sensibility. *Askesis*, which means training in the renunciation of objects, is on the point of becoming an accepted first step toward theory.

Only a few years ago, students in Oldenburg and Penn State were amazed when Professor Illich announced his Science, Technology, and Society course on the tradition of *nepsis*, the guarding of the senses, the *philakia* of the gaze, and spoke of high fidelity and worry about immune systems rather than of buttocks and breasts. I believe that the longing, which I just mentioned, is due to a widespread awareness of the desertification of the amorous faculties. Some speak of erotic expression or the experience of agape, but I prefer the term, *philia*. This yearning struggles for words, and is obscurely related to a faulty stance toward things that have mutated from instruments into systems. Students' readiness allows me to formulate the plea for philosophical attention to artifacts: Chisels and statues, text and layout, communications and systems condition the working of our senses and thereby impinge on the habitual practice of virtue. One can then begin to bring up to date the rules for the discernment of spirits, enabling one to recognize the unprecedented influence of the 1996 artifact.

The nature of an epoch's modal object, the nature of the "thing" from which only the embrace of Lady Poverty can wrench me, cannot be understood at all times by the same categories. In retrospect, this seems to be eminently so in the case of technology or instrumentality. Techology as instrument has dominated western perception for 600 years, but is now fading away. Why? Simply because the moment I perceive the *ob-jectum* as a system I operate, it loses the otherness decisive for the character of tool.

My interest in mediation by tools was awakened by an innovation in the sacramental theology of the thirteenth century. Up to that time, *organon* had meant both hand and hammer and, at times, both of them in conjunction. Only in the time of Aquinas was the "tool" disembedded as *causa instrumentalis*, an abstraction which, arguably, became the labeling category for the macro epoch into which all of us were born.

I first thought that this discovery of a novel mode of causation around 1240 was the result of *fides quaerens intellectum*. It initially seemed that the mental construct, technology, was just one more instance of *corruptio optimi*, one more example of the secularization of a conception that had been coined to interpret the faith. Here, the contemporary horrors of technological manipulation could be traced to a corruption of sacramental theology.

I was wrong. The artifact was an instrument par excellence, not to understand the infallible action of God in the sacraments, but to explain the fallible action of angels governing heavenly spheres. Being spiritual beings, they needed heavenly bodies to govern the world below the realm of the moon. For the scholastics, one of the principal characteristics of the universe was order. It was thought not fitting that God act directly on the world. It seemed appropriate——in proportion——that angels, pure spirits, guide these heavenly bodies. Instrumental causality thus entered philosophical thought.

In the study of theology, ecclesiology was my preferred subject; and, within this discipline, liturgy. Liturgy, like ecclesiology, is concerned with sociogenesis. It inquires into the continued embodiment of the Word through rituals. Necessarily, these rituals often center on objects like tables, tombs, and chalices. So, my interest in these so-called *sacra* led me to the theory of instrumentally used objects. I pursued the nature of the artifact in the belief that understanding would deepen my insight into virtue in our epoch, especially the virtue of charity. Therefore, the love of friendship, *philia*, as practicable under the social and symbolic conditions engendered by modern artifacts, has been the constant subject of my teaching.

II

The technological eye and the ethics of the gaze: To illustrate what I am talking about, let me take the *libido videndi Deum*, and examine how this desire has changed historically in the West. Increasingly, looking has been more and more affected by technology, by what technology *says*.

My historical survey is carried out *sub specie boni*; I explore the possibilities of seeing in the perspective of the good. In what ways is this action ethical? The question arises when I consider the necessity of defending the integrity and clarity of my senses——my sense experience—— against the insistent encroachments of multimedia from cyberspace.

When I pick up "image," and think of it as a mountaineer's rope to climb down from today's paradigm-screen back to Plato's ideas, I notice that the fiber that runs through my hand changes from epoch to epoch. The name by which the image goes, the power it holds, the respect it commands, changes in each iconic regime. The more I study the history of the image, however, the clearer I see how its function and place have changed, and the more forcefully I have been led to three intuitions:

First, the polemical status of the image is a distinguishing characteristic of western history.

Second, dissension about the nature of images has until very recently been experienced as an ethical issue.

Third, the current age of interface, the image that has been a subject of dispute gives way before something new that I call a "show." It is the historian's task to find and weigh the evidence establishing whether show is heterogeneous with what has been called image in the past. This historically distanced view of show is in two ways fundamental for an ethics of the gaze: It is necessary to insert ourselves into the tradition of ethical iconology, and to recognize the totally new ethical challenge that has come into being with the age of show.

When I speak of ethics, I mean disciplined reflection on my actions insofar as these are the source of my habits—my *hexis*, the Greeks would have said. What pedagogues call growing up and psychologists call development, the ethicist understands as the formation of a personal stance, attitude, or propensity which can be made up basically of virtues or of vices.

The ethics of the gaze is important because the *hexis* or "total character" of the person is dependent on the way that person acts. About this I ask two questions:

When did the image become an essential element of the gaze?

How does the "picture-world" of show affect the image within the gaze?

Dealing with image and show, I regard them as a challenge primarily to the viewer, rather than to their creator. I am concerned with ethical rather than political iconology, the formation of habit rather than of milieu. I ask, "What can I do to survive in the midst of the show?" not, "How do I improve show business?"

My theme is a narrow issue; I focus my attention on one major obstacle to the recognition of the gaze as a subject of ethics. As a matter of course, many assume a bond between gaze and image. I think I can show that this bond has a historical genesis; that the "image in the eye" is not a fact of nature but, rather, a constitutive characteristic of one particular stage in western culture. Only by recognizing the historical nature of the step by step bonding of image and gaze that originated in Byzantine iconoclasm can we appraise the moral consequences of vision reduced to an interface between show and gaze. There is an ethics of icons; that of the gaze is a much broader subject.

During the last few years, the literature on the history of scopic regimes has been growing. Those who contribute to it must not be confused, a) with those who study the history of optics, the science that deals with the generation, propagation, and recording of electromagnetic waves that are longer than x-rays and shorter than microwaves; b) with historians of physiology, neurology, and cognitive psychology; c) with historians of philosophical doctrines who take the metaphors of light, sight, and eye as their themes.

Historians of scopic regimes are people who concentrate their attention on the ethology of sense activities in different cultures and epochs. If I had to choose a name for their discipline, I would call it "historical opsis" to distinguish it from the history of optics. *Opsis* is the Greek work for gaze, and is a verbal noun. It bespeaks a human activity analogous to speaking, walking, eating, and listening. To gaze is a widespread action. While only some people swim or hunt, all look and see. Like other activities that can become the subject of a historian's attention, the gaze is the outcome of a natural endowment, the result of just growing up. In some times and places, the gaze is also the subject of reflection and training. When I speak of the asceticism or training of the eye, I mean much more than the apprenticeship of Zen archers, skeet-shooters, mystical navel-gazers, or the downcast eye of Victorian spinsters. Prussian civil servants, no matter their profession, had to pass tests in calligraphy and draftsmanship. Even in my own childhood, drawing was still part of the distinguishing skills; it trained the eye as music the ear and dancing the gait. Under the tutorship of a widow from Bremen, I had to paint flowers and views to improve my attention. Each age, craft, and milieu places its own demands on ocular techniques.

Not only the acuity, but also the moral quality of the gaze was trained. One was admonished not to stare. At puberty, Jesuits taught us to guard our eyes. It was part of good behavior to know how a boy may look at a lady and when he is supposed to look away. More than that: As Catholics, we were trained to experience some looks as defiling, and to mention them in the confessional. Even today, I feel guilty if I find my attention distracted from a medieval Latin text by the afterglow of the MTV to which I exposed my eyes. Until quite recently, the guard of the eyes was not looked upon as a fad, nor written off as internalized repression. Today, things have changed. The shameless gaze is in, but I am not speaking of leering at soft porn or sado-masochism.

The information age incarnates itself in the eye. Speed reading, pattern recognition, symbol management are part of elite skills. All this information grabbing and coding is only faintly related to the ethical cultivation of the gaze. The contemporary paradigm is instrumental: The eye is trained to compete with Word Perfect's search command. The eye is entrapped in an interface with Microsoft Windows icons, and modern eye training cuts the gaze down to a form of scanning.

Dozens of words for shades of perception have lost their meaning. Your glances can still be called leery, dirty, or kind; but hardly in textbooks of physiology. The words that qualify the gaze are now taken as metaphors. Formerly, a penetrating, dark, luminous, menacing, kind gaze had distinct powers. And some people in Mexico are still fearful of the *mal de ojo*, the evil eye. I have difficulties explaining why Medusa, the Gorgon, with her empty eyes should be taken as symbol for interface.

I acknowledge that most people take images as a natural given. They do not distinguish the interocular product of digital programs from the image formation solicited by a painter of old. Informaticians share this naïveté with semioticists, cognitive scientists, and a considerable number of philosophers. It is the main obstacle preventing one from following the route on which the image mutated to the point of becoming a trap for the gaze. I argue that this entrapment has a history, beginning in a complex adventure and now reaching the stage of a *ménage à trois*: At times our gaze is still solicited by images, but at other times it is

mesmerized by show. An ethics of vision would suggest that the user of
TV, VCR, MacIntosh, and graphs protect his imagination from over-
whelming distraction, possibly leading to addiction. There can be rules
for exposure to visually appropriating pictures; and exposure to show
may demand a reasoned stance of resistance.

For several hundred years, "to see" has meant to visualize. The act
of "making oneself a picture" in the eye of the brain has been taken as
a neurological given. This identification of vision with inward visualization must be recognized as a crucial achievement of European modernity. Understanding this, one can see that the replicative interiorization
that results from interface with a show is something quite unique.

The distinction between image and show in the act of vision, though
subtle, is fundamental for any critical examination of the sensual
"I-Thou" relationship. To ask how I, in this age and time, can still see
you face-to-face without a medium, the image, is something different
from asking how I can deal with the disembodying experience of "your"
photographs and telephone calls, once I have accepted reality sand-
wiched between shows. I argue that show stands for the transducer or
program that enables the interface between systems, while image has
been used for an entity brought forth by the imagination. Show stands
for the momentary state of a cybernetic program, while image always
implies *poiesis*. Used in this way, image and show are the labels for two
heterogeneous categories of mediation.

The radiating gaze: Euclid's *Tà optiká* (300 B.C.) can be read as the
ethical complement to his much better known geometry. The book deals
with rays emitted by the eye. These rays are something for which both
words and sense have been lost. I cannot avoid dealing with this gaze
when I speak to my students about medieval friendship. Freud has made
it difficult for them to grasp how Sister Diana of Verona could embrace
Friar Jordan of Saxony, doing so with chaste glances. They tend to
attribute Voodoo deaths in Haiti, which are inflicted by the *bokor*'s
fulminating eyes, to hysteria, not to the striking power of the gaze. It is
hard for many people today to experience their own gaze as an offensive
touch. Modern vision is something that happens to me, not to her whom
I see. When I speak of the visual cone of antique opticians, my students
tend to misunderstand it in the likeness of a flashlight that lights up
the visitor at the gate. Almost inevitably, they think of radar. The carnal
transcendence of body limits by the antique visual cone just cannot be
reduced to these electromagnetic similes and metaphors.

From pre-Socratic Alcmaeon via Plato, Aristotle, Epicurus, Euclid,
and Ptolemy, well into the High Middle Ages, all those who deal with
optics make this ocular effluent into the subject of their study. Their
object is not light but a trans-pupillar emanation. Since Euclid, they
construe the shape of this corporeal prolongation as a cone, and vision
as an outgoing activity. What interests the opticians is the fusion of this
transcending flesh with the color in the object; they do not deal with the

light reflected from the thing and that strikes the eye.

Ópseis can be thrown; English throw glances; the French can "jeter un coup d'oeil." Looks can set objects aglow, in analogy to the sun, or like a candle into whose wax the bee has gathered the sunlight. Homer and Aeschylus compare the human eye to the sun, whose light kindles color and life. Where the sun cannot reach is Hades, where only shadows dwell. There are many images used for the way these rays work. Alexander of Aphrodisias compares them with sticks. Hipparchus compares them with fingers. They are referred to as *psycho podia*, the limbs of the soul.

In spite of the distinct theories different schools held about the way these rays work, this organic extroversion of the eye is a common assumption in all of them. For Plato, the gaze never reaches reality; it fuses with the color from the thing somewhere halfway, between the glimmering eye of the cave dweller and the blazing light of the idea. These philosophers all deal with a human deed, an activity—the glance—and not with the reception of light; it is the glance through which *visibilia* come to be.

Visibilia are as foreign to our optics as the visual ray. To explain them I rely on Aristotle. When he speaks about vision, he establishes three conditions. There must be an object that under the brightness of the sun shows colors. These colors belong to the thing, and are brought forth from within it by the light that has struck it. So light is a second condition for sight. It is not that which we see, but the solar energy that brings forth the colors from the object. The third condition is the existence of translucent media like air or water, or the crystal in the eye. These *diaphaná* are the opposite of a canvas. They are media tinged by the color that the visual ray has grasped, and allow the soul to be colored.

Given these three conditions, the glance turns into the vision of something that Aristotle calls *emphasis*. What we see are these epiphanies, the revelatory manifestations of the world in the eye. What appears are *visibilia*, those qualities of the world that correspond to the sense of vision.

Visibile and *emphasis* are strangers in our TV world. In that Greek world things themselves have a quality that corresponds to the eye. This is the opposite from the picture world that techniques, all the way from woodcuts to hypertexts, have brought into being. Greek vision presupposes a connaturality between the eye and things. Vision in this self-manifesting world is a form of contemplation. The eye is made to see everything that can show colors.

Aristotle lacks an equivalent for picture. Image, as we take it today, implies some kind of representation, facsimile, or formal equivalent. It can be like a sketch or photograph, a sign, an emblem, an isometric or perspectival illustration. But it is always a medium between the thing and sense perception. The Aristotelian *emphasis* implies no conformity in shape; it is not light that radiates from the object, or that is reflected by it. Even less has it something to do with mapping in the brain. It does not affect nor is it affected by the *diaphanon* of the air, water, or crystal through which it comes, emphatically non-instrumental media. It stands for a non-mediated appearance of the world's hues.

In its perfect Euclidean uniformity, the visual ray itself became a critical scientific instrument, a final adjudicating rule, as much for mathematical astronomy as for geometrical optics. Through its own inviolable rectilinearity, it is our one sure inductive link to the inviolable, intelligible nature behind the visible appearances in both celestial and terrestrial realms. So, by permitting us to elicit the inherent, perceptible truth behind the *visibilia*, whether cosmic or mundane, the visual ray provides the necessary warrant of certainty that legitimizes and universalizes scientific concepts, while "saving the appearances."

The image: Occasionally, image (*eikon*) could be used not just for the bust of the Emperor or the stamp of a signet ring visible to the outer eyes, but also when speaking of inner perceptions. However, even when the word is used for perception with this spiritual sense, the *eikon* designates the figment, and *phantasticòn* the real, the *emphasis* that appears with closed eyes. When Plato or the neo-Platonists use image as a technical term, they deprive it of all factual and sensual qualities and use the word to designate a relation. Pagan antiquity did not elaborate a theory about the picture.

The first well-rounded iconology we owe to the Greek Church Fathers. They needed a theory about the *eikon* to interpret key passages of Paul in his letters to the Colossians and Corinthians where Christ is said to be "the image of the invisible God" (Col. 1.15; 2 Cor. 4.4). The exegesis of such a statement forces them to reflect upon the concept of image. Christ as image is not something comparable to the technical product of a seal; nor is He the natural result of an act of generation like a son, nor the fruit of an artist's imitation of the appearance of the Emperor in wax, color, or marble. His very being in the flesh is a likeness to the splendor of the Father's glory.

And not only Christ, God-in-the-flesh, is substantive image. It is written in Genesis, the first book of Moses, that God created man "according to His image and likeness." Among learned Christians in Asia Minor, iconology becomes as fundamental for thinking about man as thinking about God. This turn toward a philosophy of the image takes place shortly after Ptolemy wrote his a-iconic optics in Alexandria. Iconology becomes a foundation for Christian ethics: The human being created in the image of God is now assigned the task to grow in the likeness of Christ. During the centuries after Ptolemy, the last classical optician, the onset of a new scopic epoch in the High Middle Ages brought two important changes: in the meaning given to light as a metaphor for truth, and in the birth of iconology.

In 726, the Byzantine Emperor, Leo III, won a battle that stopped the advance of Islam through Asia Minor. Right after this victory over the notoriously iconoclastic Muslims, he went to the Bronze Gate of his palace, removed the image of Christ enthroned above it, and replaced it by a simple cross. With this ceremony he started a fierce debate that raged, sometimes with violence and war, for several generations. Its issue: Can Christians bow and pray before images? The Emperor's party, the Iconoclasts, held that this was idolatry, worship of creatures, unworthy of the martyrs who had died refusing to light incense before the bust of earlier (pagan) emperors. The Iconodules held that the cult of images was a legitimate form of piety, a devotion and liturgy customary since the origins of the Church.

John of Damascus—the most articulate defender of Christian cult images—prevailed in the Second Ecumenical Council of Nicea (787). He distinguished the Christian icon from the pagan statue. Through the latter, the presence of a person or god is conjured. He further distinguished it from the mosaic or fresco in which an artist lets others see what he has fancied internally. An icon, so the Council says, is a form of revelation: the light of Christ's resurrected body showing itself. The icon is like a threshold beyond which the devout eye reaches into the realm of the invisible. For the believer, it provides color to the Truth which he has accepted and come to know through his act of faith in the Word of God.

The regime of the radiating object: The rise of iconological thought in the Christian Mediterranean was complemented around the year 1000 by a Muslim breakthrough in optics. Like Euclid and Ptolemy, Hakim Ibn al-Haytam wrote in Alexandria. He was a mathematician, physician, and astronomer. To be able to observe an eclipse, he transformed a tomb into a camera obscura. By reflecting upon the sun's image on the dark wall, he concluded that we see not what our gaze grasps from the object, but light reflected from the object which reaches the eye. His writings, known as the *Alhazen*, had a decisive influence on Latin Scholastics during the thirteenth century. Peckam, Bacon, and Grosseteste accepted the action of light upon the eye which al-Haytam suggested. However, and this is the crucial point in the history of vision, medieval opsis preserves the immediacy of the gaze.

Gothic light is not a painter, it does not generate an image in the eye. This appears strikingly in a treatise, *De oculo morali*, written by a physician, Pierre de Limoges, contemporary with the optical treatises of Roger Bacon and Grosseteste. Though the little book elaborates the new optical theory, it still stands in the tradition of ethical optics. Its friar author deals with the physiology of vision in order to clarify the moral duties connected with this human activity. He demonstrates how the *virtus visiva* (the power of sight) descends from the front lobe into the crystal of the eye to welcome and embrace the incoming light. He deals with the right dispositions needed to welcome true light and the need to resist temptation by illusions.

To interpret *De oculo morali*, the relationship of things to God "who is light" must be understood. This is the century suffused by the idea that the world rests in God's hands, that it is contingent on Him. This means that at every instant everything derives its existence from His continued creative act. Things radiate by virtue of their constant dependence on this creative act. The illuminated pages of medieval manuscripts show things in this way, glowing from within and casting no shadows.

Contingency is one of the few concepts that are of specifically Christian origin, even though the term is derived from a Latinization of a concept in Aristotelian logic. Contingency expresses the ontic state of a world that has been created from nothing, is destined to disappear, and is upheld in its existence through the divine will, a state that is measured by unconditional and necessary existent Being. The world's very existence assumes the nature of something gratuitous, something that is a grace.

The coming to existence of the antique cosmos was in no way dependant on the act of someone's will, neither in its genesis nor in its continuation. However, since Augustine answered the question, "Why did God create the world?" with the *quia voluit*, because he willed it, the world's existence is the result of a sovereign act. As a consequence of this conception of contingency, we then find the scholastic real distinction between essence and existence, which also indicates the structure of the whole cosmos.

The beginning of modernity coincides with the attempted breakout from a world- and self-view defined overwhelmingly by contingency. With Bruno, contingency loses its roots; an infinite cosmos becomes a correlate of the infinite God. Since Descartes, a different logic leads in the same direction. Each being now finds in its nature a reason and claim not only to existence but also to being what it is.

The third scopic epoch: Only during the early Renaissance does the gaze turn pictorial. One way to illustrate this innovation consists in following the meaning of a word, *perspectiva*. Boethius introduced it to translate *Tà optiká*, and in this instance, as in so many others, his vocabulary served as a foothold on the farther shore for medieval Latin. Even more powerfully than the Greek, the Latin neologism stresses the liveliness of the gaze. *Perspicere* means "to look with attention," to examine, to look into or through. This is not the meaning, however, which the Tuscan painters after Giotto gave to the word. The medieval *perspectiva naturalis* had been understood as the *ars bene videndi*, the art of the skilled gaze. The Florentine painters redefine it as *perspectiva pingendi*, an *artificium*, or visual technique. They no longer insist in the same way on the ascetical training of a virtuous gaze, rather, on the aesthetic skill which enables the painter to transform the three-dimensional view through a window into its framed optical facsimile on a wall or canvas.

At this point, the image was transformed from an object into a geometrical construct. In today's language, we might say that the *perspectiva artificialis* rests on a tomography of the visual cone. The image which the painter is to transform into a picture is visualized as a cut through the visual pyramid. Leon Battista Alberti creates a technical device that has had immense symbolic consequences: An optical scalpel which, arguably, has been more important for the transformation of the gaze than the telescope or microscope two centuries later. This device is a wooden frame filled with a transparent screen covered by a grid. Albrecht Dürer, in a famous woodcut, shows how it is used. The draftsman places this optical knife between his drawing table and a woman who lies on a couch in front of him. The artist's chin is fixed by a support and his left eye covered. Thus immobilized, and reduced to monocular vision, he surveys the woman's anatomy, as well as the folds of the couch's robes, square by square. In front of him, on the table, lies a sheet with a grid that corresponds to the instrument (a *velum*), and with the pencil in his right hand he traces his observations in square after square. The image that results from geometrically slicing the visual cone becomes an optical facsimile called "a picture."

Most historians ascribe the transition from the naked to the armored eye to the action of the lens in the mid-sixteenth century. I prefer to follow those who place the transition 150 years earlier, and assign it to the new *perspectiva artificialis* made possible by two techniques: monocular, linear perspective, and the practice of shading through which attention is drawn to incident light, depth, and the passage of time. Art academies train the new gaze, and the eye comes to be experienced as another piece of equipment. The notion of image is wedded to that of vision. The painter is aware of the image that his picture will generate in the viewer. Since Alberti, this image has been located at the place of the *velum*—between the eye and the object.

The screen on the farther shore: Under the regime of the picture, the paradigm of vision remained for several centuries the image on the wall. By the late eighteenth century, the Claude glass comes into fashion among tourists. This was an object that looked like a cigar box and allowed the sightseer to turn his back to the view and to observe a section of it well framed and mirrored. Finally, there comes Daguerre's camera that fabricates a picture in the absence of an image. In the obscurity of the chamber, vision arises from the union of picture and gaze. The camera then became an emblem for the eye. At first, it is truly a chamber, a darkened room into which the spectator enters. Later, the camera lucida is a device that projects the image of an object onto a surface where it can be contemplated and traced.

It is tempting to speak about the day of the first Daguerreotype as the birthday of the modern age. However, I argue that the camera was then and remains now the pivot for the survival of the third scopic regime within the fourth, and that one of the most fascinating and least explored aspects of the last two hundred years is the coexistence of two heterogeneous scopic forms.

Some art historians speak of scopic regimes that are replaced one by another, for example, when Rafael is followed by Carravaggio. To the cultural historian of the gaze, however, no scopic regime ever achieves perfect monopoly. Past forms of gazing survive; and the survival might be group-specific. The touching gaze is still with us, marginally, like a horse and buggy in the age of the car. An imageless gaze at my friend's face can be cultivated only through a continual guarding of the eyes; it has become a fought-for ideal that I can pursue only by constant training, behavior that runs counter to the surrounding images that solicit me to deliver myself to the show.

The people among whom I live, more often than not, are armed--not just with cameras but with camcorders. The symbolic effects of recording as an activity, medium, and object have given rise to a literature that is even vaster than the literature on the symbolism of cars. But only rarely is our dependance on the act of recording understood as a crippling of the gaze. The gaze comes to be understood and, further, actually sensed, as a digital process. Surreptitiously, the act of recording becomes a filter that dulls the light in the other's eyes and removes me from her fleshly presence. What I call the "invitation into the show" lets the image in the gaze fade.

In antiquity, the eye had been the criterion or mirror for truth. In the Middle Ages, it was endowed with the narrower power to extract universal essence from flimsy shapes. Since the Renaissance, it has been the model for making and interpreting pictures. In front of the screen, it has become a gateway for moving into the show.

With the transition from the age of pictures to the age of show, step by step, the viewer was taken off his feet. We were trained to do without a common ground between the registering device, the registered object, and the viewer. The replacement of the picture by the show can be traced back into the anatomical atlases of the late eighteenth century. Beginning with the fetal representations by a Dr. Soemmering, in 1798, anatomists looked for new drawing methods to eliminate from their tables the perspectival "distortion" that the natural gaze inevitably introduces into the view of reality. They attempted a re-education of the scientific gaze that would then "see" every bone as through a tele-lens. Even the tiniest organ, which in reality the anatomist has to inspect by holding it close to the eye between his two fingers, is now shown in the new type of atlas orthogonally, as seen from infinity. The specimen in the textbook is removed from the space into which fingers can reach or feet can walk, and placed on a farther shore.

The new naturalists want the object to be shown as in an architectural blueprint; they want measures, not views. They do not want a facsimile of vision, but an isometric plan of things. They want to adorn their textbooks, not with views, but with maps. They want to show a biological specimen, not as it looks when you hold and turn it in your hand, but as you can photograph it with a powerful telephoto lens.

The old image was a geometrical construct resulting from the cone of light rays reflected from the surface of the object to the lens in the eye. The new show is a technically generated display that records the measurements taken with an instrument. It is the result of a program that transforms binary columns into an arrangement that fits some prejudice in the mind of the viewer. I want to distinguish this manipulated arrangement of shades and colors from Alberti's *perspectiva artificialis*. I call it a show to distinguish it from the image.

With the nineteenth century, a new scopic will affirms itself outside of medicine, too. Not only anatomists, but geologists and zoologists also condemn the mode of illustration that fills the textbooks of the eighteenth century. A new technique, wood engraving, makes finely-drawn, cheap book illustrations possible. Draughtsmen galore are trained for blueprinting reality, leaving it to the viewer's eye to imagine the model. Thanks to the new printing techniques, the study of nature increasingly becomes the study of scientific illustrations.

In textbooks, as well as in news magazines, the graphic show first encroaches upon and then overwhelms the text. Further, the eye is trained to take in objects that in nature are invisible. Molecules smaller than the shortest visible frequency are made to appear. But, even more importantly, abstract notions are given "shapes" in tables and charts that seduce the eye toward misplaced concreteness. We are trained to be horrified, anguished, or encouraged by the graphic representation of quantitative data to which nothing corresponds that the gaze could grasp: Gross National Product, population growth, the incidence of AIDS.

In this world, to speak of a human act sounds almost anomalous. Further, to think about the good would appear outlandish. I can call for help, however, from someone recently deceased, Levinas. Throughout most of this century, Levinas resisted the dominant trends of visualization: the disembedding of vision from synaesthesis, the disembodiment of the eye by interpreting it as a built-in camcorder or an abstract sex organ, the dissociation of the gaze from love. Levinas clearly strikes through the non-sense of the show-world. A certain conservatism allows him to speak of an ethics of the gaze. Levinas, stemming from a Lithuanian Jewish family of Talmudic rather than Hassidic traditions, was raised on the interpretation of the Cabala. In his philosophical teaching and writing, however, he insisted that his point of departure was non-theological.

As a contemporary philosopher, Levinas establishes the mutual gaze of two persons as the source of personal existence. He stresses the uniquely Jewish hesitancy in front of anything iconic as a condition for ocular intercourse. Such a disciplined hesitancy—disciplined or practiced, since it is outide the historic experience of most western non-Jews—seems to be the initial necessary step for an ethics of the other among a generation consistently robbed of its eyesight by ocular integration into virtual realities.

Levinas's stress on clear sight and his analysis of the gaze are of crucial importance for me. His work suggests a new look at the *libido videndi*. The face of the other stands at the center of his life's work. He does not speak of my own face, which appears reversed in the mirror; nor of the face a psychologist would describe. For Levinas, face is that which my eye touches, what my eye caresses. Perception of the other's face is never merely optical, nor is it silent; it always speaks to me. Central in what I touch and find in the face of the other is my subjectivity: "I" cannot be except as a gift in and from the face of the other. This face of the other cannot be made the subject of a phenomenological description, and by this route given sense and meaning. What the face of the other does in its exquisite delicacy and impenetrability is to address me forever in an ethical way. As he puts it: "I cannot but hear the face of the other in spite of the profound asymmetry between our faces." Again and again, Levinas repeats, "You see and hear as you touch." He shows how the tactile gaze of the other's face is the place where I discover myself. Ethics can once again begin here building on a philosophical analysis not only of human action but also of human interaction with artifacts and of friendship.

Pennslyvania State University
  University Park, Pennsylvania

Architecture: The Confluence of Art, Technology, Politics and Nature

by Robert E. Wood

I

Since building is common to men, birds and some insects, human beings are not distinguished from other animals by the fact that they build, but by the fact that they build, beyond simple function, creatively and meaningfully. By creatively I mean that architects construct different building types and in vastly differing styles, not being tied to precedent, not having the urge to specific form "imprinted onto" their physiology as a species-habit. By 'meaningfully' I refer to the way building fits into a world of inhabitance.[1] Such a world is founded upon the bipolar structure of the field of human experience constituted, on the one hand, by the limited manifestness of a highly selective sensory field serving biological need and, on the other, by an empty reference, via the notion of Being, to the encompassing totality. The latter makes possible and necessary universal description, world interpretation and choices leading over generations to the construction of a human lifeworld, a *traditum*, delivered from out of the understandings and choices of those long dead. Such a lifeworld opens up ways of thinking, feeling and acting for those born into it. It is precisely because we are referred to the whole that we are able, in artistic work, to bring into being the new and that the sensorily encountered can become an icon of the whole, so that it is not only positivistically "there" or even functionally interpreted; it is also *symbolic*. Furthermore, because, by reason of our reference to the whole, we can back off from the purely functional outside us as well as from the organically desirous within us and can learn to appreciate the togetherness of the sensory display for its own sake, we can appreciate beautiful forms in nature and produce them in art.[2]

Art is present in various places in a given lifeworld. But the most pervasive artform of all is architecture. All the artforms are found within or in relation to buildings. From time immemorial, wherever humans dwell together we find architecture as expression of the art of building. In our everyday life it is inescapable: we live in buildings, work

79

in or between buildings, are educated and entertained in buildings, worship in buildings, make our public decisions, attend conventions and perhaps also listen to lectures on architecture in buildings. Architecture is indeed the most pervasive artform.

Architecture as a fine art not only sets the context for the arts-in-general, it also requires of the architect the aesthetic sensitivities of the other plastic artists. It requires the eye of the painter to provide an aesthetic arrangement within a given perspective. As Ruskin would have it, "a wall surface is to an architect simply what a white canvas is to a painter...."[3] Further, in relation to the wall surface, the architect is a relief sculptor, sensitive to the effect of shadow in giving form to the surface. Architecture likewise requires an eye for sculpture-in-the-round to give plastic depth and thus coherence to the indeterminate number of perspectives from which one can view the building. In addition, however, to these features which the architect shares with the painter and the sculptor, the peculiar province of the architect is the handling of enclosed space as it plays in relation to single and multiple perspectives.[4] In the twentieth century, Walter Gropius, Frank Lloyd Wright and Siegfried Giedion share that view with Antoine Pevsner.[5] Louis Kahn sees the province of architecture as light, which is that which makes the space appear in its play with the sculptural and painterly aspects of the enclosure.[6]

The aesthetic dimension here is closely linked, however, to other features. Most basically, architecture has to fit the ends for which the building is constructed so that architects have to understand the concrete operation of things human: they have to be students of human nature. Moreover, in order to fit those ends, architecture requires the know-how to construct something that will stand over time: the architect must be an engineer. It is commonplace to distinguish architecture from building insofar as the latter is satisfied in producing an enclosure which provides protection from the elements.[7] Because it is tied to function and because, by reason of the functions it deals with, it is concerned with the construction of larger-than-human objects, architecture as an artform has the greatest number of natural restrictions and thus of technical know-how. It requires geological, meteorological and engineering knowledge: knowledge of geological substructure and of general weather conditions in a given territory, knowledge of the properties of materials, of load-bearing capacities, of stresses and strains, of conductivity and insulation, of acoustical properties and the like. To that extent, as Frank Lloyd Wright and Le Corbusier observed, the architectural engineer puts us in touch with the principles of the physical cosmos.[8] Although we must add that there is a difference between *using* physical principles and *showing* them.[9] Engineering knowledge is a necessary, not a sufficient condition for architecture which brings building into the arena of the fine arts.

Bridging the division between the fine and the useful arts, architecture is able, within the limits of structural stability, to elaborate aesthetic form in tandem with suitability to the ends it serves. Thus three fundamental architectural principles were enunciated by Vitruvius, the basic source of our knowledge of classical architectural theory: *firmitas* or stability, *convenientia* or utility and *venustas* or decorousness.[10]

The Vitruvian triangle expresses abstractly features whose factual functioning rests upon an historical context which defines social-political functions and upon a tradition of accumulated engineering skills and architectural styles. Further, these abstract features appear within the givenness of the lifeworld. (We will return to the latter in our last section.) Architectural historian Peter Collins claims[11] that none of these three features can be rejected entirely, though deconstructive architecture has taken up the challenge and produced deliberately dysfunctional, unintegrated and even—at least to an extent—unstable forms.[12] And, of course, there is the ever-present warehouse or toolshed which can scarcely claim decorousness. The history of architecture in modern times is in part determined by emphasis upon how *venustas* relates to *stabilitas* and *convenientia*.

In the sections which follow we will first lay out certain components of architecture by looking to a kind of ideal genesis of building. We will then offer some preliminary considerations of the development of building types in terms of the articulation of social-political functions, in terms of the symbolic character of building linked to those functions, and in terms of the way beauty of structure and ornamentation play a role. This will involve consideration of how the often invoked adage, "Form follows function," plays out in relation to the Vitruvian triangle. We will conclude with some suggestions for assimilating certain Heideggerian themes regarding the inhabitance of a lifeworld involved in architecture which sets the larger context for the Vitruvian features.

II

In a sense, architecture as building begins with an imitation of nature serving our animal needs, reproducing the naturally protective character of the cave through the assembly of durable materials. In the earliest phase of its development it seems to have taken three basic simple forms: the cone and its cognates represented by the Indian teepee on the one hand and the pyramid on the other (though the pyramid was more monumental sculpture than architecture); then there is the half-sphere and its variations represented by the Indian lodge; and finally we find the quadrangle in post and lintel construction. The pillar developed from the tree trunk used to support a roof. The cylindrical form this exhibits was later used for towers. Thus in the middle of the eighteenth century, searching for new architectural forms that would suit the modern world, Laugier went back to the primitive hut to recover the basic elements and Fournay sought a regeneration through geometry and its elementary forms.[13] Le Corbusier, paralleling Cezanne's observations in painting, claimed that the elements of architecture, abstractly considered, are the sphere, cube, and cylinder, the horizontal, vertical and oblique.[14] The building, formed out of variations on these geometrical forms, is related to the earth upon which, within which or over which it is set and out of which it is made. It is related to the sky as the spatial surround into which it reaches. (We will return to earth and sky more specifically in our final section.)

The act of erecting a building not only protects what happens inside the building from the outside, but also, from the indeterminate surround of space it carves out an interior, and from the indifference of empty space it gathers, it charges. As carving out interior space, it allows things to take place—an expression suggesting both a spatial and a temporal feature.[15]

Allowing things to take place by establishing an enclosure for protection, a building requires entrance and exit. The primary entrance/exit provides further spatial orientation, so that we have up-down, back-front, and—linked up to human bodily orientation—right-left as primary directions in otherwise indeterminate space. The entrance/exit establishes a face for the building and much of architectural art has been devoted to the articulation of that face, rendering it both expressive and beautiful.

The relation between inner and outer may be more or less open. Less open, a building requires inner, artificial illumination; more open, it has its walls penetrated by fenestration of a smaller or larger character to allow less or more of natural light. There may also be an interior relation to the natural exterior when the interior surrounds a space open to the sky establishing a courtyard, an interior within the interior which is also an exterior. The courtyard is open upwards but not outwards.

In its external relation to the natural surround, a building requires a path or paths and thus becomes a center establishing direction. In actuality, cities originate more by clustering buildings along paths formed by those traveling toward some navigational center—most frequently a center formed by the confluence of river and sea. The paths wind along rivers and streams flowing between hills and mountains and pass through the lowest gap in high elevations. Buildings are oriented with respect to those paths and that general surround which constitutes the *genius loci*, to which ancient architects were especially sensitive.[16]

In relation to other buildings the erection of a new building establishes a certain charged interspace and allows public things to take place. The public equivalent of the courtyard would be the square or piazza created by the arrangement of buildings which define the open space. In this case the space is also penetrated by streets or walkways which provide entrance and exit.

Relation to the surround has a different character depending upon the character of the natural environment—be it flat or hilly or mountainous, be it rich or poor in flora. The quality of light in a region combined with technological development suggest different types of fenestration. The character of the seasons also makes a difference. The latter, for example, makes the flat roof functional in drier and warmer climates and the pitched roof functional in wetter and colder climates—the colder the climate the more pitched the roofs to let the weight of accumulated snow run off diagonally rather than bearing down directly upon the roof.

As cities develop, the relation to the natural surround diminishes when houses immediately rise at the edge of streets and walkways. Then, nature tends to return privately in the form of courtyards with flora and publicly in the form of parks and boulevards. The rich develop walled-in gardens or country estates with grass, trees, shrubs and flowers. In modern America especially the suburban house stands back from the street, separated by grass and surrounded by shrubs, trees and flowers, with a place to grow vegetables as well. Natural and man-made form a synthesis to establish location.[17]

III

What we have considered thus far sets the most general natural context for the building function and the relation of the building to its surround. It says nothing of the different social-political functions which develop over time and of the different building forms correspondent with them. It says nothing of the symbolic character of building nor of beauty, whether structural or ornamental, in which architecture as a fine art culminates. Let us fold in these considerations and establish thereby the role of technological development in effecting architectural style.

The articulation of different common functions over time, working in tandem with the development of technology, required the introduction of different building types. We might consider here two of those developments. One of the major foci of architectural art throughout the ages has been religious architecture. In the High Middle Ages the development of the rib-vault combined with the flying buttress made possible the virtual elimination of load-bearing walls in the upper portions of the medieval cathedral. This invention met with the development of the leaded anchoring of stained glass segments to permit larger masses of glass—virtual glass walls—through which abundance of light could stream. This, in turn, opened up expressive possibilities that were tied to a certain understanding of the place of human beings in the cosmos. The cathedral made possible the gathering of large numbers of people for worship. As the Bible of the illiterate, through statuary and frescoes as well as through its general form and decorative motifs, it taught people their place in the scheme of things and set the dispositional tone for responding to that instruction. Contrary to the Greek temple whose open porticos allowed for a viewer to see from the outside the statue of the god within, and whose dominant horizontality emphasized belonging to the earth, the medieval cathedral, through the recession and decoration of its doorways, invited the worshippers within and closed off the interior from the exterior. By means of the rib-vaults, the eye was directed upward toward the soaring heights above and culminated prayerfully in the pointed arches. Through the clerestory windows light, transformed by the stained glass, shown from above. The cruciform shape of the groundplan whose dominant axis was underscored by the interior walkways drew the worshippers forward towards the altar as the termination of a journey. The gathering of the whole together in a rhythmic and proportionate manner brought engineering skill, religious function and artistic expression together into a symbolically powerful whole.[18]

A whole world separates the cathedral from the modern skyscraper—our second example of the impact of technology upon building structure and function. Louis Sullivan, one of the fathers of the modern skyscraper, approached its construction with the adage that has since become the watchword of the so-called International Style: "Form follows function."[19] As Sullivan viewed it, the form a building takes should be determined by the social function it is meant to serve. In the case of the coming to be of the modern skyscraper, congestion and high real estate values provided the socio-economic context for the high-rise building. The social functions dictated the form it would take. Technology, however, would set the limits within which everything could occur. In the late nineteenth century we find several technological developments at hand: the development of steel construction, ferro-concrete and plate glass, electric lighting, central heating (and eventually also air-conditioning), along with the invention of the elevator, all converging by reason of the need for handling the concentration of large numbers of people in relatively confined land masses. Without steel construction and within confined land conditions, the thickness of foundations being in direct proportion to height, there was a certain natural limit to the height of buildings. Until the use of steel, the high-rise building, under the limiting lot conditions of a modern city, could only rise to some ten stories, with the walls at the base twelve feet thick. But since in earlier times they did not have to arise within the cramped confines of the modern city, the dome-and-pillar construction allowed St. Peter's in Rome to rise to a height of over 450 feet [held together with an iron chain], which could easily include a 30-story sky-scraper, and any of the Gothic steeples. With steel-girder construction, greater heights could be achieved within relatively narrow boundaries without unduly encroaching upon the space available on the street level. At the same time, elevators made possible rapid access to the upper floors. Steel and glass construction established new open relations between inside and outside. The ability of steel I-beams or steel reinforced concrete beams to span larger areas led to the development of interior and exterior non-load-bearing walls. This provided flexibly adaptable interior space through the removal or addition of walls, so that the form allowed an indefinite variety of possible functions. It also involved a relation of openness between inside and outside hitherto virtually impossible and thus changed the relation of a building to space. The convergence of these technological developments with social need brought into being the modern skyscraper.[20]

Specific functions dictated the overall form the buildings would take. A basement contained boilers and the like; the first and second floors would service customers walking in from the street; the floors above would contain offices, the top floor reroutings for the heating (and later air-conditioning) system, the elevator and the like. A major entrance gave the building a face and orientation. The first two floors were often "set upon the earth" by the convention of rusticated stone. The first floor would present a light and airy welcome by higher ceilings and large plate-glass windows, while the second floor would be readily accessible to customers by stairs. The floors above, since they all served the same office function, would show an identical exterior. The top floor would present a different form and cap off in an elegant way the building's relation to the sky.

Next came the question of the proportions of each element in relation to the whole and on top of that the question of decoration. From Vitruvius through Alberti to Le Corbusier, proportions were established by selecting a module as a basic unit of measure derived from the measurements of the human body, and putting it through various manipulations of halving, quartering, doubling, etc. Taking the human body as the basis for the module established a feature of the overall form which followed the general function of serving the human being.[21]

After the question of proportion there is the question of ornamentation. One who expects in Sullivan's architecture samples of the way the International Style understood his "form follows function" will be startled at the way ornament covers Sullivan's buildings.[22] His own ornamentation was based upon a loving study of vegetative forms, so that those who viewed his architecture would be reminded that human functions take place within living nature. In this he hearkens back to the medieval cathedral with stylized vegetative motifs constrained to follow the lines of the building and the ordering rhythms imposed by the architect.[23]

Nonetheless the fanatical rejection of ornamentation, proposed by Adolf Loos in his *Ornament and Crime*[24] and turned into dogma by the International Style, was born when Sullivan publicly bemoaned the stylistic eclecticism exhibited by his contemporaries. The architects of the time employed ornamental forms taken from copybooks of historical precedents in such a way as to obscure rather than enhance overall structure and to pervert completely the symbolic meanings that adhered to such forms in the past. Sullivan himself had bemoaned especially banks and libraries made to look like Greek temples. He suggested refraining from ornament entirely for a period of years so architects could concentrate upon well-formed buildings, as it were, in the nude. They would thus relearn the values as well as the limitations of mass and proportion. Only then could they introduce ornament in a way that would complement and enhance rather than efface the dominant structure, like harmony added to melody as in the movement from plain chant to the polyphony. For Sullivan as a poetic architect, however, ornament should never be superadded. Both structural proportions and ornamentation should spring from the same emotional tonality. Ornament would thus be an organic part of the original conception, like a flower amidst the leaves and branches formed by a kind of logic of growth.[25]

The International Style fathered by Mies, Gropius and Le Corbusier and based on Sullivan's "form follows function" formula, repudiated all ornamentation and any elaboration of aesthetic form—such as that focused upon by the contemporary de Styl movement[26]—that did not flow directly out of the engineering functions serving the social-political functions of the building.[27] This was also linked to the deliberate repudiation of the practice of borrowing from past forms—"quotation" as it is called—because of the perceived need to develop distinctively modern forms—a perception that goes back at least to the middle of the eighteenth century. In the past, ornamentation was tied to the articulation of social function, underscoring aspects of deeper meaningfulness, as in religious and political architecture, or of hierarchical rank, as in the constuction of palaces and mansions.[28] Too narrow an understanding of the notion of form following function in much of Modern architecture left out that whole dimension of meaningfulness.

Nevertheless, modernism in architecture, enjoying its heyday immediately before and after World War II when it became the International Style, was eventually judged to be sterile, inhuman—indeed, boring. Enter architectural Post-modernism.[29] It rejects Modernism's rejection of historical styles and reintroduces "quotation." We end up once more, however, with a stylistic jumble of elements derived from previous architectural periods against which Modernism revolted. In Mark Jarzombek's felicitous phrase, we are confronted with "One Liner Historicism," quoting without understanding anything of the historical context that made the quotation meaningful.[30] Borrowing a contrast from literary critic Murray Abram's *The Mirror and the Lamp*,[31] we could say that the Post-modernists view architecture as mirroring society as a jumble of incoherent elements rather than giving it illuminating direction. Michael Graves, grounded in a particular and well-articulated architectural theory, ended up designing Disneyworld hotels as a way of making people comfortable with their existence.[32]

There is another dimension to Postmodernism in architecture: the deconstructive attempt. Following out the mirroring rather than illuminating view of general architectural function, the Vitruvian theoretical triangle of *firmitas, utilitas* and *venustas* is subverted on each of its three corners by the postmodernist architecture of Derridian origin. Bernard Tschumi and Peter Eisenman even aim at dysfunctionality and at making the inhabitants of their designed homes uncomfortable! Eisenman's Wexler Center for the Arts at Ohio State University is a model of dysfunctionality. And Tshumi's Villet Park Project in Paris is meant to be constantly subjected to transformation.[33] Tschumi introduced Eisenman to Derrida and commissioned them to design one of the buildings for this Project. Derrida describes the beginnings of his cooperative planning with Eisenman as a matter of free association of words connected in an odd way with the term *chora* —place or space as the receptacle and nurse of becoming—about which he was at that time reading in Plato's *Timaeus*.[34] They added the letter 'L' to *chora* and off they went playing with L-shapes and meaning-associations. Part of the associative significance of the L was the Hebrew word "El" as in *El-Shadai* or in *Gabri-El* or *Rapha-El*. Place and creation come together in the manufactured word "Choral." Characteristically, their cooperative plan was deferred indefinitely.[35]

IV

I want at this point to place the Vitruvian framework which exposes correct, verifiable aspects in architectural work within the more encompassing framework of the lifeworld within which unconcealment of the whole happens.[36] *Stabilitas, convenientia* and *venustas* are actually features set within that larger framework exposed by Heidegger. "Function" is broadened to include relation to the total environment as a relation of inhabitance—a sense of orientation, a feeling for space. These sensibilities transcend a simple formal aesthetic and are related to an appreciative awareness of a full way of life.[37] Lifeworld involves a mode of appearance in which the correlativity of the human being and what appears is constitutive. The mode of appearance is determined, not by objects separate from each other and from the subject, but by the encompassing of subject and object in comprehensive manifestness. In "The Origin of the Work of Art," Heidegger presented the Greek temple as well as Van Gogh's painting and C. F. Meyer's poem as appearing in the tension between Earth and World.[38] Here 'world' is world of inhabitance and 'earth' is correlative to it as native soil. Heidegger's "native soil" is not simply the object of chemical analysis; it is the correlate of inhabitance and plays in relation to a world of lived meaning. A temple functions in the world of the Greeks as grounding their felt understanding of how humans fit within the whole by expressing the human relation to the gods or, as he later put it, the mortals' relation to the immortals. The temple establishes a felt relation to the gods, allowing us to draw near to them, and that precisely as it configures sensuousness. The temple opens up the world of meaning as set upon the earth.

Earth is only derivatively a correctly verifiable scientific object as a peculiar location in the solar system and as a chemical mass. In order to so appear it has to emerge within a human lifeworld. In its lifeworld function, the notion of earth has several components, all of them a function of their manifestness and thus their correlativity to humanness as the locus of that manifestness. In a general sense, buildings are made of earthy materials whose *Verlässlichkeit* or reliability furnishes the

stability that permits their functionality.[39] Reliability concerns what
Heidegger calls the "sheltering" feature of earth—correlative to a sense
of being cradled and thus belonging. The sense of belonging is caught in
the expression "native soil."[40] Another of earth's properties is sensuousness, its rising up to manifestness as its own fullness is sheltered in
darkness.[41] This rising up occurs in perceivers who are themselves
made of earth, the *humanum* from 'humus' which enters into the
determination of our essential mortality. This provides the sensuous
features of materials: light, color, shadow, texture and the qualitative
relations between them as well as the perspectival appearance of quantitative relations in proportions and in scale.

There are various ways in which belonging to the earth has been
articulated in the architectural tradition. Frank Lloyd Wright's organic
architecture stressed belonging to a given natural environment by
selecting materials found in that environment for his country homes.[42]
The use of native materials as a convenience—for example, red tiles
made out of the local clays in Italy or local limestone as mandated
building material in Jerusalem—gives a certain aesthetic unity to the
villages and cities and underscores their belonging to the earth as *this*
peculiar locale—which is linked to what Heidegger means by earth as
"native soil." Other ways to emphasize a building's belonging to the earth
have appeared. There is the rustication of lower stories in Renaissance
works.[43] More recently, transparency achieved through steel and glass
construction of homes in a wooded setting was intended to let the house
melt into the natural setting from without and display panoramically
the natural setting from within.[44] Most recently, in the Portland Civic
Center, Michael Graves (the creator of Disneyland hotels) colored the
lower stories green to emphasize relation to vegetative forms. Even the
development of ornamentation composed of stylized organic forms—for
example, in the medieval cathedral, in Art Nouveau, in Sullivan's
general approach to ornament and in Lloyd Wright's stained glass
designs—was intended to emphasize our belonging to the earth. We
might add that the contemporary use of mirror-glass on the exterior of
skyscrapers adds a new dimension of relation to the environment.
During the day, they mirror in a surrealist, distorted manner, the
buildings which surround them as well as the clouds and sky. At dawn
and dusk they glow with the color of the rising and setting sun.[45]

In Heidegger's later analyses, the notion of earth plays in relation to
that of the sky: things rise up into the field of awareness on the earth
and under the sky. "Under the sky" does not simply refer to the Now of
spatial encompassment, but also to the alternations of night and day,
cloud and sun, spring and fall, winter and summer which furnish
fundamental measures of human existence.[46]

What stands there is unmoved and exhibits proportionate relations
between the parts. This led Schlegel to claim that architecture is frozen
music, a symphony in stone.[47] As in music, the relation between the
parts could be described in terms of mathematical ratios. Nonetheless,

the play of light and shadow from dawn to dusk, from cloudy to clear and from season to season, but also under conditions of artificial illumination—whether ancient candle-light or modern electrical light—creates another kind of symphony, not the frozen music of static space but the dynamic music of an ever-changing exhibition of textures, colors and forms. In the frozen music aspect of architecture we can consider Schopenhauer's claim that the aesthetic theme of architecture is the struggle of gravity and rigidity.[48] In the dynamic music aspects we have Kahn's notion of the play of light and shadow revealing texture and form and rhythm.[49]

In relation to exterior space, a building rises toward the sky and sets itself upon the earth. It stands in the light streaming from above. Because we are set upon the earth of sensuousness as oriented toward the whole, sensuous configurations become metaphors of our belonging to the whole, the dimension Heidegger refers to as that of "the divinities, the beckoning messengers of the godhead."[50] Rising, grounding and being illuminated constitute the primary metaphors of human dwelling—metaphors made live in Plato's image of the Cave[51] and picked up in the symbolism of the medieval cathedral. Human life ascends or descends, measured by how it occupies its place in the cosmos. It has strong foundations or it shifts and collapses. It occupies its place and is well-founded insofar as it stands in the light of understanding how we humans fit into the whole and it can see further in the light insofar as it ascends higher. We dwell most fully, however, insofar as we participate in the display of the beauty of the whole, for which our beautiful products are essentially icons.

The central notion which runs through all our considerations is *function*. In a sense it extends beyond everyday utility and includes symbolism and the tuning of *ethos* through the perception of the beautiful. Both the structural and aesthetic aspects of form subserve such a wider conception of function. As is well known, neither Plato nor Aristotle—nor for that matter the medievals who followed them—distinguished useful and fine art. Perhaps it was because, imitating nature, they accepted aesthetic form as following human function and not as superadded. In fact, they operated largely along the lines suggested by Sullivan.

Plato had earlier presented a view of the arts as providing an aesthetic ambiance which ideally should be characterized as orderly, harmonic, proportionate, graceful. He focused primarily upon musically accompanied poetry, but in addition to painting and the design of clothing, furniture and utensils he also mentioned architecture which we have chosen to highlight in this paper. The total ambiance these artforms provided would stimulate psychic dispositions characterized by the same properties as the objects, establishing a fit matrix for the emergence of *nous* attuned to the recognition of such properties in the encompassing cosmos.[52] Hence art—whatever its theme—is not morally neutral in its aesthetic properties. Neither is such morality acosmic; rather it lies precisely in the dispositional, interpretative and behavioral adjustment of the human to the overall context of existence. In this Plato and Heidegger share common ground. It is our contention that in that context architecture as the most pervasive of the artforms takes the lead. Drawing upon our understanding of cosmic laws, it sets us upon the earth under the sky. But the way it articulates its forms can open us to a view of the deathless order in which we live and die.

University of Dallas
Irving, Texas

* * * * *

Endnotes

1.   Cf. Christian Norberg-Schultz, *The Concept of* Dwelling, (New York: 1985; see also his *Meanings in Architecture*, (New York: Rizzoli, 1983) henceforth MA; *Architecture, Meaning and Place*, (New York: Electra/Rizzoli, 1988), henceforth AMP; *Genius Loci: Towards a Phenomenology of Architecture*, (London: Academy Editions, 1980), henceforth GL.

2.   On the founding structures of the field of experience, see chapters 2 and 3 of my *A Path into Metaphysics: Phenomenological, Hermeneutical and Dialogical Studies*, (Albany: SUNY Press, 1990). For a general approach to the aesthetic region, see my "Recovery of Form," *Proceedings of the American Catholic Philosophical Association*, 1995, pp. 1ff (henceforth RF).

3.   John Ruskin, *The Seven Lamps of Architecture*, (New York: Dover, 1889), 83; henceforth SLA.

4.   Nicholas Pevsner, *An Outline of Eurpoean Architecture*, (New York: Penguin, 1983), 15-16. The latter point handles Roger Scruton's objection to the peculiarity of interior space as the special province of architecture in his *The Aesthetics of Architecture*, (Princeton: Princeton University Press, 1979), 43ff, henceforth AA.

5.   Walter Gropius, *The Scope of Total Architecture*, (New York: Harper, 1955), 32, henceforth STA; Frank Lloyd Wright, *The Future of Architecture*, (New York: Mentor, 1963), 245, henceforth FA; Siegfried Giedion, *Space, Time and Architecture*, (Cambridge: Harvard University Press, 1980), 30ff.

6.   Louis Kahn in John Lobell (ed.), *Between Silence and Light: Spirit in the Architecture of Louis Kahn*, (Boulder: Shambhala, 1979), 50, 47, 34.

7.   Ruskin, SLA, 8-9.

8.   Wright, FA, 43; Le Corbusier, *Towards a New Architecture*, trans. F. Etchells (New York: Praeger, 1960), 8 and 23, henceforth TNA.

9.   Such expressivity might be seen in two examples. In the Doric pillar, the vertical emphasis of the fluting, the entasis or swelling in the middle of the pillar, and the cushion on the Doric capital give expression (as Schopenhauer would have it) to the tension between gravity and rigidity, displaying the impression of elasticity and strength resisting the weight of the entablature and roof (See Schopenhauer, *The World as Will and Representation*, vol. II, trans. E.

Payne, (New York: Dover, 1966), 411-18. Of course it is only an impression, since the pillar is not bulging under the weight nor is its capital cushioning. But it is one way of understanding, at the engineering level, the dogma "Form follows function." The form expresses the function of weight-bearing, but here only in the mode of "as if."

One could also create an even more external display of engineering function by means of a set of conventional signs. Consider, for example, the forty story Firstar Bank Building, the tallest building in Wisconsin. It is one step beyond the International Style, a modern gleaming white steel and clear glass construction comprised of a repetitive grid pattern. The severe verticality of its tower is cut across by three horizontal rows, one at the top, another some three-fifths of the way down and the third at the bottom. The insertion of such rows composed of diagonals alternating directions creates an ambiguous Gestalt of arrows pointing upward and downward, so that the direction of the eye is constantly altered upwards and downwards as one naturally runs along the overpowering verticality of the building. This establishes a set of signs indicating from the outside and conventionally the downward and upward thrusts indicated more naturally by the imitation of natural forms in the case of the Doric pillar.

10. Vitruvius, *Ten Books on Architecture*, trans. M. H. Morgan (New York: Dover, 1960), I, IV, 2, p. 17; henceforth TBA.

11. *Changing Ideals in Modern Architecture, 1790-1950* (Kingston and Montreal: McGill-Queen's University Press, 1984), 22, henceforth CIMA.

12. Mark Johnson, *Disfigurings* (Chicago: University of Chicago Press, 1992), 230-67.

13. Norberg-Schulz, MIWA, 166.

14. TNA, 20 and 31. For Cezanne, see Sam Hunter and John Jacobus, *Modern Art: Painting/Sculpture/Architecture* (Englewood Cliffs, NJ: Prentice-Hall, 1985), 30a.

15. For a summary of the approach we are taking here see Norberg-Schulz, MA, 224ff.

16. Norberg-Schulz, GL.

17. GL, 170

18. On the development of the symbolic interpretation of architecture, see G. W. F. Hegel, *Philosophy of Fine Art*, trans. F. Omaston, (London: G. Bell, 1920), vol. III, 91-106; (henceforth PFA). For particular considerations of both technological and symbolic developments in medieval cathedral construction, see Vincent Scully, *Architecture: The Natural and the Manmade*, (New York: St. Martin's Press, 1991), 155-81, Norberg-Schulz, MIWA, 92-112, and especially Otto von Simson, *The Gothic Cathedral*, (Princton: Princeton University Press, 1988).

19. *Kindergarten Chats and Other Writings* (New York: Dover, 1979), 42-49, 170, 208 (henceforth KC).

20. See Sullivan, "The Tall Office Building Artistically Considered," in KC, 202-13.

21. See Vitruvius, TBA, III, I, 3, p. 73; Le Corbusier, TNA, 219.

22. Still standing are, for example, the Wainwright Building in St. Louis, the Guaranty trust in Buffalo and Roosevelt University's Auditorium Building

in Chicago. (See Nancy Freazier, *Louis Sullivan and the Chicago School*, (New York: Crescent Books, 1991), 40-45, 50-51, 56-61.) Sullivan's work has been considered the consummation of Ruskinism which viewed architecture as focused upon ornamentation. (Collins, CIMA, 115-16.)

23. Sullivan, "Ornament in Architecture, KC, 187-90.

24. See Mark Johnson, *Disfigurings*, 125-28 and Karsten Harries, "The Death of Ornament" in *The Bavarian Roccoco Church* (New Haven: Yale University Press, 1983), 247ff, henceforth BRC.

25. "Ornament," 189. This integral togetherness of all the elements is what Wright meant by "organic architecture," which includes relation to the native environment and to the character of materials (FA, 15-27. This is basically Aristotle's notion of a well-made tragedy which can be carried over to all the arts. See my RF, 3.

26. On de Styl, cf. *Disfigurings*, 114-19.

27. This claim goes back to J. N. L. Durand at the end of the eighteenth century. See Collins, CIMA, 25.

28. See Harries, BRC, especially 245-46. The entire concluding discussion, "The Death of Ornament" deserves careful attention.

29. Postmodernism is said to begin with Robert Venturi's *Complexity and Contradiction in Architecture*, (New York: 1966).

30. "Post-Modern Historicism: The Historian's Dilemma," in *Restructuring Architectural Theory*, M. Drani and C. Ingraham (eds.) (Evanston: Northwestern University Press, 1988), 89a, henceforth RAT.

31. M. H. Abram, *The Mirror and the Lamp: Romantic Theory and the Critical Tradition.* (New York: Oxford University Press, 1953).

32. *Disfigurings*, 222.

33. Ibid., 242ff.

34. Plato, *Timaeus*, 49b ff.

35. Jacques Derrida, "Why Peter Eisenman Writes Such Good Books," RAT, 99-105. As in every other region of human experience, this approach *by itself* leads us into a dead end. I emphasize "by itself," since deconstruction can be helpful in pulling out submerged strands of meaning; but it fails to help us—indeed positively hinders us—in attempting to grasp d thus learn how to produce integral wholes, for all the limitations factually involved in that attempt.

36. On the distinction between the correct and the unconcealed, see Heidegger, "On the Essence of Truth," *Martin Heidegger: Basic Writings*, ed. D. Krell (New York: Harper and Row, 1971), 79; henceforth PLT.

37. In fact, for Heidegger art perishes in the aesthetic. See "Origin of the Work of Art," (henceforth OWA), p. 79 in *Poetry, Language and Thought*, trans. A. Hofstadter (New York: Harper and Row, 1971), 79; henceforth PLT.

38. OWA, Van Gogh's painting, 32-36; Meier's poem, 37; the Greek temple,

41ff.
39.   OWA, 34; cf. also "The Thing," PLT, 167.
40.   OWA, 42.
41.   OWA, 47.
42.   FA, 94ff.
43.   Serlio considered rustication as "symbol of the original forces of earth," reported in Norberg-Schulz, GL, 54.
44.   See Mies van der Rohe's "Farnsworth House" and Philip Johnson's "Glass House" in New Canaan, Connecticut in Paul Heyer, *Architects on Architecture: New Directions in American Architecture*, 31-33 and 207ff.
45.   Tom Wolfe sees them as boring reflections of other glassboxes in, *From Bauhaus to Our House* (New York: Farrar Straus Giroux, 1981), 6. Robert Romanyshyn, *Psychological Life* (Austin: University of Texas Press, 198XXXX, sees them as narcissistic.
46.   "Building, Dwelling, Thinking," PLT, 149-50 (henceforth BDT); "The Thing," PLT, 178.
47.   This is attributed to Friedrich von Schlegel by Hegel, PFA, III, 65.
48.   CF. supra, n. 9.
49.   See Norberg-Schulz, GL, 12; see also Gropius, STA, 35 on the dynamics established by the play of light.
50.   BDT, 150.
51.   *Republic*, VI.
52.   Ibid., III, 401a.

Electric Technology and Poetic Mimesis

by Daniel McInerny

> To have gathered from the air a live tradition
> or from a fine old eye the unconquered flame
> this is not vanity.

Ezra Pound, Canto 81

The influence of technological media upon poetic mimesis did not begin with the dawn of chirographic culture, much less with Gutenberg and his Bible. The Homeric Muse was the poetic art's original media. This becomes evident when we understand such words as 'technology' and 'media' as connoting something more than iron parts, internal combustion, electricity, or micro-chips. As Hugh Kenner has observed, a piece of technology or 'machine' is any large coordinated enterprise designed to concentrate energy, reduce effort, and annul duplication.[1] This definition easily applies to the hardware of our personal computers, but also, though less apparently, to such an artifact as the *Oxford English Dictionary*, which brings myriad resources representing numerous technologies and thousands of work hours under the covers of one, albeit enormous, book. The Homeric Muse, like the *O.E.D.*, coordinates the lore of an entire civilization into one 40,000-plus line hexameter poem.

The special character of the Muse as machine is brought out by consideration of the very first line of Homer's *Iliad*: "Sing O Muse the anger of Peleus's son Achilleus." Readers of the *Iliad* will notice there is no discernable break in the ensuing lines between the bard's invocation and the Muse's narration. That is, the Muse never 'takes over' from the bard; invocation and narration are fused from the outset and remain so for the duration of the poem.[2] Thus the *Iliad* is not a presentation of Homer's static 'point of view,' his 'take' on certain events in the twilight of the Trojan War. No; the *Iliad* is the presentation by a Muse, who represents not only Homer but an entire tradition of bards who with instrument in hand and prodigiousness of memory entertained and instructed a civilization. It is because the *Iliad* is more a digest of songs than Homer's personal creation that we can regard it as a kind of

95

machine.

In an oral culture such as Homer's this machinery is necessary to keep the culture alive. Father Walter Ong, S.J., who has done some of the most extensive work on the differences in storytelling between oral and literate cultures, educes at least two reasons why the machinery of epic narrative is especially effective in the maintenance of an oral tradition.[3] First, an oral culture is not capable of managing knowledge in scientifically abstract categories, and so a long story is needed to collate, organize, and disseminate the 'tale of the tribe.' It is what Ong calls the "roominess" of epic narrative, its size and complexity of scenes and actions, which allows it to serve in this way. Second, Ong cites the durability of epic narrative over other forms. While an oral culture may achieve durable repetition by means of maxims, riddles, and proverbs, these forms are nevertheless too slight to represent a significant complex of actions. Other longer forms, such as genealogies, fail in presenting only highly specialized information or, like orations, in not being normally repeated.

It is because the purpose of epic is to propagate a tradition that the roominess and durability of its narrative machinery take precedence over chronologically-ordered 'plot.' Father Ong points out that when Horace describes the epic poet as plunging the hearer *in medias res*, he refers chiefly to the epic poet's disregard for temporal sequence.[4] Horace's *res*, at least for the oral poet, is not a chronological ordering, an Aristotelian beginning, middle, and end which the poet then artfully rearranges so that the hearer may more immediately confront the action of the story. As Ong notes, this is how Horace's phrase was misinterpreted by the writers of subsequent generations, among them, notably, John Milton in Book I of *Paradise Lost*. After proposing 'in brief the whole subject' of his poem, Milton proceeds to trace 'the prime cause' of Adam's fall before "the Poem hasts into the midst of things." "Milton's words here," Father Ong writes, "show that he had from the start a control of his subject and of the causes powering its action that no oral poet could command. Milton has in mind a highly organized plot ... in a sequence corresponding temporally to that of the events he was reporting."[5] The oral poet was not capable of this structure, Ong concludes, because such interest in temporal sequence was the product, or rather the concomitant attribute, of a technology which radically distinguishes the culture of the blind Milton from that of the blind Homer, namely writing. It is due to the development of writing, ultimately, that Aristotle prescribes for the composer of tragedies a working out of a plot in terms of *temporal* beginning, middle, and climactic end. For insofar as writing could comfortably account for the epic concerns for roominess and durability through the manipulation of letters on a page, narratives comprised of stricter temporal sequence and greater formal unity quickly became more prominent. Ong reminds us that Greek tragedy, while orally performed is essentially a written genre.

Thus at the dawn of Western storytelling the technology of the Muse, a machine unconcerned with chronological development (the overthrow of Chronos by Zeus is not an event *in time*) or tight formal unity, was supplanted, in the chirographic culture which succeeded Homer's, by what Aristotle calls *mythos*, the form or *eidos* of tragedy. In the two and a half millennia since then, poetic mimesis has continued to be influenced both by the technologies which it employs and the technologies which comprise the cultural environments of its makers. Eric Auerbach's masterful work of criticism, *Mimesis,* along with Walter Ong's work among that of many others, have gone a long way in tracing these influences both in general and in the context of critiques of particular works of Western literature. In what follows I would like to concentrate on the influences of technology on twentieth-century poetic mimesis, and offer some speculations regarding the very possibility of this mimesis in our technological environment. But to understand where we are is to understand where we've come from, and so I must preface my remarks with a sketch, in the broadest of strokes, of how poetic mimesis got to the point where it is today. The very possibility of this sketch rests upon technology itself. For part of the logic of a technological environment, as Marshall McLuhan has taught us, is its ability to reveal the nature of its antecedent environment.

Pictorial and Aural Space: It has been over thirty years since the heyday of McLuhan's investigations into the media of our electric age, but reflection on these investigations show them to be more prescient by the hour. The medium is indeed still the message. This oracular utterance, as McLuhan once explained it, means "that any new structure for codifying experience and of moving information ... has the power of imposing its structural character and assumptions upon all levels of our private and social lives, even without benefit of concepts or of conscious acceptance."[6] One of the major conclusions McLuhan generated from this axiom was the claim that our technological environment has disengaged itself from what he terms "pictorial space." In my next section I will discuss what kind of 'space' our technological environment defines for us today. Here let us concentrate on what our environment reveals of this antecedent, pictorial space, especially in regard to its influence on poetic mimesis.

I have been using 'space' undefinedly. Following McLuhan I take it as a metaphor for our lived-world, including the physical, psychological, and cultural components with which, from which, and in which we think and act. McLuhan's axiom, once again, is that technology possesses the powerful ability to help create the 'space' of our lived-world, and it is precisely the technologies of literate culture which helped create the space McLuhan calls pictorial. Pictorial space is made possible by an alphabet, a machine which allows for the translation of sound into marks on a page. Communication thus comes under the bailiwick of two senses rather than one, and with the advent of movable type and printing the word is ever more disassociated from speech and hearing. Punctuation marks underscore this disassociation. Those who have seen an autograph text of Aquinas will note that even in the manuscript culture of the Middle Ages such marks were minimal or non-existent. They are a typographer's invention; a self-conscious manipulation of metal bits and ink over two-dimensional space.

With writing and type thus arises the prominence of the visual sense. As McLuhan has remarked, this prominence "created for Western man an environment built on the visual assumptions of uniformity, continuity, and connectedness. Such spatial assumptions scarcely existed in cultures based on acoustic patterns...."[7] The space which Homer inhabited was so acoustically patterned, and might be described as 'aural' space in contrast to the pictorial space of chirographic culture. In aural space, hearing and memory hold sway, and roominess and durability, as noted earlier, are the strongest influences upon poetic mimesis. Temporal sequence is of little concern to the poet relying upon memory alone. He will begin the story with whatever persists in the topmost drawer of consciousness and loop associatively backward and forward in time until the story is laid out. The epic poet has no other choice. As Ong notes: "Having heard perhaps scores of singers singing hundreds of songs of variable lengths about the Trojan War, Homer had a huge repertoire of episodes to string together but, without writing, absolutely no way to organize them in strict chronological order."[8] Only careful selectivity produces the tight pyramidal plot in which a sequence of scenes are joined together by 'probability and necessity' and build to a climax, *peripeteia*, and *denouement*. This level of selectivity, however, Ong further notes, "is implemented as never before by the distance that writing establishes between expression and real life."[9]

By "distance" here Ong means to refer to the heightened consciousness which is both the product and the producer of the climactic, Aristotelian 'plot-line.' With writing comes reflectiveness and with reflectiveness comes heightened consciousness and even self-consciousness.[10] Thus follows a greater selectivity on the part of the poet in the making of his story. He will beware of stringing scenes together with only the most tenuous of connecting filaments. It will occur to him, as it does to Aristotle, that the *Iliad* appears to be a mere 'heap.' Such heightened attention to unity among difference, to the uniformity, continuity, and connectedness of which McLuhan speaks, is best compared to the sense of sight. This is because the eye, as Aristotle observes at the outset of the *Metaphysics*, is more successful than the other senses in making the discriminations which delight us. Accordingly, Aristotle speaks of the *mythos* as the *eidos* of the tragedy. We should understand that *eidos* here means form, but also, in its more literal sense, a seeing or gazing. The poet's plot is his unifying gaze over the disparate elements of men in action.[11] Twenty-three hundred years after Aristotle we find perhaps the greatest novelist of the 19th century still comfortably ensconced in pictorial space. In the preface to his novel, *Portrait of a Lady* (nota bene: a *portrait* of a lady), we find Henry James referring to the House of Fiction, a particularly pictorial metaphor for a particularly pictorial fiction. It is an apt metaphor, as Hugh Kenner has pointed out, for in pictorial space a novel is more like a building than a statement.[12] We speak of a novel's *structure*, its *surfaces* and *depths* and *insight*. We speak of *background* and *foreground* and *levels of meaning*. Moreover, because a novel is like a building or a picture, it must be approached as buildings and pictures are, from a *point of view*. The novel is thus no longer a story but a narrative, a temporal series of events looked at from a definitive perspective or set of perspectives. No major novelist since James has been more assiduous in maintaining consistent, narrative point of view. This is because, with the advent of new technologies, some of which were developed in James's own lifetime, the psychological space of chirographic culture ceased to be predominantly pictorial, so that poetic mimesis itself underwent a revolution.

In the Timeless Air: With the innovations in this century of electric technology, which in turn make manifest the chirographic culture which they enclose, poetic mimesis is once again under the influence of a new environment, an environment which is obviously still very much in formation. What is the specific nature of this new environment? What kind of space is created? The new environment is not exclusively aural or pictorial, though it retains features, or encloses, both of these antecedent environments. It might be thought of as a combination of these two. Radio, television, the newspaper, the personal computer, all these technologies retain aspects of aural and pictorial space. Yet it would be a mistake simply to think of our new environment simply as a mixture of the aural and pictorial.

"Electric technology," McLuhan has written, "simply because it is all at once, is also discontinuous."[13] Here McLuhan highlights the essential feature of the space contemporary technology helps create. In contrast to pictorial space our contemporary space is not defined by the uniform, the continuous, the connected. It seems to be defined, rather, by the opposite of these qualities, but in a way infinitely more complex and pictorial than the discontinuous quality of aural space. McLuhan chooses the *mosaic* as the metaphor to depict our contemporary space. A mosaic, unlike a picture, resembles "a total field of energy or relationships.... It is an all-at-once mythical structure in which beginning, middle, and end are simultaneously present."[14] Our contemporary technologies thus assist in creating a space in which no one point of view, no fixed temporal sequence, is dominant. For this reason they resemble cubist painters who sought to 'flatten' perspective so that all perspectives appear in one simultaneous and luminous moment. An analogous operation has taken place in the physics of this century as a new space-time relation has replaced Newtonian space. But more mundane examples are ready-to-hand. McLuhan cites the mosaic of the front page of a newspaper, where items are arranged in such a way so as to present not a picture of the world but something more like an X-ray in depth. How much more does the news on the radio, or the Internet, or disks for our CD-ROMS serve as X-rays rather than pictures of our world.

Yet the discontinuity of mosaical space has discombobulating side-effects. For what unites the items on the front page of the newspaper other than the dateline at the top of the page? McLuhan understands mosaical space as creating exterior situations that have all the structural characteristics, that is, the all-at-onceness, of the human consciousness. In contrast to chirographic book culture, which turned human communication 'inward,' bringing man into silent conversation with his thoughts as his eyes scanned symbols on a page, electric technology projects that silent conversation (whether conscious or unconscious—that is not the salient distinction) 'outward' into the domain of art. This projection is quite evident in the interior monologues of Joyce's *Ulysses*, and in the first principle of Joyce's aesthetic, the epiphany, the moment of 'insight' in which consciousness is heightened. Still, McLuhan recognizes that to the observer trained to look for unity among difference, for connectedness, "the new situation presents an extreme form of the irrational."[15] It is less irrational if we understand it as something of an historical reprise of the aural space of oral culture. What we live in now is a new kind of mythological world, permeated with the fragments of both aural and pictorial understanding. Hence the poets and writers at the threshold of this century turned naturally to ancient myths—not as mere poetical devices or tropes, not in order to invoke gone modes of thinking and behavior—but as neighbors in the all-at-onceness of mosaical space. "In the timeless air" as Ezra Pound puts it, all voices speak together. Hearing is once again the eminent, if not preeminent, metaphor to depict the space in which we live. Arguably the greatest poem of the century, T.S. Eliot's "The Waste Land," is a vivid mosaic of pictorial fragments, mythical voices and their echoes. Significantly, the original title of this poem was a phrase out of Dickens: "He Do The Police in Different Voices." Meanwhile the book, that emblem of chirographic culture, yields at least in its fictional varieties to the all-at-onceness of mosaical space. Joyce's *Ulysses* is Homer's *Odyssey* folded over turn-of-the century Dublin. *Finnegan's Wake* is the repellant babble of all voices of all cultures at all times speaking concurrently—over and over again. Similar mosaical structures can be found in writers such as Proust and Faulkner.

Ars Imitatur Naturam?: We have just touched on two salient features of the environment in which modern poetry arose. The first is that consciousness is put, as it were, on the 'outside.' The second is the disruption of sequential time, causing a disruption of the traditional narrative structures dependent on it. These two characteristics manifest the subjective turn toward lived experience that is universally acknowledged as the defining feature of modern poetry.

What does this subjectivity imply for poetic mimesis? That is, to what extent does it maintain nature as an objective, public referent? To be sure, this reflexive turn does not necessarily imply subjectivism or solipsism, though in some varieties modern poetry has run to these

excesses. Quite to the contrary, in fact, modern poetic subjectivity tends
to emphasize non-subjective structures, whether of language or images,
tradition or history. It was Eliot who famously stressed the *imperson-
ality* of the poet. Nevertheless, these structures are always explicitly
pursued within the ambit of subjectivity. Pound described his *Cantos* as
a poem including history, but it was history as correlated and re-inter-
preted by Pound's personal history.   The question is, does modern
poetry, in its subjective turn, maintain a mimetic relationship to nature?
I think it does maintain such a relationship, but, in order to underscore
how, it would be instructive first to consider two tendencies modern
poetry has, two temptations, toward utter subjectivism.

In *The Sources of the Self,* Charles Taylor attempted to explain
Modernism's peculiar subjectivity by depicting Modernism as a transformation of Romanticism, a transformation which roughly corresponds
to that between pictorial and mosaical space.[16] Taylor's account helps
us see two ways in which Modernist subjectivity lapses into subjectivism. The first is the dissolution of the unitary self. A crucial aspect of
the nature to which the Romantics sought to attune themselves was an
alignment between instinct and feeling on the one hand, and the creative
imagination and reason on the other. A shift, however, in the notion of
nature at the dawn of Modernism, from benevolent source to great
reservoir of amoral power, led to a disruption of this alignment. The
result was a bifurcation of human identity, typified by the split in the
Nietzschean self between its Dionysian and Apollonian components.

Subjectivism also arises in the way Modernism sees the relationship
between language and truth. Taylor describes the epiphanies of Romantic poetry as 'epiphanies of being' in that they point beyond, or express,
one or more features of benevolent nature. The epiphanies of Modernism, by contrast, do not point beyond themselves at all. In the artful
manner of their juxtaposition Pound's images in the *Cantos,* or Eliot's
in *The Waste Land,* serve, to use Pound's metaphor, as a kind of receiving
set to radio signals. The Modernist epiphany occurs in the energy
created by the juxtaposition of images and words. But what exactly is
being manifested in such an epiphany is the question. Hugh Kenner, for
example, has written of the Pyrrhonism of Joyce's characters in *Ulysses,*
a book, at least for Kenner, in which the only substance is style.[17] Taylor
is ambiguous on this point. He writes, "in the post-Enlightenment world,
the epiphanic power of words cannot be treated as a mere fact about the
order of things which holds unmediated by the works of the creative
imagination."[18] The rub here is what Taylor understands by the 'mediation' of reality by language. At one point, he disturbingly links this
mediation with the rise of neo-Kantianism in the philosophy of this
period, also referring to Nietzsche's remark about "the formless unfor-
mulable world of the chaos of sensations." He then concludes, "In order
to live at all in this world, we have to impose some order on it."[19] To be
fair to Taylor, however, he also talks of the Modernist mediation of the
world by language in tones other than Kantian. In glossing Pound's

statement that bad art is that which makes false reports, Taylor says, "The reality we are meant to report accurately on is not the bare scene, but the scene transfigured by emotion. And the emotion, in turn, is not simply personal or subjective; it is a response to a pattern in things which rightly commands this feeling."[20]

I have been identifying two tendencies in Modernism towards subjectivism. Yet these are only tendencies, or as I said earlier, temptations. I would like to close by briefly suggesting how we might understand the peculiar subjectivity of modern poetry in a way which defends against these tendencies. I take my cue here from Jacques Maritain whose book *Creative Intuition in Art and Poetry* remains a much under-valued resource for the understanding of modern art. It is true for Maritain that 'poetry,' in contrast to 'art,' is necessarily born out of subjectivity. "The essential need of the poet," he writes, "is to create; but he cannot do so without passing through the door of the knowing, as obscure as it may be, of his own subjectivity."[21] Yet, Maritain's Thomism tells him that the substance of man is obscure to himself. It can only be reached "through a repercussion of his knowledge of the world of things."[22] In other words, "the primary requirement of poetry, which is the obscure knowing, by the poet, of his own subjectivity, is inseparable from, is one with another requirement—the grasping, by the poet, of the objective reality of the outer and inner world."[23] This grasping of self and thing, of thing as inseparable from self, is understood by Maritain as a union of affective connaturality which is born in the spiritual unconscious and which fructifies in the work. Without attempting to explicate this concept fully, we can at least note the ways in which it defends against the subjectivist tendencies outlined above. First, subjectivity for Maritain is only possible after a grasp of "the objective reality of the outer and inner world." There is no question of subjectivity constituting a world. Reality is only mediated by language in the sense that language tries to articulate the soul's affective grasp of the real. The epiphany is of reality, not the self.[24] Second, Maritain's understanding of subjectivity, contrary to Taylor's, keeps the subject unified. This is because Maritain understands the conscious and pre-conscious life of the soul's powers, whether of intellect, memory, imagination, or sense, as emanating from a common root.

For these reasons nature is not only available but is required for a genuine subjectivity, and Modernism in its better moments is as mimetic of nature as classical art ever was. Indeed, one of the most attractive features of modern poetry is that in helping manifest the real it brings the mimesis of classical art into its presence, gathering from the air a live tradition. In his best known essay, "Tradition and the Individual Talent," T.S. Eliot exhorts the poet to recognize himself and his work as advancing the cause of a live tradition. But tradition involves, Eliot writes, first of all a historical sense,

and the historical sense involves a perception, not only of the

pastness of the past, but of its presence; the historical sense
compels a man to write not merely with his own generation
in his bones, but with a feeling that the whole of the literature
of Europe from Homer and within it the whole of the literature of his own country has a simultaneous existence and
composes a simultaneous order.[25]

The simultaneity of voices in a tradition is a profound cultural achievement, and to the extent that our electric technology and the book culture
it encloses helps us cultivate such synergy we should be grateful. Indeed,
any article of the *Summa theologiae* evinces the attempt to bring order
to the living voices of a tradition. That this kind of work is facilitated
for us by electric technology is obviously to our benefit. The danger, for
the poet or any citizen of our environment, is to see the timelessness and
discontinuity of the environment as a mere mythological ritual which
obliterates nature's order. For as Eliot was to perceive by the time he
wrote "Four Quartets," the timeless moment of history is timeless only
insofar as it reflects an ordering governed by that which is outside of
time.

University of St. Thomas/Center for Thomistic Studies
  Houston, Texas

* * * * *

Endnotes

1.   Hugh Kenner, *The Mechanic Muse* (New York: Oxford University Press,
1987), 54.
2.   Kenner makes this point in his essay "Beyond Objectivity" in *Joyce's
Voices* (Los Angeles, CA: University of California Press, 1978).
3.   My argument is generally indebted to his *Orality and Literacy: The
Technologizing of the Word* (New York: Routledge, 1982), especially Chapter 6.
4.   Ibid., 142. Horace's description can be found in his *Ars Poetica*, lines
148-49.
5.   Ibid.
6.   From "New Media and the New Education" in *Christianity and Culture*,
ed. J. Stanley Murphy, C.S.B. (Baltimore: Helicon Press, 1960), 188. I am
generally indebted to McLuhan's work throughout my argument.
7.   "Environment as Programmed Happening" in *Knowledge and the Future of Man*, ed. Rev. Walter Ong, S.J. (New York: Holt, Rinehart, & Winston,
1968), 115.
8.   Ong, *Orality and Literacy*, 143.
9.   Ibid., 148.
10.   This is not to mention modern man's concern with the unconscious and
pre-conscious life of the intellect. For the effect of these concerns on poetic
mimesis see Jacques Maritain's *Creative Intuition in Art and Poetry* (New York:

Meridian Books, 1957).

11. On *eidos* and its relation to mimesis in Aristotle's *Poetics* see Thomas Prufer's "Providence and Imitation: Sophocles's *Oedipus* and Aristotle's *Poetics*" in his volume of essays *Recapitulations* (Washington, D.C.: The Catholic University of America Press, 1992).

12. See Hugh Kenner, *The Pound Era* (Berkeley: University of California Press, 1971), 27, and in general, 23-40.

13. "Environment as Programmed Happening," 114.

14. Ibid.

15. Ibid.

16. *The Sources of the Self* (Cambridge,MA: Harvard University Press, 1989). See especially Chapter 24, "The Epiphanies of Modernism." Another excellent discussion of the nature of Romantic poetry,and the influence of technology upon it, is Chapter 2 of Walter Ong's *Rhetoric, Romance, and Technology* (Ithaca, NY: Cornell University Press, 1971).

17. See the essay "Myth and Pyrrhonism" in his book *Joyce's Voices* (Los Angeles, CA: University of California Press, 1978).

18. *The Sources of the Self*, 481.

19. Ibid., 472.

20. Ibid., 475. Alasdair MacIntyre has also critiqued Taylor's reading of Pound, Joyce, and Eliot on the issue of their subjectivity. His Thomist view corresponds with the one to follow, but not in drawing from Maritain. See his essay "Critical Remarks on *The Sources of the Self* by Charles Taylor" in *Philosophy and Phenomenological Research*, vol. 54 (March, 1994): 187-90.

21. *Creative Intuition in Art and Poetry*, 82.

22. Ibid., 83.

23. Ibid., 83.

24. Maritain, I believe, would have an interesting interpretation of the energy that is created in a Modernist poem in the juxtaposition of images. He might equate it with the light of the Illuminating Intellect filtering through particulars in the spiritual pre-conscious. See Chapter 3 of *Creative Intuition*, "The Preconscious Life of the Intellect."

25. T.S. Eliot, "Tradition and the Individual Talent" in *Selected Essays* (New York: Harcourt, Brace and Company, 1950).

From Absurdity to Decision: The Challenge of Responsibility in a Technological Society

by Daryl J. Wennemann

I

Introduction: It is a commonplace that modern technological developments bring with them a greater responsibility in their use. This is increasingly the topic of philosophical reflection and debate. In his essay, "Responsibility and Technology: The Expanding Relationship," Carl Mitcham argues that the concept of responsibility evolved in tandem with modern technological developments. Mitcham's analysis of the areas of law, science, engineering, religion, and philosophy indicates that with the growth of technological power there has been a concomitant growth in the spheres of responsibility that are recognized by experts in these various fields, a recognition that is sometimes institutionalized in professional codes of ethics (especially in science and engineering). In Mitcham's words,

> The promotion of the abstract noun "responsibility" to linguistic and cultural prominence—even while the reality to which it refers may not have been wholly without premodern recognition—is thus a phenomenon easily associated with issues of power and readily correlated with the rise of technology to social and historical dominance. In their earliest forms the notion of what we now call responsibility and the technical activity of making and using artifacts were seldom related. But in association with the dreams of modern technics there emerge images of a new strength of will, "the greatest responsibility" of Friedrich Nietzsche's superior man.[1]

Perhaps a more immediate and accessible source of confirmation of this correlation of responsibility and technical power can be found in Karl Mannheim's sociology of knowledge. In *Ideology and Utopia*, Mannheim points to the rise of a stage of social existence in which responsibility becomes the appropriate correlate in ethical thought to

105

the stage of self-consciousness made possible by modern social science. In this regard, Mannheim sets out three broad stages of development of ethical consciousness, the ethics of fatality, the ethics of conscience, and the ethics of responsibility.

Mannheim argues that prior to the development of social science the historically original social condition of human beings was such that they had to accept existing social conditions as unalterable. This led to an ethics of fatalism. The ethic of fatalism was characterized by "submission to higher and inscrutable powers."[2] This ethical stance was overcome, in Mannheim's view, with the development of an ethics of conscience. Here human freedom is exercised with respect to external social conditions but not with respect to the formation of the conscience itself and the decisions arising from the conditioning of the conscience. It is also interesting to note that, according to Mannheim's account, the ethics of conscience did not extend its concern to the consequences of actions. Its focus was on the causal power of the agent in initiating events. As Mannheim states,

> The first break in this fatalistic outlook occurred in the emergence of the ethics of conscience in which man set his self over against the destiny inherent in the course of social events. He reserved his personal freedom, on the one hand, in the sense of retaining the ability through his own actions to set new causal sequences going in the world (even though he renounced the ability of controlling the consequences of these acts) and, on the other hand through the belief in the indeterminateness of his own decisions.[3]

The ethical principle of responsibility arises with the predictability of social events that follows upon the control of some social relations. According to Mannheim, there are two imperatives associated with the principle of responsibility. First, that actions should take into account their *foreseeable* possible consequences, as well as being done in accord with the dictates of conscience. Second, the conscience itself must be subjected to critical self-examination so as to eliminate, as far as possible, any unconscious factors determining the will.[4] These are the underlying conditions for a science of politics, which is the locus of Mannheim's concern.

> His [Max Weber's] ideas and researches reflect the stage in ethics and politics in which blind fate seems to be at least partially in the course of disappearance in the social process, and the knowledge of everything knowable becomes the obligation of the acting person. It is at this point, if at any, that politics can become a science, since on the one hand the structure of the historical realm, which is to be controlled, has become transparent, and on the other hand out of the new

ethics a point of view emerges which regards knowledge not as a passive contemplation but as critical self-examination, and in this sense prepares the road for political action.[5]

What is most interesting about Mannheim's account is that this sort of knowledge itself must be seen as the result of the unconscious striving of a social stratum which may or may not exhibit utopian tendencies.[6]

Another significant point to consider from the perspective of Mannheim's sociology is the correlation between the rise of the idea of responsibility with a class or social stratum within a historical period as well as the social effectiveness of the idea in relation to the social forces present within the society in its dynamic development. This is the kind of issue Ellul raises when he asks *who* is to control the technical ensemble, apart from the question of how. The issue, however, of the utopian character of the idea of responsibility raises a philosophical problem regarding technology and responsibility.

A general philosophical issue that I will consider is whether modern technological developments give rise to moral responsibilities that modern persons are incapable of fulfilling, what Coolen calls a pragmatic paradox.[7] In my view, the question of responsibility in a technological society gives rise to the question of the limits of pure technical rationality (to put it in quasi-Kantian terms). My argument, however, will be that pure technical reason does not give rise to antinomies of pure reason, but rather, the extension of pure technical rationality beyond the limits of human comprehension gives rise to an absurd field of action. In this context, moral demands become impossible to apply in a consistent manner. One of the paradoxes of modern technique is that it produces a situation in which it is impossible to act in a responsible way at the very time that it expands moral responsibility. It gives rise to moral demands we are incapable of fulfilling. Hans Jonas has seen this paradox very clearly,

> And here is where I come to a standstill, where we all come to a standstill. For the very same movement which put us in possession of the powers that have now to be regulated by norms—the movement of modern knowledge called science—has by a necessary complementarity eroded the foundations from which norms could be derived; it has destroyed the very idea of norm as such.... Now we shiver in the nakedness of a nihilism in which near-omnipotence is paired with near-emptiness, greatest capacity with knowing least for what ends to use it.[8]

The question Mannheim did not consider is whether the critical self-examination that he thought would prepare the road for political action might not lead to paralysis. It is my view that this question requires that we carry out a sort of critique of pure technical rationality in order to determine the condition for the possibility of responsibility in a technological society.

II

Hans Jonas and the Principle of Responsibility: An important point of departure for an evaluation of the challenge of responsibility in a technological society is the work of Hans Jonas, *The Imperative of Responsibility*. This is especially the case given the link between ethical theory and political policy making that Jonas draws. His analysis seems to agree with that of Mannheim in viewing ethical thought as preparing the ground for political action. Of course, the important and difficult practical problem here lies in the movement from ethical principle to political action.

Jonas argues that the problem of responsibility in a technological society is grounded in the unprecedented extension of possible deleterious effects of human action to which modern technological power gives rise. This represents a new ethical condition that takes human beings beyond the traditional ethics of contemporaneity and immediacy toward an ethics of the future. That is to say, modern technology gives rise to a new ethical imperative that possible future effects must serve as a basis for ethical judgments. The principle of responsibility requires that we act in a way that ensures a future for generations to come.

It is interesting that Jonas frames the issue of responsibility in a technological society in terms of collective action. As Jonas remarks,

> To be sure, the old prescriptions of the "neighbor" ethics—of justice, charity, honesty, and so on—still hold in their intimate immediacy for the nearest, day-by-day sphere of human interaction. But this sphere is overshadowed by a growing realm of collective action where doer, deed, and effect are no longer the same as they were in the proximate sphere, and which by the enormity of its powers forces upon ethics a new dimension of responsibility.[9]

Jonas goes on to make the point that the cumulative effects of modern technologies are irreversible and their aggregate magnitude is such that it may lead to a situation in which future generations must make choices within an increasingly narrow range of possibilities.[10] As such, the ethical condition of future generations may be undermined by a series of preceding decisions over which they had no influence. In Jonas's words,

> [T]he cumulative self-propagation of the technological change of the world constantly overtakes the conditions of its contributing acts and moves through none but unprecedented situations, for which the lessons of experience are powerless.

And not even content with changing its beginning to the point
of unrecognizability, the cumulation as such may consume
the basis of the whole series, the very condition of itself. All
this would have to be cointended in the will of the single
action if this is to be a morally responsible one.[11]

Here we have an ethical demand that arises from the power of
modern technology to affect future conditions in a way that seems to
make the fulfillment of the demand humanly impossible. Jonas's account of the situation is essentially in terms of the problem of the uneven
development of techniques and human faculties.[12] First, there is a
disproportion in the development of techniques of scientific prediction
with respect to the capacity of other techniques. According to Jonas, "...
the required extrapolation demands an exponentially higher degree of
science than is already present in the technology from which it is to be
extrapolated."[13]

In addition, the faculty of technical intelligence has outpaced the
development of the moral faculties necessary to make rational moral
decisions. Jonas rightly points out that this situation requires an
expansion of the effective knowledge of future effects of technological
action. It is theoretically possible that technological development itself
might expand our foresight so that we could make practical judgments
based upon the foreseeable consequences of our actions. This would
provide a link between ethical principles and political decision making.
Nevertheless, the present uncertainty of our prognostic ability weakens
the possibility of political action. The danger inherent in the kind of
self-examination called for by Mannheim within an ethic of responsibility is paralysis in the face of uncertainty.

[I]t must be admitted now that this same uncertainty of all
long-term projections becomes a grievous weakness when
they have to serve as prognoses by which to mold behavior—
that is in the practical-political *application* of whatever principles were apprehended with the help of the heuristic
casuistry.... Being so much in the dark, why not trust our
luck including that of posterity? But in this way, all the gains
of our hypothetical heuristics are kept from timely application by the inconclusiveness of the prognostics, and the finest
principles must lie fallow until it is, perhaps, too late.[14]

To summarize Jonas's position we may say that in lieu of a technological expansion that might provide the analytical tools for making
informed decisions based upon the prediction of future effects of technological actions, that is, a futurology, Jonas calls for what Henri Bergson
described as an expansion of soul.[15] Jonas holds that we must train the
soul to respond in fear to the remote possibilities of evil effects our
actions might have on future generations, based on a casuistry of the

imagination.[16] This is a spiritual fear that is cultivated by an act of the will (as a spontaneity?). Such an expansion of soul will allow us to exercise the self-restraint that is demanded by the principle of responsibility:

> ... we must educate our soul to a willingness to *let* itself be affected by the mere thought of possible fortunes and calamities of future generations, so that the projections of futurology will not remain mere food for idle curiosity or equally idle pessimism. Therefore, bringing ourselves to this emotional readiness, developing an attitude open to the stirrings of fear in the face of merely conjectural and distant forecasts concerning man's destiny—a new kind of education sentimentale—is the second, preliminary duty of the ethic we are seeking, subsequent to the first duty to bring about the mere thought itself. Informed by this thinking, we are obliged to lay ourselves open to the appropriate fear.[17]

There are two more points I would like to emphasize regarding Jonas's ethic of responsibility. First, it should be noticed that the imaginative prediction of catastrophic future effects is to lead to a decision now. When viewed from the perspective of prediction, however, the uncertainty of future outcomes seems to undermine any claim the future might have on us. As Jonas notes,

> [T]he envisaged distant outcome should lead its beholder back to a decision on what to do or abstain from now: and one demands, not unreasonably, a considerable certainty of prediction when asked to renounce a desired and certain near-effect because of an alleged distant effect, which anyway will no longer touch ourselves.[18]

The fear produced by this exercise of the imagination is supposed to give rise to an action that has the effect of disproving the imagined prediction of catastrophe. It might be fruitful to remember, however, the existentialist's insight that fear can be distinguished from anxiety by the fact that fear always has a definite object, whereas anxiety is directed toward the possibilities that lie in the future. It would seem that the heuristic Jonas proposes should lead to a state of anxiety, not fear. As a result, the problem of the political effectiveness of Jonas's ethic of responsibility remains. The anxiety we experience in the face of the uncertainty of future possibilities may result in paralysis.

The second point I would like to emphasize is that Jonas's ethic of responsibility is supposed to be based on an objective imperative.[19] This is essentially a duty to ensure the future. It goes without saying that there are many difficulties involved in establishing an objective imperative that can ground an ethical responsibility to future generations.

Hans Küng has noted, in this regard, that Jonas really has two separate arguments, one based on theological grounds and the other based strictly on philosophical grounds.[20] Küng holds that the theological argument is more compelling. In a sense, Küng finds the philosophical argument to be so complicated and artificial, so to speak, that it is not able to motivate us to be responsible in the new way required by a technological environment.

My view is that if such an imperative exists and it lies beyond our capacity to carry it out, then we are driven to a condition of absurdity by the very technological means that gave rise to the new imperative.

Ronald Green has shown that Kant's reflection on the burden of the moral law in *Religion Within the Bounds of Reason Alone*, led Kierkegaard to see the need to transcend the ethical sphere of existence altogether.[21] Jonas's supposition that an ethics of the future might require a sort of religious faith is especially cogent here.[22] Kierkegaard's argument was that the perfection of the moral law is such that no one can possibly live up to its demands. The result is despair. We must consider whether an ethic of responsibility is finally absurd because it is not self-imposed but is imposed by an objective condition (the technological system or the collective action of persons in a technological society). Jonas's view of the inadequacy of traditional ethics may also apply to his own reconstruction of an ethic of responsibility.

> Living now constantly in the shadow of unwanted, built-in automatic utopianism [of technical progress], we are constantly confronted with issues whose positive choice requires supreme wisdom—an impossible situation for man in general, because he does not possess that wisdom, and in particular for contemporary man, because he denies the very existence of its object, namely objective value and truth. We need wisdom most when we believe in it least.[23]

III

Jacques Ellul's Existential Critique: Jacques Ellul's analysis of modern technique shows that when technical rationality extends beyond a humanly definable limit the result is an existential condition that is characterized by absurdity. In his study, "The Latest Developments in Technology and the Philosophy of the Absurd," Ellul argues that modern technique has become a sponsor of absurdity, producing absurd social situations. It is obvious that in such a social context it becomes extremely difficult for individuals to act in a rational way. But Ellul's analysis is deeper. He indicates that the irrational reactions of individuals within a technological environment support the rigid rationalized structure of the technical system. As Ellul remarks,

> We live in a society which is highly ordered, coercive (even if

morally lax), demanding, etc., and it is necessary to try to break out of its constraints. It is indispensable to revalue chaos and to deal with excessive order, to counter-balance order with disorder. That I well understand. Still, this may mean that the absurd is introduced as a positive form of disorder.[24]

It seems, therefore, to be a dynamic of technique to place us in absurd situations; and the absurd behavior elicited by technique is one of the factors that maintains the technical system. Absurdity is functionally rational, to use Mannheim's terminology.[25] Thus, we are led toward absurdity by the dynamics of technique itself or, I would say, by the extension of technical rationality beyond any humanly bearable limits.

Ellul provides a number of examples of absurd situations that reflect a "need" that arises from a technical impulse to produce various technical devices or objects. First, however, he points to the general compulsion we have for technical growth which, in itself, is absurd.

Things which nobody needs, which correspond to no use, are produced because the technical possibility is there, and this technical possibility must be exploited. We must move, inexorably and absurdly, in one direction. In like manner, products that fill no need are used in the same absurd and ridiculous way.[26]

Ellul argues that one example of this situation can be found in the development of the telephone system in France. First, Ellul points out, it seems to be a part of the national character of the French people, so to speak, that they simply do not use telephones very frequently. A 1982 statistic indicated that the French tended to use their phone only 1.3 times a day. This is a very low rate of use. It is, however, a general technical requirement of such an electronic system of communication that the efficiency of the system is increased per unit phone with higher levels of use. The situation in France in the early '80's was such that there were just too many phones for the level of use at that time. As a result, the French telephone company, Teletel, launched a public relations campaign to increase the use of telephones among the French people. This was justified, in part, by appealing to the superiority of a telecommunications system linking telephones, computers, and televisions. Here, the technical possibility of establishing a communications network gave rise to an absurd situation in which the behavior of people must be molded to conform to the technical possibility.

Thanks to this system, a telephone call can get you the number of a pen-pal, train and airplane schedules, merchandise catalogues, TV and movie programs, etc.—all of which *forces* the use of the system. It is quite seriously proposed to cease printing telephone directories, train schedules, etc., so that users will have to phone to get such information. At this point, average telephone use is going to rise and the inevitable technical progress will have been justified. Here we have behavior absurdly dictated by the imperatives of sophisticated technologies that no one needs.[27]

A similar, but perhaps, more significant example for our purposes involves the development of the French nuclear energy program. This example is important because it is based on the *prediction* of energy use. Ellul notes that after the second world war France began constructing hydroelectric plants on a massive scale. By 1955 it was found that there was simply too much electricity, based on the generating capacity of the existing plants in relation to demand. Again, a campaign was begun to induce the French people to use more electricity, including housing projects that used electric heating units and a regressive tax system making the use of electricity cheaper. Of course, this led to an enormous growth in consumption of electricity, requiring the building of nuclear electric plants. Now, Ellul points to a study done in 1971 that concluded that the price of electricity generated by a nuclear plant would be much higher than conventional plants. It also projected that the nuclear program would exceed demand by 1985. Of course, some responded to this report by suggesting that the demand for electricity must be increased.

Ellul sees a fundamental absurdity in the fact that even though such projections are inaccurate, there is an ineluctable need for them inasmuch as we seem to be dependent on them to act. A meaningless use, then, must be created in order to justify the technical development. The absurdity, therefore, lies in the prediction itself.[28]

Ellul shows that a broad range of human situations and experiences are characterized by absurdity in a technological society. The rise of the philosophy of the absurd mimics this social situation in the form of plays and films that promote absurdity. For example, many modern plays and films include scenes in which the dialogue is drowned out by background noise which is now, of course, in the foreground.[29]

The economy produces "superfluous gadgetry" for public consumption.[30] Waste is built into the economy in the form of planned obsolescence. The productive power of the economy builds up waste, polluting the environment from which we derive the resources necessary to maintain production (for a time).[31] Indeed, the very size of modern economies leads to absurdity as governments attempt to manage markets that are so large as to be beyond human comprehension.[32] Ellul remarks,

It is quite remarkable how excessive technique leads everywhere toward absurdity, toward an unprecedented situation which we do not have the faintest idea how to escape. There

is no exit. We are up against the wall. The facts outstrip the
possibilities for human comprehension.[33]

Beyond the problem of rational comprehension there lies the existential dimension of the absurdity of human experience in a technological
society. Our experience of time and space is altered by the technological
environment in the simple fact that the speed of life is so drastically
increased. Ellul points out the absurdity of the fact that this distortion
of experience has given rise to an entire tourist culture that is global in
extent.[34] Tourism has the function of a compensatory mechanism,
making life in a complex technological environment psychologically
bearable, but it does so by creating a completely artificial culture that
merely adjusts the individual to the technological environment.

There are other distorting influences in our experience of reality.
The virtual reality of many games based on the most sophisticated of
technical advances leads persons into an unreality in which computer
generated space and time sets the rules for human (non)action. Persons
become entranced by such projected images. Their influence is similar
to that of an intoxicating drug.[35] Here Ellul asks,

> What is the current effect of technology on children and
> adults, and what kind of person is being created by the
> millions, right now, without the least genetic intervention? I
> would describe this person, as I have encountered him in
> others and in myself, as a captivated, deluded, and distracted
> individual. In our society, having once been obsessed with
> work, we have now become fascinated by the multiplication
> of images, the intensity of noise, and the spread of informa-
> tion. In each area, technology impinges on anybody and
> everybody, even those who are not enamored of its achieve-
> ment, in television and mass entertainment. There is no
> exit.[36]

Furthermore, to add to the absurdity, everybody also has an increased
burden of responsibility.

So much of modern music has become sheer noise. Modern art
reflects the vertigo of the lived experience of persons living in a technological milieu. Perhaps all the more ominous, contemporary science has
become fascinated with chaos, catastrophe, and noise as objects of
rational analysis. Ellul reflects on the response given by Nobel Prize
winning scientists who were asked to comment on the goals of recent
developments in genetic engineering and other sciences.

> Apart from banalities like "to make man better" or "more
> intelligent," they were incapable of saying what human goal
> they desired. In actual fact, people do not know what they
> mean when they say that man is going to become more

human. Nobody really knows what to do with these wonderful and prodigiously effective technologies. This does not at all mean that Frankenstein monsters are going to be created, but rather that *anything at all* may be made.[37]

At last we arrive at the point at which we can surmise how it is that technique has taken away our responsibility (at the same time that it increases it).[38] In a technological society anything and everything is possible. This was Lewis Mumford's remarkable conclusion of his massive study, *Technics and Civilization*. According to Mumford, "[H]owever far modern science and technics have fallen short of their inherent possibilities, they have taught mankind at least one lesson: Nothing is impossible."[39]

I would like to suggest that if there is a sort of antinomy of pure technical rationality it is that the technological imagination can hold opposite possibilities for technical development simultaneously. Just as Capital, according to Marx's analysis, produces both an accumulation of wealth and impoverishment, modern technique produces a condition in which opposite possibilities become realized within the same social milieu. Techniques of genetic manipulation, for example, produce the possibility of overcoming infertility, treating disease, and increasing food production; but they also contain the possibility for eugenics, organ farming, and biological warfare. Ellul's analysis suggests that we cannot simply choose the good uses of this technical capacity and avoid the evil ones. The dynamic of technique ineluctably produces all of the possible manifestations of the technology. An absurd situation that makes ethical decision-making all but impossible.

IV

Existential Decision: Cutting the Gordian Knot: The absurdity of the technological environment, the lack of limits and the concomitant lack of points of reference for ethical orientation, leads us beyond the conventional ethical sphere of existence (perhaps to the religious sphere). For, a condition in which anything is possible is essentially what Kierkegaard called "the despair of possibility."

It is remarkable that Ellul is able to apply a Kierkegaardian critique to the problem of technological universalism. He argues that if anything is possible then everything becomes necessary, what Kierkegaard called "the despair of necessity." Kierkegaard's insight is that possibility gains its significance from its dialectical relation to necessity. As long as the possibilities for human-decision making are limited by some necessity, a degree of orientation is still possible. The power of technique, in Ellul's view, severs the relation between possibility and necessity. As Ellul states, quoting Kierkegaard from *Sickness unto Death*,

When technology makes everything possible, then it becomes

itself the absolute necessity. Necessity which was once the
mother of invention, has created an inventive process which
is the mother of a new necessity. 'The loss of possibility
signifies: either that everything has become necessary ... or
that everything has become trivial.'(p. 173). In fact, with
modern technology, both happen at once.[40]

Technological universalism thus leads to a technological determination of human life. As a result, whatever it is technically possible to do
must be done. Here we can say that technique is beyond good and evil.
It is no longer meaningful to speak of ethical action which is responsible.
For, responsible action is meaningful only within a limited field of
action. Here, Ellul shows how Kierkegaard analyzed the constitution
of the self as a dynamic relation between possibility and necessity.[41] In
breaking the dialectical relation between possibility and necessity technique undermines the self and its responsibility, that is, its ability to
respond.

Ellul's response to this entire situation is to appeal to a fundamental
existential decision, a decision at the level of lived experience. Such a
decision is the condition for the possibility of ethical responsibility
because it establishes a limit within which ethical responsibility is
meaningful. As Ellul points out, "A decision is never the simple solution
to some problem (something a computer can figure out); it is always a
'decision' that cuts through some Gordian knot."[42] At the existential
level a decision represents a fundamental stance the self takes with
respect to the world, establishing its responsibility.[43] This is the condition for ethical action in a technological milieu in which there are no
limits established by the technical environment itself.

Ellul argues that within a technological milieu we must choose
non-power as a guiding principle. The reason for this decision is that
now it is with respect to the technological system that human action
must be oriented. For technique is the specific environment within
which human life must be lived out.

Ellul conceives of technique as an unlimited power that has as its
only goal to augment its power; and unlimited power excludes moral
values.[44] The ethic of non-power opposes this necessity with the freedom of a decision not to exploit technological means to the fullest. Ellul's
ethical intuition is thus very close to that of Jonas. In my view, the
existential dimension of Ellul's ethic provides for the needed motiva-
tional quality that Küng found absent in Jonas's philosophical account.
It is not fear that leads to an ethic of self-restraint but anxiety (perhaps
despair). Ellul's existential insight is that the choice of non-power is
tantamount to the choice of selfhood itself.

Ellul's ethic is characterized by non-power, freedom, conflict, and
transgression.[45] These are characteristics of human action that oppose
technique and found selfhood. As such, the ethic of non-power is the
condition for the possibility of responsibility in a technological society.[46]

For, within the limits of non-power human action is definable and significant. As Ellul asserts,

> What is at stake is a vital principle (I simplify here to make a point) of setting limits: given that the almost unlimited means at our disposal permit almost unlimited action, *we must choose, a priori, non-intervention each time there is uncertainty about the global and long-term effects of whatever actions are to be undertaken.*[47]

The peculiarity of the present ethical situation that both Jonas and Ellul recognize, is the fact that technical action has such momentous consequences that now our ethical responsibility extends to the unforeseeable consequences of our actions. This is one reason that it is so difficult to find ethical orientation in a technological society.

The question our age puts to us is whether we can bear the burden of responsibility introduced by modern technique. If our response is to be meaningful, we must take up the burden of our freedom and live it out in face of the unlimited power of technique within the bounds of an ethic of non-power. Then we can exercise a responsibility that we freely choose, not one imposed on us by the technical system. This would be a responsibility guided by hope, with its positive orientation to the future, as opposed to the imaginative prediction of catastrophe.

> There is no other choice before us but suicide—or the reinvention, both individual and communal, of the ethics of responsibility. Once that happens—but not before and nowhere else—everything will become possible once more.[48]

University of Scranton
Scranton, Pennsylvania

* * * * *

Endnotes

1. Carl Mitcham, "Responsibility and Technology: The Expanding Relationship," 3, *Technology and Responsibility*, vol. 3, Society for Philosophy and Technology, ed. Paul T. Durbin, 1987.

2. Karl Mannheim, *Ideology and Utopia*, trans. Louis Wirth and Edward Shils, (Harcourt, Brace & World, Inc., 1936), 191.

3. Ibid. This is apparently the stage of thought Elizabeth Dodson Gray criticizes in "Parenting Technology," in *Research in Philosophy and Technology*, ed., George Allan, vol. 14, 1994, 59-68. Gray's model of parenting new technologies would seem to fit the third stage of moral development in Mannheim's

model.

4.   Karl Mannheim, *Ideology and Utopia*, 191.

5.   Ibid.

6.   Within the context of Mannheim's study, *Ideology and Utopia*, utopian ideas are ones that are effective in bringing about social change.

7.   See T. M. T. Coolen, "Philosophical Anthropology and the Problem of Responsibility in Technology," in *Responsibility and Technology*, ed., Paul T. Durbin, 1987, 41. Coolen's explicit concern in this article is with the concept of responsibility with regard to technology and not with "practical matters that may arise in actually exercising responsibility," 42. This latter concern is the focus of my study.

8.   Hans Jonas, *The Imperative of Responsibility*, (The University of Chicago Press), 22-23.

9.   Ibid., 6.

10.   Ibid., 7.

11.   Ibid. Jonas maintains that the traditional morality did not take into account the problem of cumulative effects because the effects of human actions in traditional societies were so negligible. This is the basis of what he calls the ethic of immediacy. It should be remembered, in this regard, that Kierkegaard dealt with the problem of cumulative effects in his analysis of original sin.

12.   See my "The Contemporaneity of the NonContemporaneous, or the Problem of Uneven Technological Development," in *Research in Philosophy and Technology*, Frederick Ferre, ed., vol. 13, 1993, 253-63.

13.   Hans Jonas, *The Imperative of Responsibility*, 29.

14.   Ibid., 30.

15.   See Henri Bergson, *The Two Sources of Morality and Religion*, trans. R. Ashley Audra & Cloudesley Brereton with the assistance of W. Horsfall Carter (University of Notre Dame Press, 1977).

16.   Hans Jonas, *The Imperative of Responsibility*, 30.

17.   Ibid., 28.

18.   Ibid., 30.

19.   Ibid.. x, 25.

20.   Hans Küng, *Global Responsibility*, (The Crossroad Publishing Company, 1991), 151, n. 78. See 52, "Even a 'duty for humankind to survive' can hardly be demonstrated conclusively in a rational way ... Is an appeal to reason by means of which it is so often possible to justify one thing or its opposite enough?"

21.   See Ronald Green, *Kierkegaard and Kant: The Hidden Debt*. (Suny Press, 1993).

22.   Jonas, *The Imperative of Responsibility*, 12, "to underpin this proposition theoretically [that we have a moral responsibility to future generations] is by no means easy and without religion perhaps impossible."

23.   Ibid., 21.

24.   Jacques Ellul, "Recent Developments in Technology and the Philosophy of the Absurd," *Research in Philosophy and Technology*, ed. Frederick Ferre, vol. 7, 1984, 81. Ellul develops the same theme in *The Technological Bluff*, trans. Geoffrey W. Bromiley (William B. Eerdmans Publishing Company, 1990),

Part III, "The Triumph of the Absurd".

25.   See my "The Contemporaneity of the NonContemporaneous, Or The Problem of Uneven Technological Development," in *Research in Philosophy and Technology*, Frederick Ferre ed., vol. 13, 1993, 253-64.

26.   Jacques Ellul, "Recent Developments in Technology and the Philosophy of the Absurd," 82.

27.   Ibid., 82-83.

28.   Ibid., 84.

29.   Ibid., 82. See Don DeLillo's, *White Noise* (New York: Penguin, 1986).

30.   Jacques Ellul, "Recent Developments in Technology and the Philosophy of the Absurd," 86.

31.   Ibid., 87.

32.   Ibid., 88.

33.   Ibid.

34.   Ibid.

35.   Ibid., 92.

36.   Ibid., 89.

37.   Ibid.

38.   See Henryk Skolimowski, "Freedom, Responsibility and the Information Society," Vital Speeches 50, no. 16 (June 1, 1984). Cited by Carl Mitcham in "Responsibility and Technology: The Expanding Relationship," *Technology and Responsibility*, ed. Paul T. Durbin, Society for Philosophy & Technology, vol. 3, 1987.

39.   Lewis Mumford, *Technics and Civilization*, (Harcourt Brace Jovanovich, 1963), 435.

40.   Jacques Ellul, "Recent Developments in Technology and the Philosophy of the Absurd," 95.

41.   Ibid., 94.

42.   Ibid., 91.

43.   Ellul actually holds that the self is made responsible by the word of God which elicits a response by putting questions to us, "Where is your brother?" The Bible is thus not a book of answers but a book of questions.

44.   See Jacques Ellul, "The Power of Technique and the Ethics of Non-Power," in *The Myths of Information: Technology and Post-Industrial Culture*, ed., Kathleen Woodward, (Coda Press, Inc., 1980), 244. See also Ellul, "Lust For Power," *Katallagete*, (Fall, 1979), 30-33.

45.   Jacques Ellul, "The Power of Technique and the Ethics of Non-Power," in *The Myths of Information: Technology and Post-Industrial Culture*, ed. Kathleen Woodward, (Coda Press, Inc. 1980), 245.

46.   See Gilbert Hottois, "Technoscience: Nihilistic Power versus a New Ethical Consciousness," in *Technology and Responsibility*, ed., Paul T. Durbin, 1987, 81, "This is why, further, any counterweight to the nihilistic power of technoscience cannot come from one or another moral system but only from a new burst of moral consciousness in general, from the originating source of every possible moral system. That is how deep the level of the encounter is between ethics and technology ... And here we are at the heart of ethics itself, not of some particular expression of it conditioned by a need to respond to the technical

context."
    47.   Jacques Ellul, "The Power of Technique and the Ethics of Non-Power," 245.
    48.   Jacques Ellul, *Living Faith*, 60.

Rationality and Responsibility in Heidegger's and Husserl's View of Technology

by R. Philip Buckley

The analysis of technology is perhaps the most phenomenological component of Heidegger's later thought. His description of this complex phenomenon called technology, however, produces certain tensions that are frequently resolved through one-sided readings. Some commentators emphasize the critical component of his reflections (often making Heidegger out to be a progenitor of "green" philosophy or "deep ecology"). Other philosophers highlight the call for a new, non-technological type of thinking (here Heidegger is often transfigured into a type of "new age" thinker). Some readers simply proclaim the "neutrality" of Heidegger's description (i.e., he is "neither for nor against technology"). A problem with the first type of interpretation is that it often re-inscribes a form of "calculative" thinking that Heidegger is clearly criticizing. The difficulty of the second view is that it sometimes goes in the direction of an "irrationalism" of which Heidegger is often accused but of which I believe he is not guilty. The third standpoint might be called a form of "phenomenological correctness" - clinging to the neutrality of the phenomenological project so as to offend nobody. Ultimately, accepting such a "distant" standpoint not only strips Heidegger's philosophy of some of its vigour, but is also rather "un-Heideggerian" in that his philosophy undermines the very notion of the philosopher as a calm, detached, neutral spectator. The aim of this paper is to chart a different course of interpretation through Husserl's earliest work; a course which doesn't take all of the bite out of Heidegger's critique of technology. Following a summary of the salient features of Heidegger's view of technology, I indicate in the second part of this paper the Husserlian foundation of much of that view. In the concluding part, I attempt to develop the sense of rational responsibility for technology which arises out of the juxtaposition of Husserl and Heidegger.

I

At the root of Heidegger's understanding of technology is the funda-

121

mental distinction between "calculative" thought (rechnendes Denken) and "contemplative" thought (besinnliches Denken).[1] The word "calculative" is connected to a type of thinking which finds its most powerful expression in modern science and which is motivated by measurement, by the search for results. "Calculative" also connotes how this thinking aims to manipulate and control. Just as a "calculating person" is someone who seeks to gain advantage, so too the thinking of science aims not just to observe a situation, but to make predictions, to plan for the future, to quantify in the sense of "taking stock" and thereby to keep everything in order. This thinking betrays for Heidegger a fundamental need for certainty and security: it wants to know exactly where "things" are and precisely what "they" might be doing.[2]

"Contemplative" thought, to the contrary, seeks neither to measure nor to control things, but to uncover their meaning (Sinn), above all, to question the meaning of things. It is a thinking which is fundamental and it is linked to Heidegger's vision of authentic philosophy. Though Heidegger is far from consistent with his terminology, contemplative thinking as authentic philosophy is often just called "thought" in his later works, and the word "philosophy" itself is frequently reserved for the philosophical tradition. Thus "thought" is at times severely contrasted with "philosophy"–that is, with the philosophy of the tradition.

The link between the philosophy of the tradition and the calculative thought of modern science is made through the introduction of yet another type of thinking: "representational thinking" (vorstellendes Denken). This thinking takes the world as something that can be "placed before" (vor-gestellt) the subject, just as one places a picture before oneself and hence representational thought treats the world or reality itself as if it were a picture (Bild). For Heidegger, the appearance of the "subject" and the world becoming a "picture" are two "interwoven events" which mark the beginning of the modern age dominated by science, the age of the "world-picture."[3] The calculative thinking which characterizes modern science is itself only possible on the basis of having a subject that can calculate and a "world" which is "placed before" it, a world that is easily manipulated, controlled and contained. For Heidegger, there would be no science without philosophy and its representational thinking.[4]

What does this "opposition" between calculative and contemplative thinking amount to? First, it is crucial to note that the thinking which Heidegger describes as taking place in science is not a "lesser" form that could be "upgraded" to a contemplative form of thought. The calculative thought of science is constitutionally incapable of being contemplative thought, and hence Heidegger's oft-quoted assertation that "science does not think."[5] Certainly scientists can reflect on their own field, on its methods, procedures and so forth. But this sort of self-interrogation aimed at improvement is part of calculative thinking in the first place. Calculative thought turned in on itself remains calculative thought. This implies a "distance" between the calculative and contemplative forms of thought, or an unbridgeable "gap" (*Kluft*).[6] The difference between these two types of thinking is one of kind and not degree.

This "gap" does not mean that calculative thought is somehow "bad," or that contemplative thinking is "better." To judge contemplative thought as superior to calculative thought is to think calculatively, and hence cannot be the task of authentic philosophy. Neither is Heidegger claiming that the nature of modern science as calculative is to be viewed as negative. It is the good "fortune" of science that it cannot "think" in the contemplative, deliberative or recollective sense.[7] The problem, it seems, occurs when calculative thought pushes aside other forms of thinking.

Heidegger wants to undermine the exclusivity of calculative thinking without denigrating it. He desires to open a space for other forms of thinking. A first step away from the domination of calculative thinking consists in uncovering the presuppositions which underlie it, in seeing that calculation is not the only possibility of human "thought." It may well be that the realm of contemplative thought can only be approached by means of this method which ultimately might be characterized as a *via negativa*. Nonetheless, the description of calculative thought and its representational character does tell us something about the nature of contemplative thought. Contemplative thought is extremely difficult to attain because, by its very nature, it cannot be "attained." To *want* to *have* a contemplative style of thought is to remain in the clutch of the basically possessive calculative style of thinking.[8] Contemplative thought is hence marked by a fundamental "passivity,"[9] it consists of a certain "letting-go" of all "attitudes," of any "picturing" of the world. Put in terms which are even more expressive of passivity, contemplative thought is a "releasement" from the dominating style of calculative thought. Both "letting-go" and "releasement" are plausible translations of Heidegger's basic characterization of contemplative thought as *Gelassenheit*.

The fundamental "passivity" of contemplative thought is a great distance from the notion of placing everything in a picture and maintaining everything in order. This distance becomes even more marked when the central notion of Heidegger's analysis of modern technology is encountered–*das Gestell*, translated here as "framework."[10] Modern technology "enframes" nature, that is, attempts to "capture" it so that it can be used, or as Heidegger puts it, transforms nature into a "standing-supply" or a "stock" (*Bestand*) which can be stored up and distributed.[11] As Heidegger says cryptically, in view of the "framework" which is modern technology, "nature becomes a gigantic gasoline station," and in principle, all relationships become technical relationships.[12]

Modern technology would then seem to be the application of calculative thought in the practical realm. However, this assumes a distinction which Heidegger does not accept, namely, between "pure" science and "applied" science, between calculative thought and its physical instantiation in technology. For Heidegger, "pure" science is not as pure as it would have us believe. In the first place, this is not meant negatively, but simply expresses Heidegger's belief that theory encroaches upon reality.[13] Moreover, modern technology is not the mere application in one way or another of modern science, but in fact it manifests the hidden nature of modern science.[14] This essence lies in the nature of calculative thought, to control, order, and organize the world, to put it in a picture that is secure in its place and can be found whenever we "need" it. Pure science is as much possessed by what Marcuse calls the "logic of domination"[15] as is technology.

Heidegger's description of the nature of calculative thought, modern science and technology, and his call for contemplative thought leaves one with very much the same feeling that one has after reflecting on the role of authenticity and inauthenticity in his earlier writings. Though continually claiming that nothing "negative" is meant by either his diatribes against life in the "They," or his description of modern science, one cannot help but feel that some sort of judgement is being passed here by Heidegger. To be sure, at least once in virtually every piece that touches upon technology, Heidegger reminds us that technology is not to be seen as "the work of the devil."[16] Yet in his various descriptions of the working of the sciences and technology, in his portrait of the on-going "busyness" of scientific research (Beitrieb) and the way we are "lost" in Beitrieb and Machenschaft,[17] the sense does emerge that something "negative" is indeed going on here. One cannot help but feel that the essence of science revealed in technology is somehow de-humanizing, cuts humanity away from its true self and its world. Even though Heidegger denies envisioning an "idyllic past" when talking about the nature of "handicraft,"[18] his exultation of simple labour and his denunciation of the effect which the typewriter has on the "handicraft" which is writing[19] certainly make it difficult to see modern technology as a "neutral" phenomenon. Even more "negativity" is felt in certain "romantic" pieces such as Der Feldweg: "Man tries in vain to bring the globe in order through planning, when he is not in tune with the voice of the country-lane."[20]

Heidegger's reflections on technology are two-sided, and it is important to maintain that ambiguity, for he sees "technology" itself as being ambiguous in its very essence.[21] But how are we to understand this ambiguity? One way is to take seriously what Heidegger says about epistemé, techné, and aletheia. His beloved Greeks also recognized and nurtured the relationship between knowledge and the "technical" ability to produce something beautiful. Hence, "handicraft," the arts and even poetry belonged to techné. Indeed, techné was related to the "bringing-forth" of poiésis, it had essentially a "poetic" function. Both epistemé and techné are ways of making something present, of bringing something into the open, of revealing. Hence, techné is related as well to that act of "unconcealment" which is designated by Heidegger as "truth" (aletheia).

For Heidegger, this notion of technology as revealing, as being part of the ambiguity of truth (for revealing includes concealment), continues to be characteristic of modern technology. What seems to have changed for Heidegger is the *episteme* intrinsically connected to *techne* which is now calculative thought with its representational character; a type of thinking which Heidegger claims was totally foreign and impossible for the original Greeks.[22] Put simply, technology has lost its truly "poetic" quality and now "reveals" the world only as a commodity or stock-pile. The revealing of modern technology is not the bringing-forth of ancient *techne* (which is not willed creation, but letting a world approach us), but rather an aggressive challenging[23] of the world to produce that which can be stored up and manipulated. In issuing this challenge, modern technology has become mere "technique," mere functioning.

Modern technology as a form of revealing that consists only in the ordering, control, and distribution of resources not only conceals the notion of revealing at work in ancient *techne*, but obstructs other possible forms of revealing as well. Modern technology is thus a type of revealing which banishes every other possible form of revealing.[24] In bringing to completion the essence of calculative thought, modern technology also adds to the hegemony of this thinking. This technology not only dominates the earth, but it dominates humanity to the extent that it becomes virtually impossible to conceive of humanity or the world in a non-technical way. At times, one has the impression in reading Heidegger that the remarkable feature of the modern age is not so much that it is dominated by calculative thinking, but that it is dominated by one form of thinking throughout the world. This "planetary age" of calculative thinking extended by technology is for Heidegger a form of revelation. It reveals the essence of calculative thought to dominate everything.[25]

In reviewing the ambiguity of Heidegger's approach to technology, one certainly comprehends why Heidegger once described his approach as saying "yes" and "no" to technology.[26] But is this yes and no merely an attempt to place limits on technology without condemning it; or is something more at stake in Heidegger's reflections than simply reaching a liveable balance between the benefits and drawbacks of modern technology? Comparison with Husserl's own "yes" and "no" to science provides a worthwhile approach to these questions.

II

Husserl's early analysis of calculative rationality in the *Philosophy of Arithmetic* is quite stunning in its similarity to Heidegger's. Calculation within the framework of Husserl's analysis of number refers to the use of symbolic or inauthentic number concepts, and it is opposed to "authentic" counting which is rooted in intuition. The ability to calculate is for Husserl a great power of our intellect. This is particularly true when we consider that one of the theses attached by Husserl to his "Habilitation" in 1887 is that "in an authentic sense, one can hardly count beyond three."[27] Most of what is normally called counting in ordinary language is in fact calculation or "inauthentic" counting according to Husserl. Without this power to calculate, the human mind would be capable of little arithmetical progress. Moreover, symbolic or inauthentic thought tends to be very "efficient." It allows for the greatest result from the least amount of effort, and hence is representative of the principle of "economy of thought" of Avenarius and Mach to which Husserl devotes substantial commentary in the *Logical Investigations*.

Calculation plays a central role in the rapid progress which science in general is able to make. However, there is a negative aspect to this power. Though economical, calculation for Husserl also implies a certain "blindness." To put it rather paradoxically, it allows for a powerful "thinking" without much true thought. The "thoughtless thinking" of calculation is not wholly negative for Husserl: at one point he declares that all cultures which have lifted themselves out of barbarism have developed a symbolic number-system.[28] However, the blind functioning or "thoughtlessness" of calculation can "get out of control." What Husserl describes in his widely read *The Crisis of the European Sciences and Transcendental Phenomenology* regarding the mathematization of nature by Galileo and the forgetting of the "life-world" from which science arises is really an extension to European culture of his analysis of calculation from more than 40 years earlier.

Husserl, like Heidegger, is ambivalent in his understanding of calculation. On the one hand, inauthentic thought seems almost intrinsic to science, and indicates an advanced culture. On the other hand, when inauthentic or symbolic thought is not linked to authentic insights and awareness, it loses the right to be called "science" and becomes for Husserl mere *technique*. This in no way prevents science from "working"–it even allows it to become ever more active. But Husserl calls this inauthentic functioning "activity in passivity." It is acting without real insight into what one is doing, without a thorough awareness of the meaning of one's action, without true understanding. Husserl became more and more convinced that despite all the proclamations of mastery which science made and all of the advanced technical manipulation evident everywhere, science was actually becoming impoverished in terms of its self-awareness and enslaved in such "activity in passivity."

The ambivalent sense of both Husserl's and Heidegger's descriptions of a calculative thinking within a science that had become just "technique" is remarkable. The blindness and narrowness of science, its "activity in passivity" with its anonymous function, its forgetfulness of its own origin and its inability to gain insight into itself, all these Husserlian themes seem even more evident in the description of *das Gestell*. The consequences of one-sided thinking become more dramatic, and the domination of one sort of one-sided thinking over the entire world threatens to remove any possibility of reaching a standpoint where that one-sided thinking can be contemplated. And yet while both seem to attack a certain imperialism evident in calculative thought, neither wishes to destroy this thought, replace it, nor do they imply that such a desire would be even possible to fulfill.

This ambiguity is often resolved vis-à-vis Husserl by stressing that while he admits of some benefits to calculative thought, the goal of authentic philosophical life is ultimately to ground this thought in genuine insight. If inauthenticity is marked by "activity in passivity," then authenticity for Husserl might be described as "activity in activity." Inauthenticity is always able to be overcome through an active and brave willing not to live in passivity, through an active seeking of insight into what one does and why one does what one does, through what Husserl calls at one point a "heroism of reason."[29] According to Husserl, authentic human life consists of pulling oneself out of the life of simply living along in a current of unreflected "tendencies" and pre-given validities, and the formation of a life of critical choice.[30]

This "activity in activity," this development of a continually questioning attitude, is described as a true *habitus* of critique. The acquisition of this *habitus* to critique, understand and justify all of one's positions, to constantly inquire into the unreflected habits and "passive" actions of everyday life, can only take place on the basis of a willed effort, of a choice to direct one's life in a new way.[31] A certain solemn decision must be made to consciously and consistently escape from the passive tendencies and thoughtlessness which characterize inauthentic life. For this reason, the best description of authenticity is self-responsibility, and the ultimate characterization of Husserl's philosophy is one of *absolute self-responsibility*.

From this interpretive standpoint, the apparent convergence between Husserl and Heidegger in their descriptions of technical-calculative thought diverges regarding the significance of the phenomena they describe. For Husserl, the blindness of modern science is due to objectivistic and naturalistic prejudices, and these prejudices in fact indicate that it is not true science. For Heidegger, this "blindness" belongs to the essence of science as science. According to Husserl, for science to be true science it must escape from the pressing demands of everyday praxis; for Heidegger, the entanglement of theory and praxis shows that for finite, human *Dasein* there is no such thing as *pure* theory. Whereas Husserl interprets the narrowing of science to mere calculation as a loss of the original ideal of science, Heidegger holds that the dominance of calculative thought reveals the very essence of science. For Husserl, the forgetfulness of science is an avoidable weakness on the part of the scientist, while for Heidegger it is inevitable and even necessary for the functioning of science. Where Husserl speaks of correcting all these short-comings of science by a willful assuming of another attitude, Heidegger alludes to a completely other form of thought which can neither be willed, sought after, nor mastered.

Such an interpretation would explain why Heidegger, in what appear to be his most explicit comments on the *Crisis*-text itself,[32] launches a strong attack on all those who talk about a crisis of the sciences, for indeed all these "crisis-makers are in fundamental agreement with the hitherto existing science and embrace it, even become its best advocates,..."[33] For Heidegger, the group of "crisis-makers" was a fairly large club, but Husserl would clearly belong to its charter-members. Even though Husserl attacks the "old" science, Heidegger would see him as remaining fully in its grips, even belonging to the moment of its fulfillment. The attempt to grasp all by a masterful and self-mastered subjectivity is taken by Heidegger as a sign of a fundamentally calculative thinking, the same sort of thinking that underlies the very sciences said to be in crisis. From this viewpoint, Husserl's prescription was itself rooted in the disease it sought to heal.

III

The reading just worked out stresses the difference between Husserl and Heidegger; but such a reading that disarms the ambiguity of Husserl's phenomenological analysis of calculative thinking, seems itself rather calculative. Is it possible to think about the ambiguities in Husserl's and Heidegger's description of technology without resolving them and without having to make a calculative decision about which view is "better?" Such questions can be approached by picking up on a word that Heidegger uses at the same time as Husserl's *Crisis*-text, namely "responsibility." Certainly, no word is more Husserlian–the dominance of calculative thought is ultimately for Husserl the result of lapsed responsibility, of not being willing to "answer" both to myself and to others for what I believe and what I do. With Heidegger, it appears in the context of his own discourse on the destitution of spirit. What is this responsibility in an age of destitution? What could it mean in the context of a thinking that sees the domination of calculation and technology in the modern age not as primarily rooted in the "irresponsibility" of the scientist and the philosopher, but as belonging to the "fate" (*Geschick*) of our times?

It is not easy to find the proper locus for the notion of responsibility in Heidegger's thought. There is a sense in Heidegger's work that "nothing can be done," or that "we should do nothing else than wait."[34] After all, responsibility seems to entail answering for a decision, and deciding seems a highly calculative act, which by its very nature, "excludes." Heidegger's later writings really do seem a questionable place to search for a notion of responsibility. But even the posing of this question yields a potential point of departure: it is responsibility for nothing else than questioning. It is not a responsibility that is to be conceived in terms of the answers sought by calculative thought. Nor must we view responsibility in terms of a moral lapse, a mistake.[35] The point of departure for Heidegger's philosophy is a forgetfulness of the question of Being, but this is a forgetfulness which belongs to the nature of the question and the presencing and withdrawal of Being itself; it is a necessary forgetfulness which ought not to be seen as a moral lapse.

And yet, the question of Being is that for which we are called to assume responsibility. Responsibility for the question of Being is a possibility for *Dasein*, and part of the recollection of the question of Being is the recollection of the question of *Dasein*'s responsibility for the question. There are two aspects to the recollection of responsibility–the "response" and the "ability." These aspects fit well within the framework of the active-passive interplay so crucial to Heidegger's thought in its entirety. At the outset, there is clearly an emphasis on the "ability" of *Dasein*, of the question of Being as a possibility of *Dasein*, and indeed, the "ownmost" possibility of this being. Because this being is the focus of attention, the highpoint of Heidegger's early work is the revelation of the ontological meaning of this being as temporality.

The shift in the later Heidegger is to the presencing of Being in terms of time's own self-extending.[36] This shift is no indication of the failure of *Being and Time*, but the attempt by Heidegger to recollect the question from a different angle. This new angle is the giving of Being itself, the "es gibt." Being gives itself in all sorts of entities, these entities are its gifts. That Being gives "itself" does not imply that Being is an entity. "It" is its own giving, and time itself is the "es" of "es gibt."[37] Being is given finitely, historically, epochally in time, and in "our" time this giving is manifested in technology. Humanity is there to receive this revelation, though as we have pointed out earlier, it is a form of revelation which conceals its revealing and other forms of revealing as well.

In the later Heidegger, *Dasein* appears more passive, a greater emphasis is placed on *Dasein*'s openness to Being. It could be said that the emphasis here is on the "response" of *Dasein* to this giving of Being. Yet, Heidegger is still able to speak of the responsibility of *Dasein*, of the "ability" to "respond" to this gift. There is still an active comportment. Perhaps the priority in the later Heidegger has shifted from the active side of the active-passive interplay to the passive side; but only in the entirety of Heidegger's work can the movement between activity and passivity be grasped in total balance.

If *Dasein* is called to respond to Being, and if Being's giving of itself in our times is through the revelation of technology, then in a very specific sense, *Dasein* can be said to be responsible for technology. This can certainly not be construed in the manner of technology as an instrument which lies within the control of *Dasein*, for technology is not essentially machines which are at the disposal of *Dasein*. To conceive of technology as such is not only to misunderstand it, but is also to stay firmly within the manipulative, possessive, challenging type of thinking that makes technology possible. Rather, technology is the *Gestell*, which itself enframes everything, including *Dasein*. For this reason, Heidegger is able to claim that the essence of technology is "nothing human."[38] To be sure, the "framework" is not possible without humanity, but it remains something beyond complete human control. Speaking of the revelation of technology Heidegger asks:

> Does this revealing happen somewhere beyond all human
> doing. No. But neither does it happen exclusively *in* man, or
> decisively *through* man.[39]

In being responsible for technology, for the question concerning technology, *Dasein* is responsible for that which to a certain extent is beyond *Dasein*. Being a "responsible" person in no way implies a complete or full autonomy. To the contrary, it suggests a "dependence" on what comes from the outside, an ability to be touched, a fundamental vulnerability.

Heideggerian responsibility is thus a great distance from a traditional interpretation of responsibility which has much to do with control and with calculative thinking. What is usually intended when someone is said to be responsible for their actions is that one is responsible for the results of one's actions. A responsible person is therefore the person who takes into account these possible results, who estimates and predicts the effects a particular action will have. The debate around the limits of such responsibility centers on the limits of estimation and prediction, and the absolution of responsibility usually occurs at the point where it is admitted that nobody could have possibly predicted such and such an outcome, that certain results were "unforeseen." Those who do not wish to see any diminishing of responsibility are forced to an argumentation which claims that one should have foreseen these consequences, that what is claimed to be "unforeseen" was able to be seen, and that indeed, it probably remained unseen not due to a lack of ability but due to a lack of willing to see. The defense of a traditional notion of responsibility is inevitably connected to the desire and attempt to see all.

This traditional sense of responsibility means to be in control of that which comes from us, and the efforts to extend this sense of responsibility lead to extended notions of what comes from the subject. To the contrary, the sense of responsibility found in Heidegger's thought does not mean to control or to manage, but rather to respond to what comes from afar and to assume the care for that which we can never master. The far-reaching consequences of our actions, how people react and how these actions return to affect us are certainly beyond our control. It is implausible to assume that we could ever know the full consequences of our actions; and yet, it would seem an inhuman life if we were to deny responsibility for our actions and their results. Could it not be said that authentic responsibility is precisely this–to assume responsibility for those people or things which are given to us? To claim responsibility only for that which we can predict with certainty or control with ease, or to abdicate responsibility on the basis that we cannot predict or simply cannot control, these seem less than human ways of behaviour.

Oddly enough, to remain within the active, dominating, willful type of calculative thinking which belongs to technology can actually be taken as a refusal to be "responsible." To be sure, such thinking is often closely allied with talk about responsibility, but this is a pseudo-responsibility, a responsibility for that which I can control, but not for that which might place me under its spell. The notion of responsibility in Heidegger's works thus places an emphasis very much on a "response" to that which comes from beyond. In the realm of authentic responsibility for our actions, such a response has little to do with the retroactive justification of certain actions based on calculations made at the time of decision. Far more suitable responses might be gratitude, or remorse,–resulting in the request for forgiveness. Without doubt, the great difficulty which many philosophers have with Heidegger's own involvement with National Socialism is not just the involvement itself, but Heidegger's quasi-calculative defense that nobody could foresee the course that National Socialism was going to take. Perhaps he genuinely could not. But this in no way lessens responsibility and the obligation of a correct response–which in this case could only be humility and remorse. Such a response was never forthcoming.

A notion of non-calculative responsibility as a different way of dealing with the ambiguity of Heidegger's view of technology also opens up a way to see Husserl in a less one-sided manner. To be sure, Husserl's notion of responsibility seems linked to a controlling subjectivity, to the attempt to take control of that which comes from beyond. It is also true that responsibility for Husserl can even be said to be directed not to the outside, but to the self: it is "self-responsibility." But it is important to note that Husserlian responsibility ultimately never functions within the realm of calculative thought. Husserl is never one to say that the responsible person is one who is willing to estimate well the effects of his or her actions. Self-responsibility for Husserl is linked much more with an attitude of self-critique, with the willingness to engage and question oneself, to provide the evidence for that which one believes, and to discard that for which no evidence can be provided. Responsibility for Husserl is self-justification. However, this justification is not rooted in showing how one had calculated properly, but in producing "evidence," In fact, evidence is precisely that which Husserl had seen to be missing in purely calculative thinking. The willingness to provide evidence is thus linked to the willingness to critique oneself and to assume responsibility for one's beliefs.

Evidence is something that humans have an "ability" to gather, but it is also something to which we "respond." It is something that comes from beyond, and it too can be seen as a "gift." In the *Logical Investigations*, the experience of "truth" which occurs when an "empty intention" is "fulfilled," when there is evidence for something, is not just a subjective feeling. The evidence to which one is able to point comes from the outside. For sure the reception of this evidence is rooted in a readiness, in a skill at "seeing," in a desire to have evidence, but the fulfillment of this desire is never solely of the subject's making. Evidence is not easily obtained, it must be "worked for," and the responsibility for doing this work belongs to philosophy. Responsibility as the *habitus* of critique is an element of Husserlian philosophy which can be freed from the framework of calculative and technological thinking.

The notion of responsibility which has been developed here is full of a sense of acceptance and places emphasis on the ability to respond to that which comes from beyond. Such responsibility does not attempt to control that which is uncontrollable, and it does not maintain that the recognition of limits of what can be controlled puts an end to responsibility. In being open to that which is beyond us, responsibility forms a critique of that thinking which seeks to control. The recognition of this receptivity does not spell the end of calculative thought and its fulfillment in technology, but it leaves open a space for non-technological ways of comportment towards the world and one another.

McGill University
Montreal, Canada

* * * * *

Endnotes

1. Martin Heidegger, *Gelassenheit* (Pfullingen: Günther Neske Verlag, 1959), p 15-16; *Discourse on Thinking*, trans. J. M. Anderson and E. H. Freund (New York: Harper and Row, 1966), 46. *"Besinnung"* could also be translated as "reflection," though this has a rather epistemological connotation. In the translation of *Gelassenheit*, "meditation" is used, but the Cartesian background here is difficult to avoid. Also possible are "consideration" or "deliberation" [proposed by Theodore Kisiel, "Science, Phenomenology, and the Thinking of Being," in *Phenomenology and the Natural Sciences*, ed. J. Kockelmans and T. Kisiel (Evanston: Northwestern University Press, 1970), 173], but these maintain something of calculation–one normally considers things in view of future planning or deliberates upon options in order to choose the best one.

2. In a rather playful fashion, Heidegger brings out the longing of calculative thought for security by pointing out that a crucial figure in the development of calculative thought, namely, Leibniz, was also the inventor of "life-insurance." See *SvG*, 202.

3. Martin Heidegger, "Die Zeit des Weltbildes," in *Holzwege*, ed. F. W. von Hermann, *Gesamtausgabe 5* (Frankfurt: Klostermann, 1977), 93. [Hereafter *Holzwege (Ga 5)*]; "The Age of the World-Picture," in *The Question Concerning Technology* , trans. W. Lovitt (New York: Garland, 1977), 133. [Hereafter *OCT.*]

4. Martin Heidegger, *Der Satz vom Grund* (Pfullingen: Neske, 1957) 110-111. [Hereafter *SvG*.]

5. Martin Heidegger, *Was heisst Denken?* (Tübingen: Max Niemayer, 1957), 4. [Hereafter *WHD*]; *What is Called Thinking*, trans. J. G. Gray (New York:

Harper and Row, 1968), 8. [Hereafter *WCT*.]

6. *WHD*, 4-5; *WCT*, 8.

7. *WHD*, 4; *WCT*, 8.

8. The lines from T. S. Eliot's *Four Quartets* come to mind: "In order to possess what you do not possess/You must go by the way of dispossession."

9. I use the word "passivity" with caution, for as will be shown in the discussion of responsibility in the conclusion, Heidegger wants to distance himself from traditional notions of both passivity and activity.

10. The central essay in which this notion of *Gestell* is developed is "Die Frage nach der Technik," in *Vorträge und Aufsätze* (Pfullingen: Neske, 1954, 13-44. [Hereafter *VA*]; "The Question Concerning Technology," in *The Question Concerning Technology and Other Essays*, trans. W. Lovitt (New York: Harper and Row, 1977), 3-35 [Hereafter *OCT*]. See as well the beginning of "Die Kehre," in *Die Technik und die Kehre* (Pfullingen: Günther Neske Verlag, 1962), 37-47; "The Turning," in *OCT*, 36-49.

11. Heidegger, "Die Frage nach der Technik," in *VA*, 22-27; "The Question Concerning Technology," in *OCT*, 14-19.

12. Martin Heidegger, *Gelassenheit* (Pfullingen: Neske, 1959), 20; *Discourse on Thinking*, trans. J. M. Anderson and E. J. Freund (New York: Harper and Row, 1966), 50.

13. Heidegger, "Wissenschaft und Besinnung," in *VA*, 56; "Science and Reflection," in *OCT*, 167.

14. *WHD*, 53; *WCT*, 22.

15. Herbert Marcuse, *One-Dimensional Man* (Boston: Beacon Press, 1966), 144-69.

16. Martin Heidegger, *Identität und Differenz* (Pfullingen: Neske, 1957), 33; *Identity and Difference*, trans. J. Stambough (New York: Harper and Row, 1969), 40. See also, *Gelassenheit*, 24; *Discourse on Thinking*, 53.

17. Heidegger, *Beiträge* (GA 65), 57; cf. also 126ff.

18. *WHD*, 54; *WCT*, 23.

19. Martin Heidegger, *Parmenides*, hrsg. von M. S. Frings, *Gesamtausgabe* 54 (Frankfurt a.M.: Vittorio Klostermann, 1982), 119.

20. Martin Heidegger, *Der Feldweg* (Frankfurt a.M.: Vittorio Klostermann, 1989), 21.

21. Heidegger, "Die Frage nach der Technik," in *VA*, 41; "The Question Concerning Technology," in *OCT*, 33.

22. Heidegger, "Die Zeit des Weltbildes," in *Holzwege* (GA 5), 91; "The Age of the World-Picture," in *OCT*, 131.

23. Heidegger, "Die Frage nach der Technik," in *VA*, 22; "The Question Concerning Technology," in *OCT*, 14.

24. Heidegger, "Die Frage nach der Technik," in *VA*, 34-35; "The Question Concerning Technology," in *OCT*, 26-27.

25. David Kolb has pointed out the link between *Gestell* and the military notion of "mobilization order" (*Gestellungsbefehl*). Kolb suggests that "the military connotation is useful, since the military is a realm where everything is to be set in order waiting to be used at a moment's notice. This instant and complete availability is much of what Heidegger has in mind by *Gestell*." David

Kolb, *The Critique of Pure Modernity: Hegel, Heidegger and After* (Chicago: University of Chicago Press, 1986), 145.

26. Heidegger, *Gelassenheit*, 24-25; *Discourse on Language*, 54.

27. "Im eigentlichen Sinne kann man kaum über drei hinaus zählen." Edmund Husserl, *Philosophie der Arithmetik*, hrsg. von L. Eley, *Husserliana XII* (The Hague: Martinus Nijhoff, 1970), 339. [Hereafter *Hua XII*.] Later, Husserl placed the limit at ten to twelve, in order to allow for an intuitive foundation for the decimal number system.

28. *Hua XII*, 272-273.

29. Edmund Husserl, *Die Krisis der europäischen Wissenschaften und die transzendentale Phänomenologie*, hrsg. von W. Biemel, *Husserliana VI* (The Hague: Martinus Nijhoff, 1954), 348. [Hereafter *Hua VI*]; *The Crisis of the European Sciences and Transcendental Phenomenology*, trans. D. Carr (Evanston: Northwestern, 1970), 299. [Hereafter *Crisis*.]

30. "Leben ist in sich Streben und ist als menschliches ein Aufsteigen von Passivität in Aktivatät, von einem Sich-gehen-und-tragen-lassen, von einem Leben nach Neigung, zu einem überlegten Wählen, zu einem kritischen, höher Bewertes bevorozugenden Leben." Husserl-Archive manuscript *F I 24/70b* (1923). I am grateful to Ullrich Melle for offering his insights on this manuscript.

31. *Hua VI*, 147-148; *Crisis*, 145.

32. Martin Heidegger, *Grundfragen der Philosophie: Ausgewählte "Probleme" der "Logik"*, hrsg. von F. W. von Herrmann, *Gesamtausgabe 45* (Frankfurt a.M.: Vittorio Klostermann, 1984), 53-57. [Hereafter *GA 45*.]

33. *GA 45*, 53.

34. Heidegger, *Gelassenheit*, 37; *Discourse on Thinking*, 62.

35. Heidegger, "Die Frage nach der Technik," in *VA*, 18; "The Question Concerning Technology, in *OCT*, 9.

36. Heidegger, "Zeit und Sein," in *ZD*, 16; "Time and Being," in *TB*, 10-16.

37. Heidegger, "Zeit und Sein," in *ZD*, 16; "Time and Being," in *TB*, 16.

38. *WHD*, 53; *WCT*, 22.

39. Heidegger, "Die Frage nach der Technik," in *VA*, 31; "The Question Concerning Technology," in *OCT*, 24.

The Machine That Couldn't Think Straight

by Stephen A. Dinan

Since at least the 15th century B.C., when the "singing" statue of Memnon, king of Ethiopa, was built in Egypt, human beings have been fascinated by the possibility of creating a machine in their own image.[1] After the first practical electronic computers appeared in the late 40's, this ancient fascination took root in the new discipline of Computer Science in the form of a desire to build a machine that could think. In 1950, the computer theorist Alan Turing initiated the field of artificial intelligence (AI) when he expressed the hope that "machines will eventually compete with men in all purely intellectual fields" such as chess playing and language use.[2] For over thirty years, the best known and most relentless critic of the attempts made by AI researchers to fulfill Turing's hope has been Hubert Dreyfus.[3] Throughout this period, Dreyfus has carefully analyzed and rejected the goal of AI workers to produce human-like intelligence in a computer, as well as the presuppositions they have about human intelligence which led them to formulate this goal and the programming techniques they have used to reach it. Nevertheless, even the first edition of his study of the failures of AI theories and techniques, published in 1972, concludes with an argument that artificial intelligence techniques can be used to approximate, at least to some extent, several types of intelligent behavior (Dreyfus, 291 ff). In this paper I will use Dreyfus's "critique of artificial reason" to explore the limits and possibilities of simulating human intelligence using the techniques of knowledge representation developed in artificial intelligence research.[4]

I

Formalizing Human Knowledge: According to Dreyfus, the programming techniques used by AI researchers are based on the Platonic view that human knowledge can be formalized in explicit definitions. These definitions express the criteria to be used in identifying objects and actions in the world of our everyday experience. Because he reads Platonic definitions as instructions for interpreting experience, Dreyfus

135

believes that Platonism culminates in the attempt to simulate intelligence by formalizing human knowledge as explicit facts and rules stored in a computer (Dreyfus, 67-72). Though this formalization occurs in many ways, artificial intelligence software attempts to simulate intelligent behavior by classifying data and by determining an "intelligent" (that is, rational) response to this data. The information stored in such programs usually includes some representation of the knowledge which purportedly guides humans when *they* engage in the behavior to be simulated, that is, knowledge of the objects and situations with which the computer program is intended to deal. The rules used by the program, on the other hand, prescribe how data presented to the computer is to be identified or classified, as well as how to determine an appropriate response to this data (given the purpose of the program). Whether the computer is attempting to classify data or determine an appropriate response to it, it will have to search through often large amounts of stored information to find the information it needs to resolve the problem at hand. There are two kinds of search techniques that may be used: a "brute force," "trial-and-error" search of all available information until the needed information is found, and a more limited search of only that information prescribed as relevant to the present situation by so-called "heuristic" rules. Heuristic rules are generally reliable procedures that usually work in certain kinds of situations but are not guaranteed to succeed each time they are used. Where the amount of information to be searched is limited enough that an exhaustive search for a solution is practical, such a search technique is preferred because of its reliability. In many cases, however, artificial intelligence programs use heuristic rules because an exhaustive search of all relevant information is impractical. The adequacy of these rules for classifying data or identifying an appropriate response is one of the major concerns of AI researchers. The other major concern is how to ensure the adequacy of the knowledge base used by a program. Before turning to the theoretical issues regarding the limits and possibilities of artificial intelligence, it will be useful to first examine how several common types of artificial intelligence programs use stored representations of knowledge and rules, and what problems they encounter as a result.

II

Artificial Intelligence and Its Discontents: In *chess programs and other board games*, the data presented is either a command to produce an opening move, or information about the move just made by an opponent. The strategy rules of the computer program dictate which move is an appropriate response. This requires stored information about every legal move for every piece, and general information such as the size of the board, the rules determining the order of play, and the definition of winning. The computer must also have rules to determine the probability of any one move leading to a win in any possible situation. If a "brute force" search is used, it determines the probability of winning by considering every possible move that could result from each move the computer could make in the present situation. This has always worked well in games like tic-tac-toe that have a small number of distinct moves. Because of the exponentially large number of possible moves in chess, however, most chess programs have relied heavily on heuristics to limit brute force searches to the most promising paths.[5] In 1972 Dreyfus noted that "all ... heuristics either exclude some moves masters would find or leave open the risk of exponential growth" in the number of moves considered by a brute-force search (Dreyfus, 101). Seven years later, despite the improvement in heuristics and the increased speed of computers, Dreyfus doubted that any chess program would ever achieve the level of master play.[6] Even so, he recognized that chess was "decidable in principle by counting out all possible moves and responses" if the problem of exponential growth "inevitably connected with choice mazes" could be solved (Dreyfus, 101). In the last few years, that problem *has* been substantially solved. Relying on massive brute-force searches made possible by parallel processing and much faster components, supercomputers can now compete credibly not only with chess masters but also with grand masters and even the current world champion without having to use improved heuristics.[7] In the process, however, computer scientists have abandoned the attempt to replicate the techniques of play used by master chess players. Many desktop systems still produce very good results, however, by combining sophisticated heuristics with faster brute-force searches.

Visual pattern recognition software attempts to identify data in the form of a digitized image of one or more objects—that is, an image whose distinguishable parts have been encoded as a patern of zeroes and ones and stored in the computer. Rules specify how to represent the relevant features of the digitized object and how to classify that object by searching through stored information until a match is found for its represented features. Programs such as character recognition software or fingerprint identification software are very reliable because objects such as printed characters and the lines of a fingerprint have a limited set of features which are relevant to their identification. It is much more difficult, however, for a computer to reliably identify more complex objects with many different kinds of relevant features. It has so far proven impossible, for example, to replicate perception of complex visual objects in a natural environment. To do this, a program would need stored information about all the features of previously categorized objects that might be relevant to identifying new objects. It would also need rules to isolate the image of each object from the images of other objects in complex digitized input, rules for abstracting and representing all the possibly relevant features of each object, and rules for searching through stored information about previously identified objects to find a match for this abstract representation. Given the enormity of this task, it is perhaps not surprising that successful knowledge-based pattern recognition has been limited to relatively simple objects in restricted environments.

Voice recognition software attempts to translate a digitized stream of sounds into a stream of words. The properties of the sound stream are first "sampled," or measured, and recorded in the computer as a set of binary numbers. In the traditional AI approach, rules prescribe how to isolate sound patterns for individual words, represent them in the computer, and match them against stored sound patterns derived from previously digitized words. A number of problems have plagued attempts to program such rules. For instance, it has proven extremely difficult to extract individual words from continuous speech—voice recognition systems often require that individual words be spoken with a significant delay between them. Moreover, words spoken at different times, by different people or under different conditions never sound exactly alike and often sound quite different. Regionally distinct accents, slurred words, slow or fast speech, differences in voice timbre, the location of a word in a sentence, changes in its grammatical function, and many other variables affect the way the same word will be pronounced. Because of this, voice recognition systems often require the user to "train" the computer to recognize his or her voice. Even then, a limited vocabulary must be used since the sound pattern of every word to be recognized must be input several times by the user during training. If anything affects the sound of the user's voice, such as having a cold or speaking in a room that produces echoes, the computer is unlikely to recognize the voice. Recently developed techniques have made some advances in overcoming these problems by bypassing the kinds of rules employed in the traditional AI approach to artificial intelligence, relying instead on the massive computing power of neural networks to discover statistically significant acoustic patterns in various combinations of words spoken by a variety of speakers. Even so, the results are a far cry from human voice recognition. Unlike all existing computer systems, we routinely accept virtually unlimited "input" in our "mother tongue" and understand continuous speech by a wide variety of speakers with different accents, speaking under difficult conditions. We can even understand words we have never heard before!

Language translation software tries to convert text from one natural language to another. Rules prescribe how to determine the words, syntax and meaning of the initial text, and how to produce an output text in another language that has the same meaning. *Natural language processing software* is similar to language translation software, but only has to do half as much work—its rules prescribe how to identify the words, syntax and meaning of a text, but it does not translate this meaning into another language. As in attempts to computerize other complex intelligent tasks, success in these two areas has been limited. The more limited the allowable syntax and vocabulary of input, the more successful the software has been. Success has been achieved in processing natural language over restricted domains of meaning, such as

translating common words and expressions from one language to another and responding to database queries input in natural language.
Using scripts and other devices to represent stereotypical situations, it
has even been possible to develop programs that can summarize some
news stories, answer questions about the motives of a character in a
story, and even write a story based on input information. Syntactic
analysis has improved but the determination of meaning—semantic
analysis—is inadequate for processing unrestricted language input.
The reason is not only the virtually unlimited range of possible linguistic
expressions, but the tremendous dependence of meaning on the context
of a statement as well. Whereas humans experience most everyday
language as unambiguous because of their comprehension of the context
in which language is used, computers must "disambiguate" most linguistic expressions through semantic analysis. Unrestricted natural language processing would require semantic analysis to determine every
kind of context in which every word or expression could be used. This
would require that we store and organize an adequate representation
of the greater part of our knowledge of the world about which we
speak—and no one has any idea how that could be done.

Expert system software attempts to provide the kind of advice that
would be given by a human expert, or a group of human experts, if they
were consulted about a problem in their area of expertise—selecting
wines, investing in stocks, drilling for oil, diagnosing illness, repairing
computers and automobiles, etc. Typically an expert system requires
the user to answer a series of questions about his or her situation or
problem. The answers to these questions—the data the program receives—will trigger other questions, eventually narrowing down the
problem to one of several possible solutions. These solutions are prescribed by conditional rules which match the user's description of the
problem to certain conclusions or actions to be performed. Expert
systems have been very successful because they recognize two essential
limits on the attempt to model human knowledge. First, by addressing
only particular kinds of problems, they limit the amount of human
knowledge they must represent and store. Second, they supplement,
but do not attempt to replace, the intelligence of the user of the system.
An expert system's questions force the user to observe facts, and it is
these facts which provide the basis for the final identification of a
solution. By asking an attending physician questions, for example, a
medical diagnostic expert system draws upon that user's expertise in
observing, classifying and reporting symptoms. Its success thus depends
at least as much on the intelligence and experience of the user as it does
on the rules which draw connections between the user's observations
and possible solutions to the proposed problem.

III

The Problem With Formalization: As I suggested earlier, the adequacy of the facts and rules which guide artificial intelligence programs is frequently problematic. As we might expect, the chances of a program's success in representing, identifying and responding to data is dependent upon the degree to which the program's designers can anticipate exactly what will be encountered in the problematic situation. The most successful programs are those in which the range of data is limited, the data has relevant features that can be precisely determined, and the appropriate response to that data can be precisely calculated. On the other hand, programs which deal with highly variable and ambiguous data and attempt to simulate a human response that is not simply the result of numerical calculation do not fare nearly as well, since it has proven impossible to formally represent all the knowledge humans use in dealing with complex objects and situations.

In the first edition of his book, Dreyfus complained that research purporting to replicate human intelligence used *ad hoc* methods that left "untouched the problem of how to formalize the totality of human knowledge presupposed in intelligent behavior." (Dreyfus, 226) But why should "the totality of human knowledge presupposed in intelligent behavior" *have to* be successfully formalized for use by a computer program? Why can't computer scientists be content to formalize only that knowledge needed to make particular artificial intelligence applications work? With the successes they have already achieved in chess, optical character recognition, and expert systems software, why shouldn't computer scientists look forward to a time when they will have formalized enough human knowledge to make the remaining areas of artificial intelligence applications succeed as well? The answer, I believe, is that *there is no limited subset of human knowledge that is sufficient* to allow the success of the most difficult artificial intelligence tasks such as natural language processing of unrestricted input, voice recognition of unrestricted speech, and the visual perception of unrestricted or even reasonably complex objects in a natural environment.

As Dreyfus notes, the difference between expert systems and chess programs, on the one hand, and unrestricted natural language processing programs on the other, "is precisely the difference between domain-specific knowledge and general intelligence," (Dreyfus, 33)   that is, intelligence which draws on the totality of our knowledge of the world. Domain-specific programs, like optical character recognition and fingerprint identification software, limit the amount of human knowledge they must represent by adopting techniques which have little in common with the ways human beings produce intelligent behavior. This is even true of rule-based expert systems, which use a question-and-answer technique that controls what problems will be solved and what knowledge will be needed to solve them. Even so, the most difficult part of developing expert system software is modelling the knowledge of experts in a particular field—so much of an expert's expertise has to do with perception rather than the formulation of formal rules. Areas like natural language processing of unrestricted input, voice recognition of unrestricted speech, and visual perception of unrestricted objects in a natural environment are much more general in their purpose than domain-specific programs, and it is this generality of purpose which most clearly requires the representation of vast amounts of human knowledge. For a computer, to identify the meaning of natural language is to identify that meaning *as it is perceived by human beings.* This is why it must represent not only syntax, but all the things which affect the meaning of language for human beings. For a computer to recognize natural, continuous and unlimited speech, it would have to identify the meaning of the context within which that speech takes place as well as all the cues contained in that speech which allow words to be distinguished and identified—that is, all the things which humans perceive when they listen to one another. For a computer to recognize complex objects in a natural environment, it would have to discriminate all of those same aspects and attributes of objects that human beings discriminate when they perceive. In short, to simulate any complex intelligent human behavior, the computer must have a stored representation of all the things humans know about the objects and situations such behavior involves—if it doesn't have a store of such knowledge, there is no way for it to grasp the meaning of the data as that is grasped by humans, and consequently, no way for it to simulate the behavior it is attempting to simulate.

Suppose that it is necessary for computer scientists to represent a great deal of human knowledge if the applications mentioned above are to succeed. Why should the fact that they have not yet found a way to formalize large amounts of knowledge necessarily mean that they cannot do so in the future? According to Dreyfus, AI workers have tried to "analyze human behavior in terms of rules relating atomic facts." (Dreyfus, 33) This has led them to consistently misunderstand or ignore four central forms of human knowing *which cannot be formalized,* but which enable "human subjects ... to avoid the difficulties an artificial 'subject' must confront." (Dreyfus, 100) Dreyfus calls these four forms of knowing "fringes of consciousness" or "global awareness," "ambiguity tolerance," "essential/inessential discrimination" or "insight," and "perspicuous grouping." (Dreyfus, 100-28)

What Dreyfus calls "fringes of consciousness" (William James's term) or "global awareness," I will call "background knowledge" because it seems to me a clearer and more accurate term. Dreyfus says that "rather than being explicitly taken into account," background knowledge "is constantly at work in organizing our experience...." (Dreyfus, 106) In a passage quoted by Dreyfus, Michael Polanyi points out that background knowledge "compellingly affects the way we see the object on which we are focusing," and claims that we grasp this knowledge primarily "in the appearance of the object to which we are attending."[8] Background knowledge is based on past personal experience and cultural communication. It leads us to organize and understand the present situation as we do, even though we do not avert to such knowledge explicitly. Dreyfus tells us that the background knowledge of the master chess player, for example, includes his or her past experience of the present game and other games, the strategy attributed to the opponent, and the player's own strategy—in short, what Dreyfus calls the player's "sense of the developing game." (Dreyfus, 30) This understanding of the whole game supplies the context within which the present move is understood. It structures the player's focal understanding of the present situation in such a way that he or she can "zero in" on weaknesses in the opponent's position and strengths of the player's own position.[9] In addition, the player's background knowledge includes the "apprenticeship" of thousands of games he or she has played or studied and understood in the past. These games allow the player to recognize similarities between the present situation and situations encountered in those games. Such similarities, however,

> cannot be defined as having a large number of pieces on identical squares. Two positions which are identical except for one pawn moved to an adjacent square can be totally different, while two positions can be similar although no pieces are on the same square in each. Thus similarity depends on the player's sense of the issues at stake, not merely on the position of the pieces (Dreyfus, 31).

In other words, based on their past experience, human beings can recognize analogy, the similarity of things that are in many ways dissimilar. A human being sees through the dissimilarities and grasps the similarities, extending what was learned in the past to enlighten the present situation. The context of a particular situation in a particular chess game thus implicitly includes past games as much as it does the development of the present game. In the same way, human experts in any area learn much of what they know through practice gained during their "apprenticeship" in school or elsewhere. Much of their expertise consists, not in explicit judgments they can articulate, but in know-how that is implicit in their practice—that is what makes it so difficult to extract an expert's knowledge and represent it in an expert system (Dreyfus, 28-9).

The computer does not recognize analogy, only identity and non-identity. It represents situations in terms of context-free features which allow the situation to be matched on a feature-by-feature basis with previously identified situations whose descriptions are stored in the computer. Instead of relying on background knowledge to provide the context for understanding a situation and guidance in responding to that situation, special-purpose artificial intelligence programs, like heuristics-based chess programs and expert systems programs, use what Dreyfus calls "situation→action" rules. These rules are "inference chains" which connect attributes of isolated situations to actions in a way that simultaneously determines which attributes of a situation are relevant to the prescribed action and which actions are to be considered in light of these attributes.[10] Chess programs using heuristics, for instance, analyze the position, importance and color of each piece on the board, evaluate board positions using a numeric scale, and use rules to determine which moves are most promising. They then calculate the possible consequences of each promising move to determine the move with the greatest probability of winning (Dreyfus, 30). This technique predetermines, and therefore limits, the meaning of a situation and the actions prescribed in it, based on the distinct set of features which circumscribe the situation for the computer. Dreyfus suggests that, unlike a computer, a master chess player or other expert does not grasp a situation's features prior to discovering its meaning. Instead, a global experience of the situation which associates it with past experiences allows the expert to immediately grasp the situation's meaning, and with it, those features which are relevant to that meaning. Similarly, the human expert does not act on context-free rules, but follows "maxims" that are only meaningful to someone who understands the significance of the present situation and possesses sufficient expertise to understand how the maxims are to be applied (Dreyfus, 31-32). This understanding is based on the analogous continuity of the present situation with the past. Even if the computer had enough power and memory to keep track of the features of past situations, its inability to perceive analogous similarity rather than simple identity makes it incapable of representing the knowledge of human experts in any other way than in situation→action rules. But the ability to understand the present by analogy with the past is *essential* in grasping the meaning of complex perceptual objects, unlimited natural language and unlimited continuous speech, among other things. The computer's inability to succeed well in these areas, therefore, may be directly traced to its inability to store the content of background knowledge and use it as a basis for its interpretation of problematic situations.

The presence of analogy in our background knowledge points to a characteristic of problematic situations that has been a major stumbling block to a variety of artificial intelligence programs, even though it seldom presents a problem to human knowers. That characteristic is formal ambiguity, the presence of two or more possible meanings for a word, a sound, an object of perception, or any other problem data. Our "ability to deal with situations which are ambiguous without having to transform them by substituting a precise description" Dreyfus calls "ambiguity tolerance." (Dreyfus, 107) Our background knowledge provides a "global context" within which we understand each situation. The "sense of the situation" which results, "allows us to exclude most possibilities without their ever coming up for consideration" and gives us the "ability to narrow down the spectrum of possible meanings by ignoring what, out of context, would be ambiguities...." (Dreyfus, 109) As we have seen, however, the computer must represent the situation "out of context," that is, in terms of context-free attributes. Any rules for interpreting the situation, therefore, cannot ignore most possibilities, as humans do, but must take all of them into account—even those which, for a human being, would appear absurd. Since there are a vast number of such possibilities for most situations, the computer can only follow up on and eliminate unreasonable interpretations if it has a stored representation of all the knowledge necessary to do so—which, for a program that would be able to accept unlimited input, would mean all the background knowledge that humans use in interpreting situations. A natural language processing program, for instance, would have to represent all the knowledge humans have that lets them correctly interpret the meaning of most words in most contexts. A perception program would likewise have to represent human knowledge of all possible objects of perception, which includes our knowledge of all the expected appearances and behaviors of each object under all possible circumstances, all possible relationships between objects, etc. In short, a computer could only simulate "ambiguity tolerance" by explicitly representing the entire store of so-called "common sense knowledge" gained through individual and communal experience and applying this knowledge analogously in similar (but not identical) situations. Since it can do neither of these things, the best a computer can do is to search through programmed interpretations of predefined collections of context-free features to identify the preconceived meanings of a (perhaps very wide, but not unlimited) variety of situations in domain-specific applications.

The third form of knowing mentioned by Dreyfus, "essential/inessential discrimination" or "insight," (Dreyfus, 112-20) is the ability to see what is or is not relevant to the matter at hand. Suppose I am playing a game of checkers and it is my move. My background knowledge of the game lets me immediately grasp what pieces on what squares are relevant or irrelevant to my next move. My ability to ignore every piece and every square that is not relevant to my next move, to focus on those that are relevant, and to consider only those moves that might help me win the game, is an example of "essential/inessential discrimination." Moreover, these aspects of the game are not "context free" features, that is, "they do not exist independently of the pragmatic context."[11] Similarly, when I perceive objects in my immediate environment, I focus on those few objects and features of objects that are relevant to my interest, ignoring everything else. I will notice the color of a sunset because that is essential to my aesthetic interest—even so, I will only experience these colors together, as part of the sunset, rather than as isolated features. Other things, such as the cars driving by on the road, I will totally ignore—I see and hear them, but I don't attend to them and probably couldn't tell you much about them later. Like "ambiguity tolerance," this ability to discriminate between what is and is not relevant to the matter at hand presupposes the background knowledge which structures my situation and opens up areas of interest in terms of which things have relevance. For computer programs, as for humans, the determination of relevance is a necessary first step in analyzing data and selecting a path to be pursued in solving a problem. Because they cannot represent and use background knowledge, however, computer programs lack the ability to discover relevance in the situation. All decisions concerning what features of data are or are not relevant to the program, what kinds of problems the program will attempt to solve, and what strategies are relevant to solving them must be built into the program's procedures by the programmer. In other words, it is the programmer's "essential/inessential discrimination," rather than the program's, that determines what is and is not relevant.[12]

The fourth form of knowing examined by Dreyfus, which he calls "perspicuous grouping," is the ability to recognize any kind of likeness between objects without having to make explicit the criteria for identification (Dreyfus, 120-8). The likeness recognized may be no more than a presumed likeness based on the similarity of the situations in which two objects are perceived: a formally ambiguous object may be perceived as a small child, for instance, simply because it is being held by a woman in a way that suggests a mother holding her child. Or the likeness recognized may be a partial or analogous likeness: for example, a person may be recognized as a member of a family whose other members they resemble in some indefinite way. Finally, an object may be recognized as an instance of a generic type: an object may be recognized as a pen, for example, and the thing it is resting on may be recognized as a desk. While the first two examples of "perspicuous grouping"—contextually-based associations and analogous likenesses—clearly depend on the kind of background knowledge that the computer seems unable to adequately represent, recognition of generic likeness initially seems close to what computers do in pattern recognition. Dreyfus notes several interesting characteristics of this kind of recognition, however, which seem to indicate that it, too, could not be adequately simulated by a computer. First, our recognition of an object is seldom affected by distortions in appearance or changes in orientation: I can recognize a melody, for example, in spite of the improvisations of a jazz musician, or the distortions of a scratchy record, or the presence of other sounds. Second, the recognition of an object's type does not generally require that the object's defining features become explicit. *Classifying* an object admittedly requires checking for the presence of those traits which identify the object as a member of a class. *Recognizing* an object, however, requires only that one experience that object as being of the same type as others of its kind (Dreyfus, 122). I do not have to focus on the defining characteristics of a chair, for example, to recognize a chair and use it appropriately. In fact, Dreyfus notes that there are some objects that are so complex that their distinguishing traits are too numerous for me to notice each one—I simply experience the object as a particular type of object without ever explicitly becoming aware of the traits that lead me to do so.[13] Such distinguishing traits remain on the "fringe of consciousness" while they do their work (Dreyfus, 122). Third, the traits experienced as defining an object may change as one's pragmatic interest changes: while a pen used to be a sharp, metal-tipped instrument for writing with ink from a bottle or inkwell, for example, it now may have a tip that is a metal ball or even a rigid piece of felt and carry its own supply of ink. All three of these characteristics of our recognition of an object's generic type distinguish it from the way a computer must proceed in recognizing an object. The computer must have relatively complete data on the object to be recognized—it cannot simply be suggested by its context. A computer must isolate all of the defining features of the data—they cannot remain implicit. Finally, a computer must classify the data using unchanging criteria of evaluation—indeed, such classification *is* pattern recognition for the computer.

IV

There are obviously some areas in which the computer has successfully simulated and even surpassed the results of human intelligent behavior. Some new techniques used in AI research try to avoid knowledge representation completely, or else develop computer-generated representations that do not require the programmer to foresee the kinds of representations needed. Only the future can show us how much may still be accomplished by the ingenuity of computer scientists in simulating the results of human behavior. Nonetheless, Dreyfus has provided convincing evidence that a computer program must fail anytime it attempts to simulate general human intelligence, which imposes no limit on input, in application areas such as natural language processing, voice recognition and perception. He has also provided a convincing explanation of why this is so. Dreyfus notes that "the basic problem facing workers attempting to use computers in the simulation of human intelligent behavior" is that "all alternatives must be made explicit." (Dreyfus, 129) The areas which have achieved the most promising results are those in which alternatives can be made sufficiently explicit, so that the great calculating power of the computer can "make up for the basic inability of computers to think," to quote a recent New York Times article.[14]   Human reason succeeds, however, by not having to make most things explicit in complex activities, and much of the knowledge which guides intelligent human activity cannot be made explicit because of its analogous character. To the extent that complex intelligent human activity requires such analogous knowledge—knowledge which no computer can represent—the adequate simulation of that activity will remain beyond the capacity of computers.

Mount Mary College
  Milwaukee, Wisconsin

* * * * *

Endnotes

1.   See Jasia Reichardt, *Robots: Fact, Fiction and Prediction*, (New York: Penguin Books, 1978), esp. 8-16.

2.   Alan M. Turing, "Computing Machinery and Intelligence," in George F. Luger, Ed., *Computation and Intelligence: Collected Readings*, (Cambridge, MA: MIT Press, 1995), 46.

3.   Hubert L. Dreyfus began his critique of artificial intelligence in the December, 1965 monograph, "Alchemy and Artificial intelligence," published by the RAND Corporation. This was followed by his landmark study, *What Computers Can't Do*, (New York: Harper and Row, 1972). A second edition of this work appeared in 1979, reprinting the text from the 1972 edition with only minor changes, but adding a new introduction which described and critically evaluated developments in artificial intelligence research from 1968 to 1977. A third edition of the book appeared as *What Computers Still Can't Do*, (Cambridge, MA: MIT University Press, 1992). In this latest edition, again a new introduction is added that describes and critically evaluates developments in artificial intelligence since 1977. The inclusion of the original study plus three introductions presents an interesting history of the evolution of the successes and failures of artificial intelligence from 1957 to the present decade. It also shows the development of Dreyfus's critique of the discipline. All citations to Dreyfus in this paper refer to this latest edition.

4.   Dreyfus has directed most of his criticism against what he now refers to as GOFAI—Good Old Fashioned Artificial Intelligence. In the introduction to the latest edition of his work he says that "[m]y work from 1965 on can be seen in retrospect as a repeatedly revised attempt to justify my intuition ... that the GOFAI research program would eventually fail." Dreyfus, xi. According to Dreyfus, this phrase was introduced by John Haugeland to designate the AI theories and techniques which were dominant until the mid-80's, but are gradually being supplemented by newer theories and techniques. See Dreyfus, ix. Most of the comments in the present paper are directed at this "traditional" form of AI research, since it has created the foundation for many currently used AI products as well as developed the context within which the problem of knowledge representation is understood by today's computer scientists. Developments in AI research since 1985 or so deserve their own treatment. It is worth noting, however, that the limits encountered by "traditional" AI research must form the basis for any analysis of the goals of more recently developed AI technologies.

5.   "[C]ertain games can be worked through on present-day computers with present-day techniques [that is, brute force searches]—games like nim and tic-tac-toe can be programmed so that the machine will win or draw every time. Other games, however, cannot be solved in this way on present-day computers, and yet have been successfully programmed. In checkers, for example, it turns out that there are reliable ways to determine the probable value of a move on the basis of certain parameters such as control of center position, advancement,

and so fort. This, plus the fact that there are relatively few moves since pieces block each other and captures are forced, make it possible to explore all plausible moves to a depth of as many as twenty moves, which proves sufficient for excellent play.... Chess, however, although decidable in principle by counting out all possible moves and responses, presents the problem inevitably connected with choice mazes: exponential growth....

[T]he right heuristics are supposed to ... [limit] the number of branches explored while retaining the more promising alternatives." Dreyfus, 100-01.

6. See Dreyfus, 32: "[W]e can expect ... to continue to find some experts who outperform even the most sophisticated programs."

7. Dylan Loeb McClain, "Chess Computers Press Attack with Humans," *The New York Times*, October 2, 1995, pp. C1, C4.

8. Michael Polanyi, "Experience and Perception of Pattern," in *The Modeling of Mind*, eds. Kenneth M. Sayre and Frederick J. Crosson, (South Bend, Indiana: Notre Dame University Press, 1963), 214. Quoted by Dreyfus, 103.

9. Commenting on a master chess player's report of his game, Dreyfus says: "we can conclude that our subject's familiarity with the overall chess pattern and with the past moves of this particular game enabled him to recognize the lines of force, the loci of strength and weakness, as well as specific positions. He sees that his opponent looks vulnerable in a certain area ... and zeroing in on this area he discovers the unprotected Rook. This move is seen as one step in a developing pattern." Dreyfus, 105.

10. "In special purpose programs the form of knowledge representation can be limited to situation-action rules in which the situation is defined in terms of a few parameters and indicates the conditions under which a specific heuristic rule is relevant. Again, because relevance is defined beforehand, reasoning can be by inference chains with no need for reasoning by analogy." Dreyfus, 27-28.

11. Dreyfus, 119. Dreyfus says this of the essential operations to be carried out in solving a logic problem, but it is equally true of all aspects of the game of checkers.

12. As Dreyfus says of Newell, Shaw and Simon's logic program: "The classification of the operators into essential and inessential ... is introduced by the programmers before the actual programming begins." Dreyfus, 116.

13. In a different context, Dreyfus notes how difficult it is to even specify what features allow an object to be identified—even an object as mundane as a chair. A chair can have no legs, one, two, three, four or more legs, and be any shape and color you want. A chair can be anywhere, in any part of a room, and even outside a room. It can have a different spatial relationship with other objects, including the body of the person sitting on it, if a person happens to be sitting on it at all. Even its function as something used for sitting is not simple, because there are some chairs which no one ever sits on—for example, doll chairs and ritual chairs. Because chairs may be as different as a barber's chair, a camp stool or a bar stool, there are many different ways of sitting on chairs. There are even different socially defined meanings for the way one sits on a chair—sedately, demurely, naturally, casually, sloppily, provocatively, etc. Dreyfus, 36-38; quotation, 37.

14. McClain, C4.

Grading Worlds

by *Sandra Menssen*

I

Is There a Functional Criterion of Goodness for Worlds? We commonly grade or evaluate many sorts of things in the world, sorts of things that have associated criteria of goodness: knives, for instance, and violins, and human beings. There are other sorts of things in the world that seem not to have associated criteria of goodness: rivers and jungles, minutes, pains, and numbers, for instance. No objective standard separates good rivers from bad; and the claim that a particular river is good must be made from a particular point of view–that of a trout fisherman, perhaps, or dreamer, or industrialist.

Is there a criterion of goodness for *worlds*, a standard that gives sense to the assertion that the world is good, or that determines which worlds are good and how they rank against one another? Only if there is such a standard does it make sense to say that God must (or need not) create a good world. If a world is not the kind of thing that can be either good or bad, then certain attacks on God's goodness are precluded; so are certain theodicies.[1]

In a wider range of cases, criteria of goodness for things are determined by the functions they serve. Does a world–or *the* world[2]–have a function? If so, does its function provide it with a criterion of goodness? I will examine the possibility of developing a functional criterion of goodness for worlds at some length, since the problems encountered in attempting to define a functional criterion are illustrative of problems that arise in defining other criteria of goodness for worlds.

People speak of an individual or object as performing its function in (at least) three situations: (i) An individual or object may be doing what it was designed to do, thus satisfying a designer's purpose. A knife, for example, may be cutting; a person may be worshipping God. (ii) An individual or object may be doing what its user needs or wants done, thus satisfying the user's purpose. For example, a knife may be prying the lid off a can; or a knife may be cutting. A slave may be building a pyramid. (iii) An individual or object may be doing what's necessary to

149

satisfy some role; as a special case of satisfying a role, the individual or object may be doing what's necessary to (help) keep some system going. For example, a father may be educating his daughter; a university may be educating its students; a flautist may be playing the flute; a bass-player may be balancing the violins in an orchestra; an eye may be seeing.

A world might be thought to be satisfying a function in each of the three ways just mentioned. (i) Perhaps a good world satisfies its designer's (or designers') purposes, so this world is good if it is what its designer intended it to be, or if it does what it was designed to do. One who is inclined to say that what a creature is for depends on its creator's intentions in creating it might find it reasonable to say that, likewise, a world's function depends on its creator's purposes. If an evil genius or a super-powerful, super-knowledgeable *demon* created a world, would there be any plausibility to the claim that the world's goodness depends on whether its designer's intentions were fulfilled? There would be less plausibility, certainly, than if the designer were good; and this fact seems to discredit the suggestion that a criterion of goodness for worlds is determined by the creator's intentions.

The criterion just considered seems irrelevant to the claim that a good creator will create a good world. The criterion does not lay down a sufficient condition for identifying good worlds, since when we say that a world is good we presumably intend to say more than that its designer is powerful and clever enough to fulfill his intentions. Moreover, the criterion may not lay down a necessary condition for good worlds either if we allow that a good designer's intentions may be thwarted, at least in part: recalcitrant creatures may, for example, choose not to behave as their creator wishes. If we change the criterion at issue to include reference to the creator's intentions or character--if we say, for example, that a world is good just in case it satisfies a perfectly good creator's purposes--then we move out of the realm of non-theistic criteria of goodness for worlds.

(ii) One might design an aquarium with the intention of creating an interesting, educational diversion for one's son (or oneself). One's designs may be thwarted if he doesn't take the appropriate interest in the aquarium, or if the creatures in the aquarium manage regularly to hide themselves among the rocks and coral and seaweed and castles you've put in the aquarium. But one can imagine a friend saying: "Though your designs have failed, still you've created a very good world; basically, that world is not for your son's pleasure, or for your pleasure, but rather for the benefit of the creatures who live in it; they are its prime users, and you've arranged things so that the fish are well off." This circumscribed world of the aquarium is called good not because it satisfies its designer's purposes, but because it promotes the well-being of its creatures; what they need or use the world for is what counts. So here's a function a world may have, and an associated criterion of goodness for the world.

If one says that a good world promotes its creatures' well-being, the question arises: how is this standard applied to a world where the well-being of one creature or species of creatures conflicts with the well-being of others? Is a world in which just one creature fails to flourish bad? Perhaps we can say: some creatures are better than others; it's they who should flourish, and if none of the creatures is good, none should flourish. A world without creatures will be a world to which this criterion of goodness–the criterion that a world with creatures is good if and only if it promotes its creatures' well-being–can't be applied. So the criterion has limited application. It's incomplete in another sense, too: it seems to require that we have a criterion of goodness for sorts of creatures; maybe even for individuals within a sort.

Perhaps we can avoid the requirement that we be able to grade creatures by saying that all creatures, whether good or bad, need a certain minimum of basic goods in order to flourish; and a good world will provide those goods (for all creatures). If we did have a criterion of goodness for sorts of creatures, we could articulate a different standard for grading worlds: if one world has better creatures than another– whether or not the creatures flourish–it's a better world. This does not seem to be a standard of goodness based on any obvious function of the world.

A criterion of goodness for worlds based on what its users need would seem more properly to be described as a criterion of goodness of *habitats* of creatures, rather than of goodness of *worlds*. It reminds one of Hume's discussion of evils in the *Dialogues*.

I have asked: "What is a world for?" and considered the response: "That depends on what those who inhabit the world need or use it for." I took it that under this criterion of goodness, a world will be good if it promotes its creatures' (or some of its creatures') well-being. It might be suggested that we should look in a different direction to see what creatures use their world for: it might be suggested that the function of a world is to satisfy its creatures' *desires* or *preferences*, to make them happy.

Here, I want simply to point out one problem concerning such a utilitarian criterion of goodness for worlds. If worlds should be evaluated in terms of their ability to make inhabitants happy or, more specifically, to satisfy their preferences, that would surely be because satisfaction of preferences is important. But if it is important, why isn't satisfaction of preferences outside as well as inside the world important? Why should a creator neglect his own preferences, those that are, so to speak, outside the world, to satisfy his creatures' preferences? If, being eternal and in various ways infinite, a god's capacity for satisfaction is infinite, then perhaps his preferences are more important than the preferences of all his creatures. Perhaps not: maybe there are worlds in which infinite numbers of creatures exist, either all at one time or one after another in a world without end (or both), and maybe their preferences can carry as much weight as the creator's. If a creator's preferences are more important than those of his creatures, the creator could make any world he pleased, and the world would be judged good if the criterion of goodness took into account all preferences. It may not be the case, however, than any world the creator made would be judged good: an omnipotent, omniscient, imperfectly good creator--if such a being is possible--might be torn between wicked preferences and good ones, and might bring himself to put aside his wicked desires when creating the world (and so there would be, outside the world, enormously important preferences that remained unsatisfied). These considerations show, I think, that even if we could define some meaningful criterion of goodness for worlds that was based on satisfaction of preferences, that criterion would not operate as one would expect it to if God's goodness required that he create a good world.

(iii) Recall the last sort of case where an individual or object was said to be performing its function: the individual or object was playing a role or keeping a system going. The system the individual or object sustains may be itself: for example, a man recovering from a broken heart, who is "barely functioning," doing what's necessary to maintain himself, but no more, is functioning to keep a system--himself--going. One might, in designing and setting up an aquarium, intend to provide a habitat in which a group of blue damsels--say 20 blue damsels--will thrive; one could fail to provide the right habitat because the nitrifying bacteria in the aquarium don't grow fast enough, the ammonia builds up, and 18 of the 20 fish die. But then one is left with 2 fish, which provide enough ammonia (let us assume) for the nitrifying bacteria to grow; and the nitrifying bacteria break down the ammonia from the 2 fish; and one has an ecosystem so well-balanced that a friend could say: "Even though most of the fish died you've set up a very good world." Here we have a case where the goodness of a world is judged not by how well it satisfies its designer's purposes or intentions, nor by whether the world promotes the well-being (or satisfies the preferences) of its creatures, but rather by the world's capacity to be self-sustaining, by the world's capacity to promote is own benefit, so to speak. The world is seen as a system, correctly balanced--perhaps self-balancing--and hence self-sustaining. (Plato is among those who have taken the world to function as a self-sustaining system.) Here, as before, however, there may be worlds that get counted as "bad" which wouldn't strike us as worlds God shouldn't create.

Moreover, why should this function--or, for that matter, other possible functions we've discussed--be *the* function of the world, as opposed to *a* function the world might serve? Sometimes we can identify one function of an object as most important by asking what the object characteristically or essentially does. By an essential function of an object, I mean a function such that the property of being able to carry out that function is an essential property of the object; an essential property of an object is one such that it's a necessary truth that if the object exists, then it has that property. If one were searching for the function of human beings, for what humans as humans (not as slaves, or parents, or flautists) do, one might, by asking for an essential function, come up with the answer: it is the function of human beings to reason. This answer has some plausibility. If one asks this question of individuals or objects that function in certain roles, of objects that are artifacts, one may likewise get reasonable answers (a knife is for cutting, not prying lids off cans). If the world had an essential function, we might say that it would be good just in case it performed its essential function well.

Is there a function such that the ability to perform it is a necessary condition of a *world's* existence? I don't see that there is. The world as a world doesn't *do* anything essentially. Is it *for* anything, essentially (according to the non-theist)?

If there were worlds of different sorts, then perhaps there would be, for a given world, a function such that the ability to perform it would be a necessary condition of the existence of that world *as a world of a certain sort*. For example, one might say: the sort of world that sustains oxygen-dependent creatures has an essential function-sustaining the oxygen-dependent creatures. It is not clear here just what it would be for a world of this sort to perform its function well: perhaps it would need to sustain for an appropriate stretch of time all oxygen-dependent creatures it contains, or to sustain many oxygen-dependent creatures. Worlds, however, don't seem to come naturally sorted into kinds. If one is non-arbitrarily to sort worlds into different kinds, one needs some notion of what the salient features of worlds are. One needs to view worlds from some particular vantage point.

Sometimes a description of a world may suggest a particular point of view from which to judge the world. For example, if one selects a set of science fiction books related in theme or subject and asks which of the worlds described in the books is the best world, or whether one of the worlds described is better than the actual world, one gives notice of the features of the worlds that should be examined in deciding which world is best. If one chooses books that describe worlds where people have developed striking social structures, one will be taken to be asking for an evaluation of the social structures (so one might say: Heinlein's Mars, described in *Stranger in a Strange Land,* is a better world, or a worse, than our own). One of the three worlds described in C. S. Lewis's science fiction trilogy is Malacandra, a world where most of the people are good and happy but are protected by angels who limit their independence, who limit the range of choices available to them. Thulcandra–our world–has in it people sinful by nature, beset with troubles, but who are under the protection and guidance of God alone. They are free to choose actions having dire consequences. If one asks whether Malacandra or Thulcandra is the better world, one will probably be taken to be asking people to evaluate the importance of agents who act independently, who are independent enough to choose their own destruction.

A non-Leibnizian criterion for individuating worlds might provide the vantage point we seem to need, and suggest a criterion of goodness for worlds. As an example of a non-Leibnizian criterion for individuating worlds, consider Aquinas's criterion: he apparently individuated worlds in terms of the species they contain; unlike Leibniz, he allowed that the history of this world might have been different. Aquinas's principle of individuation does not by itself tell us what the world's essential function is, nor determine a criterion of goodness for worlds, but it suggests a point of view from which one might be developed: it suggests we focus on species in a world. With such a focus, we might hold that the more species a world contains, the better the world; or perhaps, that the goodness of a world is determined in part by its ability to preserve, to sustain the species it contains.

Aquinas's focus on species is not the only way, nor the most important way, in which he gives us a vantage point for understanding the world: his commitment to theism also helps identify an Archimedean point. Aquinas holds that the world is good because it images or represents God. The world's goodness depends on the goodness of its parts (or species) and on the way in which the parts are ordered; the order of the parts depends on their relation to one another, and on the end to which they are directed, that is, God. The Thomistic criterion of goodness for worlds has recently received attention from two careful and insightful expositors and critics: in 1992 Oliva Blanchette published *The Perfection of the Universe According to Aquinas: A Teleological Cosmology*,[3] which articulates and defends Aquinas's standard for grading worlds, and in 1991 Norman Kretzmann published two essays which consider Aquinas's criterion of goodness for worlds in the course of arguing for an account of creation that at key points overlaps Aquinas's account.[4] I can't here present a detailed recapitulation or a critical evaluation of the work of these two expositors, but I hope by the end of this paper the significance of a theistic standard for grading worlds (such as the Thomistic standard) will be apparent.

Notice that a non-Leibnizian criterion for individuating worlds or possible worlds, such as the Thomistic criterion, may suggest non-standard–that is, non-Leibnizian–interpretations of the claim that God must create the best possible world. A non-standard interpretation of the claim can be illustrated by thinking of an artist working on a painting who produces successive versions of it. One could reasonably hold, for instance, that when Picasso was working on the *Guernica*, he produced different versions of *one* painting before he was finished and that the photographs that exist of the various stages of production are photographs of *one* painting. It makes sense to consider which version is best; it makes sense to hold that Picasso should have made the best version the final one, that he should have made the painting as good as it can be. In saying this one does not say that he should have made the best of all possible paintings. Likewise, in saying that God must create the

best possible world, one may mean simply that God must make whatever world he creates as good as it can be (though there could be other
better worlds). Indeed, Aquinas said this very thing: "Given the things
which actually exist, the universe cannot be better....Yet God could make
other things, or add something to the present creation, and then there
would be another and a better universe" (ST I.25.6).

II

Are There Other ("Non-Functional") Criteria of Goodness for Worlds?:
I claimed above that very often (if not always) when we try to ascertain
the criterion of goodness for a thing, we ask about its function. Here I
want to comment briefly on several other methods that have been (or
might be) proposed for determining a criterion of goodness for worlds.[5]

Philosophers who implicitly or explicitly suggest criteria of goodness
for worlds sometimes focus on the balance in a world between happiness
or pleasure, and unhappiness or pain. Hume, in the *Dialogues*, and
Voltaire, in *Candide*, may have had some rough criterion of this sort in
mind. Peter Geach reports[6] that Berkeley and Paley held that "God can
and does make...calculations (for the best possible consequences), and
his commands are to be justified as setting up practices that will lead
to the greatest happiness of the greatest number of his rational creatures." Leibniz sometimes appears to take a similar position. He
explains why we cannot be act-utilitarians, though God can be:

> One will not approve the action of a queen who, under the
> pretext of saving the State, commits or even permits a crime.
> The crime is certain and the evil for the State is open to
> question....But in relation to God nothing is open to question,
> nothing can be opposed to *the rule of the best*....God being
> inclined to produce as much good as possible, and having all
> the knowledge and all the power necessary for that, it is
> impossible that in him there be fault, or guilt, or sin....[7]

Elsewhere, Leibniz indicates that the "good is what contributes to
pleasure."[8] As we will soon see, Leibniz gives an alternative criterion of
goodness for worlds, one that, he believes, will rank possible worlds in
goodness in the same order that the utilitarian criterion ranks them. I
seriously doubt that there is a total amount of happiness or pleasure or
unhappiness or pain in the world; hence these utilitarian criteria of
goodness for worlds don't seem to me to be meaningful.

Another criterion of goodness for worlds some philosophers have
suggested focuses on the organic unity or harmony in a world. Plato,
for instance, explicitly presented a conception of good-making (or best-
making) characteristics of worlds, a conception in which harmony plays
a prominent role. Certain that this world is the best, in the *Timaeus*
Plato uses his conception of best-making characteristics of worlds to

"deduce" that this world will be proportional (31c) and uniform (33b). Such attributes mark the world as well-ordered, as harmonious. Nevertheless, the best world has, in his view, other attributes that may not seem to us so clearly constitutive of harmony: the best world will be unique (31b), everlasting (32d), and self-sufficient (33d, 34a); and it will be complete (30c,d; 32c, d), though it's not clear what its completeness involves. As we indicated above, Plato sees the world as a self-sustaining system, and hence sees it as having a function. The exact connection between the world's being harmonious, and its having a function, is not clear.

Robert Nozick, a contemporary analytic philosopher, also bases a criterion of goodness of worlds on harmony or organic unity. He holds that organic unity, or the unification of diverse parts, "is the common strand to value across different realms."[9] Though he commonly speaks of a thing's value rather than its goodness, he points out that "The English word 'god' stems from a root, 'Ghedh', meaning 'to unite, join, fit, to bring together'."[10] Among the things he mentions that may have some degree of organic unity, of value, are worlds (or universes). The value of a world is equal to its degree of organic unity, plus the sums of the values of its parts. Our world has in it things of very different sorts--spiders and symphonies and suns and scientific theories, for instance--and since Nozick holds that "some things from different realms will be incomparable in value," it isn't clear how he would sum the values of the parts of the world. Furthermore, it is difficult to state the unifying relationships that (partially) determine a world's organic unity. (That is, it is difficult to do so if one is committed to remaining neutral about the truth of theism. Aquinas didn't have any problem in stating the unifying relationship in the world.)

We find varying conceptions of what the harmony of *this* world consists in if we turn to the writings of the ancient Greeks. Their conceptions of cosmic justice provide a basis for extrapolation. Gregory Vlastos suggests an intriguing way of sorting out the different conceptions: he divides those that take harmony as a relationship among unequals from those that regard equality of related parts as essential to harmony. Intriguing as this is, and interesting as it might be to consider the extent to which these ancient conceptions of harmony differ from, say, Nozick's, one gets the feeling one would be inventing the criteria as much as analyzing them--they are not very specific.

Just as we found when we considered functional criteria of goodness for worlds, these additional criteria do not seem to give the results we would expect when we hear people say that God is obliged to create a good world if he creates.

We've already considered the difficulty in fixing on a utilitarian criterion of goodness for worlds, where the good to be maximized is happiness or satisfaction of preferences. If satisfaction of preferences inside the world is important, why isn't satisfaction of preferences "outside" the world at least equally important? Unless we beg the question about God's goodness, it isn't at all clear the preferences a creator would have would match up with our own intuitive conception of a good world.

However, it's not clear why a world which doesn't count as good given some criterion of harmony or organic unity is a world God shouldn't create. Leibniz seems not to have thought that there exist disharmonious possible worlds. Nozick, however, suggests that value is "something that has an opposite which can be present." Perhaps any world that runs down, or any world in which marvelous civilizations crumble and decay, or in which mental and physical vigor of species declines, is a world marked by disunity. Perhaps worlds where there is only "random flux," as Plato puts it, would be disunified or disharmonious. It isn't intuitively clear what would be wrong with God's creating such worlds. So even though criteria of goodness for worlds that focus on organic unity may be meaningful, they let us find "bad" worlds a perfectly good God might, it seems, create.

III

An Unexpected Implication of the Claim that There is a Criterion of Goodness for Worlds: I have so far been arguing that is very difficult to identify a criterion of goodness for worlds that operates as we expect a criterion to operate (matching up with whatever intuitions we have about worlds too awful for God to create) unless one adopts a theistic criterion of goodness. For some criteria count as "good" worlds we intuitively judge bad, and some criteria count as "bad" worlds that we think a good God could create.

I now suggest a supplemental claim: The traditional problem of evil implies that there is a criterion of goodness for worlds which God uses or displays in choosing to create our world. What I have to say in defense of this claim is only the beginning of what needs to be said—but perhaps it will point to some fruitful lines of inquiry.

Why should a careful formulation of the problem of evil require reference to the goodness or badness of the *world*? Because a careful formulation of the problem will spell out the reasoning that lies beneath the immediate premises of the argument, and when that reasoning is spelled out there will be a reference to a lack of goodness in the world as a whole.

The problem of evil looms over individuals bothered by specific, particular evils: this child's suffering, that mother's anguish. Particular cases of apparently unredeemed suffering and affliction, experienced or witnessed, typically force the problem of evil. The problem may express itself variously: it can cripple the life, cripple the intellect. If it expresses itself intellectually, it results (at this level) in questions about particular evils: How, one may wonder, can a perfectly good, omnipotent and omniscient creator allow my child to suffer this pain?

At this level the intellectual problem of evil can be solved (or so one might plausibly argue), though the "pastoral" problem of evil, as Plantinga has called it, can be insuperably difficult to solve. Particular evils may be explained either by causal connections those evils have to over-balancing goods; or, if this seems unpersuasive in some certain case, it seems possible to explain particular evils by eschatological considerations. That is, one might argue, particular instances of suffering can be defeated or redeemed in the afterlife of the individual sufferer, and our awareness of this fact is sufficient to handle the (intellectual) problem of evil at the level of specific, particular evils.

I do not mean to suggest that appeal to the possibility of an afterlife can be used to "explain away" any conceivable kind or amount of suffering and wickedness. One consideration that exhibits the limits of such an appeal is as follows. The argument from design (among other things) tells us that human existence as we experience it was intended, was something worth intending. If, however, human beings generally experience pain or suffering of a sort that can only be defeated in an afterlife, then it seems that generally speaking life (this life) is not worth living; and human existence as we experience it was not worth intending. That is not a conclusion the theist will embrace. Still, eschatological considerations are useful (and may well be necessary) in explaining the occurrence of any particular evil.

If any particular evil can be explained by eschatology, however, the critic of theism needs to move to a more general level to plausibly formulate the problem of evil: the critic needs to refer either to certain sorts of evil, or to the amount of evil in the world.

Anyway, one who objects to some particular evil presumably objects to a general feature the evil instantiates; and that by itself is enough to move us to the more general level at which we consider the justifiability of certain sorts of evil.

Nevertheless, as soon as we object to sorts or kinds of evil, we seem to be objecting not to certain things that happen within the world–we are, rather, at least on some intuitive level, objecting to the sort of world we have, to the design of the world. To object to the fact that human beings suffer and are afflicted (an objection framed in terms of quite general "sorts") is to protest the design of the world; to object to the fact that human beings are capable of heinousness seems likewise to object to structure. An objection to the design or structure of the world is plausibly understood as an objection to the world's goodness.

Unless–until–we have some specific criterion of goodness for worlds, the point I'm making here will remain on an intuitive level. I think it is reasonable (at this intuitive level) to think that one taken by the problem of evil is committed to saying either that the world is not good, or that the world would be better if it were different.

Indeed, some who pose the problem of evil explicitly object to a system in which evils are (said to be) causally necessary to good. In *The Brothers Karamazov*, for instance, Ivan objects not merely to the suffering of children, but to a system which brings a greater good or harmony out of that suffering. Philosophical discussions of the problem of evil fairly frequently take Ivan's impassioned denunciation of the system as paradigmatic of the convictions that generate the problem of evil.

It might seem that we could avoid objecting to the system or the design of the world by taking a further step away from the particulars and objecting to all evils, objecting to evil in general. Indeed, tradition makes use of this formulation (following Epicurus). Even if objecting to all evils is objecting to a mass of particular events within the world, however, rather than to the world's structure, there is still a problem: all "evils" are not "intuitively" odious; they don't all require explanation. It's not fair to formulate the problem of evil as if they did. At least for the ordinary person, there are evils that are not problematical. The death of a plant is an evil to that plant, but not an evil that keeps us up nights.

So I'm inclined to think that a careful and complete formulation of the problem of evil implies that there is a criterion of goodness for worlds. If I'm right about this, and also in my earlier argument that it's not possible to formulate an adequate criterion of goodness for words from a non-theistic perspective, then we are led to the conclusion that atheistic arguments from the problem of evil are self-defeating.

University of St. Thomas
St. Paul, Minnesota

* * * * *

Endnotes

1. There are other reasons for considering whether there is a criterion of goods for worlds. The question is worth raising in part because it transects and connects fascinating issues in metaphysics and ethics. What is a world? Are there different kinds of worlds? Is the criterion of goodness for a thing determined by its kind's form or function? Does the world have a function? These are among the intriguing issues encountered.

Another reason for asking whether there is a criterion of goodness for worlds is that although there is Scriptural warrant for calling our world good--indeed, very good--it is not obvious to human reason (at least not to the modern mind) that there is a standard of goodness that undergirds the claim. Contemporary philosophy, with its relativistic emphasis on perspective, on world-view, makes it easy to deny sense to the claim that the world itself is good.

2. The term "world" is used in different ways by different people. "The world is everything, all that is," one person might say; "God created the world," another person (or maybe even the same person) says. A theist will typically deny that God created himself. Similarly, as we'll see, people have had different things in mind in speaking of "possible worlds."

It's not so easy to define the world. We can't point to it as we might point to a violin or a knife or a human being, offering ostensive definitions of these objects. If we point out, or up, and with a sweep of the hand indicate "all that," or "everything that is," it's still not clear we pick out a referent. We don't, it seems, pick out any particular number when we try to refer to "the number of things in this room"; one can't count things in a room as one can count people or pencils. One can only count the number of things of some sort. Since one can use various schemes–an unlimited number, some would argue–for classifying sorts, and since even in any one scheme there may be no determinate number of things in a sort, it's hard to imagine a counting procedure that uses "things" as a place holder. So how can we pick out a referent with the word "everything"? Ludwig Wittgenstein opens the *Tractatus* with the contention that "The world is the totality of facts, not of things"; but there are also problems with taking this as a definition of the world, problems that parallel the difficulties in trying to define the world as "all things."

Sometimes when philosophers have spoken of the world as "all things" or have said (as Aquinas did) that "all things belong to one world (ST I.47.3) they have simply been identifying their position in the ancient debate on whether there is or can be more than one world. Aquinas emphasized that the world is a unity, and not simply a collection of things, since the things that comprise it, created by God, "have relation of order to each other, and to God himself": because all things are related, the world is one. (Aquinas also had a scheme of classification that would allow him to use "things" as a place holder for certain sorts of things–probably species.)

G. E. M. Anscombe offers a definition (or characterization) of the world which I find as helpful as any, and will be my working definition: she suggests that "the world is the totality of bodies and physical processes, together with whatever is contained in any manner within the compass of that whole" (see "Times, Beginnings and Causes," a Philosophical Lecture read to The British Academy, 22 May 1974). Of course there are concepts here that could use clarification–as much could be said about any definition–but unless we encounter a specific need for clarification at some point, we won't worry about it. Under Anscombe's characterization, whether or not Aquinas (and Aristotle, and others) are right in seeing an intrinsic unity in all things, there cannot be more than one world, though there could conceivably be different "sub-worlds" which do not, cannot, causally interact.

Since the time of Leibniz, philosophers have tended to have a technical understanding of the phrase "possible world." It was his view that the world could not be different in any way without being a different possible world. Every complete possible history of events constitutes a different possible world, he held; if the actual world were in any way different, it would be the actualization of a different possible world, and hence would be a different world.

To the non-philosopher, it is likely to seem more natural to say that this world could be different without being a different world; that there is (to use the contemporary philosophers' language) a set of possible worlds similar to this one, and close enough to permit utterances such as "if this world had gone differently yesterday, today I might be in London." Before Leibniz introduced the technical notion of a possible world currently used by contemporary philosophers, philosophers seemed to have in effect a conception of "possible worlds" more closely akin to the "non-philosophical" one just mentioned, a way of regarding them which admits the possibility of the world being different in some way without being a different world. Aquinas, as we'll see, finds it sensible to imagine that the history of *this* world might have been different. It is no small matter–not "merely a question of semantics"–what one takes a possible world to be if one endorses the claim that God must create the best possible world. If there can be different versions of a possible world, then as Aquinas suggests, when one says that God must create the best possible world if he creates a world one may mean that he must make the world he creates as good as *it* can be. But his doing so is consistent with there being other (and better) worlds.

3.   Oliva Blanchette, *The Perfection of the Universe According to Aquinas: A Teleological Cosmology* (University of Pennsylvania Press, 1992).

4.   "A General Problem of Creation," and "A Particular Problem of Creation," in *Being and Goodness*, ed. Scott MacDonald (Cornell University Press, 1991).

5.   Alternatives such as the ones considered in this section merit more discussion than I have room for here. I discuss some of the alternatives at greater length in my doctoral dissertation, *Foundations of Theodicy: Is There a Criterion of Goodness for Worlds?*, University of Minnesota, 1984.

6.   Peter Geach, *The Virtues*, ch. 5, "Prudence."

7.   G. W. Leibniz, *Theodicy: Essays on the Goodness of God, the Freedom of Man, and the Origin of Evil*, ed. Austin Farrar, trans. E. M. Huggard (New Haven: Yale University Press, 1952), 138.

8.   Leibniz, from Grua's edition of his writings, quoted by John Hostler, *Leibniz's Moral Philosophy* (London: Duckworth, 1975).

9.   Robert Nozick, *Philosophical Explanations*, 418.

10.   Ibid.

Saint Anselm: An Ethics of *Caritas* for a Relativist Age[1]

by Graham McAleer

And if you salute only your brethren, what more are you doing than others? Do not even the gentiles do the same? You, therefore, must be perfect, as your heavenly Father is perfect.[2]

I

Relativism is both loved and hated. For some, it is pure liberation, a chance for a heady and playful encounter with a manifold of long-silenced meanings. For others, it bespeaks a crisis, an all too dangerous confusion about what is good and what is evil. Very often, what is loved most by those who wholeheartedly embrace relativism is its iconoclasm. Enthusiasts take it to shatter entire edifices of systematic thought and to render obsolete distinctions which have been nuanced over centuries. Better still, the argument of relativism is said to be of such a radical nature that it is able to put into the same boat such arch enemies as Christianity, Marxism and Liberalism, and undercut their superficial dissensions by isolating the naive ontology that is their common foundation.

To the contrary, however, this essay shall show that the putative radicalness of relativism only demonstrates the radical nature of the positively ancient theological virtue of *caritas*. The argument of this paper shall be that Saint Anselm of Canterbury, writing nearly one thousand years ago, understood charity to be an ethical practice engaged in under the very same ontological conditions that relativism demands ethics must now confront.

The kind of relativism that is the object of this study is taken as a starting point for the original social philosophy of Castoriadis; as a thesis worthy of constant critique by Habermas; and as a cause of optimism for thinkers like Caputo. These thinkers are all interested in a form of relativism that is said to show that the tradition of ethical thought has not confronted the reality of an ontology of difference.[3]

Philosophical giants of relativism, like Foucault and Heidegger, insist that truth cannot be the foundation of ethical systems because truth is relative: values cannot be justified, or critiqued, by an appeal to certain truths which hold now, have held in the past, and will hold in the future. It is time to admit, these authors argue, that there no longer are singular truths about persons or about things but rather, that the truth about what things actually are is quite relative to various cultural and historical systems of language, education, scientific investigation, government bureaucracy, and social organization.[4] Truth is not the foundation of scientific and social order but is rather a production of the different scientific methods employed and cultural values invoked at different times.[5] In consequence, no claim to possess the truth, to have access to the ontology, respecting persons or things is legitimate.[6] To insist that the categories of truth, being and value that are dominant amongst us now are eternal truths is simply to ignore the history of thought and civilization: history and the social sciences show all too clearly that there have been different ontologies and values respecting persons and things in a multitude of places and ages.[7] Ethics, these relativists conclude, must learn to live with an ontology of difference and must abandon its past naive assumptions. Ethics must at last explore the complexities, frustrations, pleasures and adventures of difference among competing systems of thought and manners of living.[8]

There can be no doubt that authors like Foucault and Heidegger have picked their target well. It is a time-honoured principle in the tradition of Western ethical thought that truth is the foundation of value. Marx, to give but one example, founds his critique of capitalism on what he takes to be a universal truth about the person: labour as self-activity belongs to the very essence of a person and thus any form of labour which alienates the person from her self-activity is immoral.[9] Christianity shares the assumption that truth founds value as fully as does Marxism.[10] Yet, it can be shown, I think, that the argumentation of relativism is not radical enough to unhinge all prior traditions of ethics. The ontological relativism of Foucault and Heidegger, it will be seen, condemns one to a particularity: a manner of life, including a method of thought, with values peculiar to that manner of life. According to Anselm, however, the structure of charity shows that particularity, and thus an acceptance of difference, is its very condition of expression; one is called to charity in just those situations where there are no shared truths or values; in those situations in which one is called to be a neighbour.[11] I will describe Anselm's account of the ontological conditions of charity (largely) through a consideration of his *De casu diaboli*. Saint Anselm explicitly discusses why the Devil fell in terms of an ontology of difference. Anselm tells us that Lucifer fell because he refused to accept his particularity as a being *ex se* (created being) and the otherness of God as a being *a se* (the uncreated source of created being). This text of Anselm shows that charity holds the possibility of an ethical practice which embraces particularity[12] and therewith difference and otherness.

The argument of this essay, however, is not merely that relativism has not successfully challenged all traditional conceptions of ethics, but further, that charity can address the most chilling predicament revealed by relativism. Many of those who are impressed by the philosophical power of relativism are yet deeply disturbed by the realization that ontologies and values are relative. These thinkers appreciate that it is one thing to accept that there are historical and cultural differences in truth and value, but it is quite another to know what to do when one of these different truths about what a person is, and its attendant value on how persons are to be treated, provokes real disgust and horror.[13] The problem that faces so many philosophers and theologians is how to orient to those different truths about persons and things that are found to be horrifying, and which cannot possibly be accepted as one's own, in such a way that does not simply entail the eradication of those truths and values, with all the institutions and manners of life that sustain them. How does one accept truths, practices and policies, which disgust and horrify, without effacing or denying one's own deepest held convictions about persons and things?

The theological virtue of charity is peculiarly well placed to address this question. It encourages us to embrace and love precisely those with whom we cannot agree or those who offend us the most. To accept the difference between who we are and those who are quite different from us is exactly what makes the concrete exercise of this virtue so profoundly difficult for most of us. Charity provokes a manner of action which does not efface our cherished beliefs, that does not lead us to abandon who we are; nor does it seek to eradicate the difference which exists between our particularity and that of the other. Rather, this difference in particularity is to be embraced and sometimes personally undergone *usque ad sanguinem*.

In Part I I would like to briefly detail the relativist ontologies of Heidegger and Foucault. I shall not critique these authors, but rather accept their conclusions, and yet I do not thereby mean to deny that there are excellent arguments against relativism. When showing in Part II that Anselm's concept of charity operates over the same ontological conditions of ethical practice as demanded by these authors, my point is only to demonstrate the better that the Christian tradition can meet the ontological strictures of relativism and more, that in charity it possesses an ethical practice which can transcend relativism's own limitations.

II

The relativistic claims about truth and value that have so caught the attention of writers like Castoriadis, Habermas and Caputo have issued, in the main, from ontologies developed by Heidegger and Foucault. I will treat Foucault first as his philosophy is more readily associated with

this kind of relativism, especially in the social sciences, and because I
agree with Christopher Norris that the texts of Foucault still provide
the most radical formulation of a relativism founded upon an ontology
of difference.[14] Moreover, unlike Heidegger, who was quite sure he
could offer a solution to avoid the worst consequences of relativism,
Foucault seems to have felt himself trapped by his commitment to
relativism. He admits himself that he never came to believe that any
one kind of ethical practice could be given priority over any other sort
once the full force of relativism was genuinely acknowledged. Foucault
explicitly states in his comments on political practice in *Power/Knowl-
edge* (the section on "Power and Strategies") that his thoughts on
whether it could ever be said that one kind of ethical practice was more
legitimate than another kind remain provisional and problematic.[15]

Through historical analysis, Foucault sought to show that there is
no enduring ontological truth about persons or things. He sought to
demonstrate that there has been no progress towards an absolute
knowledge of an absolute truth; but rather, a discontinuous series of
knowledges leading to very different ontologies of persons and things.[16]
According to Foucault, history is not merely epochal but historical
orders[17] are active ordering principles[18] which inject into a material
reality (this reality being insignificantly true)[19] epochal forms or meanings that produce different realities at different times.[20] In his analysis
of homosexuality, for example, Foucault showed that the 18th century
witnessed the birth of a new species. No longer were acts of sodomy
committed by persons but rather, there entered into history a new
species of person, the homosexual.[21] The 18th century, it was argued,
witnessed the dissolution of certain kinds of species, the litany of
monsters so dear to the terratology of the Middle Ages and the Renaissance,[22] and the creation of other natural forms, like the homosexual.
The 18th century thus established an order of truth peculiar to the
period, and radically different from that of the Middle Ages, that
dismissed some species as fables and established the reality of others.
Foucault therefore concluded that since truth is of an order, distinctions
like truth and falsity, sane and mad, constitute a violence of exclusion
(DL, 218)[23] for each epochal circumscription institutes a reality which
is arbitrary:[24] a reality is produced which is said to be 'within the true,'
but so too is an unreality, that which is 'outside of the true' (DL, 224).
Foucault found every order to be the same in this regard, and all equally
at violent fault.

With respect to ethics Foucault's argument is very dark, indeed. All
ethical practice is based upon an order of truth; every order of truth is
merely an imposition of one kind of truth in opposition to another kind
of truth; therefore, all ethical practice is necessarily violent. In consequence, no ethical practice can invoke any genuine claim to realize the
Good and all must be judged as illegitimate by classical standards. For
Foucault, every ethical practice suffers from the same illegitimacy—
each is based on an arbitary and violent order of truth—and herein lies

the origin of his relativism. As Foucault acknowledges in *Power/Knowledge*, it is no longer possible to say that one practice is more ethical than another for none can claim to realize the Good. Thus, ethics is condemned to living as best it can with an ontology of difference.

Foucault's analysis certainly owes much to the thought of Nietzsche[25] but it surely owes a debt to Heidegger as well. For Heidegger also concluded that history is epochal and productive; that human subjectivity always operates within an horizon sent by history.[26] He argued that all epochs are alike in that they make us forget that we are in an order of truth since we necessarily must be in-sistent, that is, absorbed in and busy with an order if we are to live and satisfy our needs. We cannot help being absorbed in what is produced by an order and thus failing to see the order itself; failing to see that the truths which we live by are but truths of an historical order of truth (EP, 390; QCT, 306-07).[27] We thus inevitably (that is, ontologically) live in the "illusion" that the produced beings are the only ones possible and hence come to believe in a static ontology that fosters the mistaken notion of enduring truths.[28]

As is well-known, Heidegger was particularly dissatisfied with the order in which the contemporary subject must live,[29] an epoch he described as dominated by subjectivist-constitution philosophies.[30] Such kinds of philosophy, he argues, remove any possibility for the subject to appreciate her finite nature, as the being to which meaning is granted (QCT, 299). Heidegger's concern is to ensure that the subject appreciate that she is sent or destined meanings, that she is a creature of particularity.[31] Thus, although Heidegger articulated a relativist ontology he tried at the same time to salvage the ethical practice that the subject guard against what fosters the illusion of human subjectivity as infinite (QCT, 308). One commentator, John Caputo, has called Heidegger's account of ethical practice, to ponder on our openness and particularity (QCT, 308-310), "an eschatological story in ethics."[32] For Caputo, such a story reduces ethical practitioners to persons left "waiting for a new dawn" (RH, 255).

Although Caputo is sympathetic to Heidegger, he stresses, along with others who are less sympathetic,[33] that Heidegger insists that human subjectivity "belongs to" the epochal meanings which are destined;[34] that subjectivity is the captive of the event of Being. The subject can only accept these meanings and relate to them, pro or contra, in the particular historical or cultural terms destined in the epoch. Should an epoch itself be found unsatisfactory for some reason, ethical practitioners can only hope that things might change and another more satisfactory epoch come into being. The problem is obvious for a case in which a particular order is found to be abhorrent: it cannot be overthrown or rejected. Commentators as different from one another as Caputo and Habermas argue that, at bottom, this is the story that Heidegger is offering, and I think they rightly find this model for ethical practice unacceptable. Heidegger has not found a satisfactory answer to the relativist's own worst problem.

It ought to be noted that Heidegger's position echoes, albeit faintly, the Christian ideal of charity: a loving acceptance of the evil done to one; as in Christ's loving acceptance, *usque ad sanguinem*, of sin. Charity is not an intellectualistic contemplation (with the hope of change) of the otherness of an order which cannot be accepted as one's own but a complete acknowledgement and acceptance of that otherness out of a love for the other. This is not to agree with what is found abhorrent or distasteful, nor is it indifference to the conditions invoked by that order. Rather, as Anselm argues, charity is something far more unnerving, an invitation to suffer that order, to have that unacceptable order painfully enter into, and effect, one's own interiority; to share the pain of an unacceptable order.

III

Anselm is one of the first of the great builders of that theological tradition which, despite its varieties and subtleties, has emphasized the virtue of charity as the acme of the person's affective and moral possibilities.[35] Anselm's thought is especially apt to contest that kind of relativism which depends on an ontology of difference because he explicity articulates his theory of charity in terms of ontological difference.[36] Although Anselm's texts, including *De casu diaboli*, have been the object of some excellent recent studies in French, my treatment of the theory of charity to be found in Anselm both extends these French studies and makes a philosophical application of the ideas of this great medieval philosopher.

Anselm argues that the subject has two broad choices—to will a happiness that can find its satisfaction in God or not. The first choice is to have love for the other, the latter to have love only for oneself. Anselm speaks of God having given to the subject the rational nature or power of discernment, "so that she might detest and shun evil and love and choose good" (*ut odisset et vitaret malum, ac amaret et eligeret bonum*).[37] Love and hate are here linked with the verbs that Anselm employs to describe the choices of the will. Indeed, the first chapter of the second book of *Cur Deus homo* stresses again and again that the choices of the will are rendered by an affectivity of attraction and repulsion, of love and hate. This identification has been noted by other commentators, most notably, by Paul Gilbert. Apart from a documentation of the passages in Anselm's works where this identification is made, Gilbert has also sought out the philosophical significance behind Anselm's statement in the *Monologion* that a person approximates most profoundly to God when the choices of her will express the image of God imprinted within her soul.[38] Gilbert argues that since Anselm understands the Persons of the Trinity to relate to one another through relations of love, indeed, since he thinks the very substance of God is love, he must also mean us to understand that love is the highest possibility of created will[39] and that charity is at the foundation of what it is to be a human subject.[40] In the pages that follow, I shall apply and extend Gilbert's comments to show how love functions ethically in the thought of Anselm.

The created subject is morally ordered to articulate herself in terms of the image of God, and this means in terms of love. For, as Gilbert explains, the Son lives in a unity with the Father through the love that is the Holy Spirit, and not through any cognitive or rational relationship:[41] "love is the name of God in his proper unity, living between the Father and the Son; the name 'Spirit' «denotes the substance of the Father and likewise the Son» (M, 69, 10)."[42] Anselm tells us that the subject was created in order that she might love,[43] indeed the purpose of the rational creature is to love the highest good;[44] to transform herself into a totality of love.[45] Crucially, Anselm argues that while reason discerns what is the subject's true good, such discernment proves useless unless the good is loved by the subject[46] and, as will be seen, the subject who loves God has a just will.

The passages upon which these various points have been based show that, according to Anselm, the will is the locus of an affective power.[47] In understanding the Anselmian will in terms of a kind of affectivity, a central but elusive concern of Anselm's can be explained quite well. Anselm's ethical theory relies on the possibility that the will can transcend willing its most immediate happiness so as to approximate a certain selfless election: he argues at some length that the will for justice (*affectio iustitiae*) is only kept when the subject relinquishes or sacrifices an immediate good out of love for God,[48] who is justice itself.[49] Anselm insists then that love is the phenomenon that can best explain this surpassing of the self. Antoine Vergote has noted of the general structure of affectivity:

> Whether in joy or terror, love or hate, guilt or peace, affectivity opens up a trajectory to something other, another locus, a future in which the subject will find himself, as it were, outside of himself, but which radically addresses his singularity.[50]

Anselm made Vergote's point long ago. When he argues that the will for personal happiness (*affectio commodi*) is "inseparable" (he describes this *affectio* as a *voluntatem naturalem*),[51] Anselm recognizes that the subject cannot absolutely transcend her will for personal happiness (her singularity or particularity). The most selfless of loves is still rooted in what is of vital importance to the subject, a certain love of self. Briancesco has neatly rendered the distinction in Saint Anselm's thought that the just will is the will that wills God for no other reason than because of who God is: that is, God or justice is to be willed *propter se* and not *propter aliud*.[52] This is quite correct, but we must note that Anselm does not argue that such willing is absolutely selfless. Not only does his insistence that the *affectio commodi* is "inseparable" show this, but so too does his claim, and that of his theological tradition, that when one is just, one is "like God."

The Devil's sin was to want to be God, to annul all difference and distance between himself and God. To will justly is for a person to comport herself to God, to become as like God as is possible for a created subject. It is not to abandon herself or to cast off that which marks her the most thoroughly, her own particularity or, as Anselm says, her *affectio commodi*. Anselm, in his characterization of what it is to will justly, does not speak of the complete abandonment of a subject's own self but instead insists upon the transformative capacity of the will (Saint Bonaventure and Henry of Ghent after him will stress this aspect of the will):[53] the just will does not cease to be the subject's and that which defines the subject the most; rather does it put the will into new relations which give a new definition to the subject.

While love always addresses the lover's particularity, its peculiar structure enables it to be an opening out towards another. Love of the other enables the subject to transform herself into the other in such a way that she is truly with the other, that the other somehow enters into her own definition of self. When the subject defines herself through love of the other, she does not force the other to be the same as herself, which would only be a self-love, but nor does she abandon her own particularity, since such an abandonment is only a means for her to love herself in the complete identification she has made of self and other. In Anselm's terms, one has an *affectio commodi* and an *affectio iustitiae*: a love of self mediated by a love of the other. Vergote also highlights this structure of love:

> There is no love that does not also entail the desire to be loved
> and the wish to confirm and broaden - through an exchange
> of gifts - the self as well as the other (GD, 132).

According to Anselm, a love of self is mediated by a love of the other when a person loves God. For to love God is to will the order of God, a just order (DCD, c. 4, 241). It is this order, an order of love and justice, which allows the other to appear as other since it demands of a person that she hold before her a difference between self and other. That the order instituted by the love of God is an order of difference is evident from Anselm's treatment of the Fall of the Devil. The Devil rejected the difference between God and himself, and ceased to love God as God, through willing *inordinate*.[54] The Devil willed *inordinate* when he sought an identification between himself and God: which is to say, when his love of self was so great that he sought to eliminate the particularity of God and thus to remove the very difference which established God as God. There is only a just will, therefore, when a person mediates her love of self through the love of another, which is to embrace the other in her different particularity. Anselm makes this very clear when he says that the will to justice is separable and can be lost.[55] A "separable" relationship with the other cannot be embraced if there is only a love of self since such love does not relate anyone to another person. Love of self does not allow the other to appear at all: in indifference, the other in her particularity simply ceases to exist, and in hate, the other is destroyed. It is only when a love of self (*affectio commodi*) is ordered through a love of the other (*affectio iustitiae*) that a difference is acknowledged in which more than one person can appear and establish a relationship which could be lost.

If Anselm's argument is that it is in the nature of charity to accept a difference between the subject and the other, a difference which characterizes love of the other as other, then it is no surprise that the Fall is taken by Anselm as paradigmatic of the ethical situation. The Angels, just as Adam and Christ later, Anselm tells us, were created in justice and happiness[56] and to be created in justice is to be created loving God (DCV, c. 8, p. 152). Understanding the ontological preeminence of God, the created will ought to respect God's status as a subjectivity *a se*.[57] Otherwise, created will shows contempt for God, sins thereby, and causes evil[58] (the definition of sin that Abelard will take up).[59] Satan fell because he willed to be the same as God: understanding that it belongs only to God, due to His nature, to have a *propriam voluntatem*, Satan nevertheless willed to have an autonomous will.[60] In abandoning the ontological difference between *a se* and *ex se*,[61] Lucifer ceased to love God and ceased to define himself through the order of love instituted by God.[62] In his love of self, Satan refused to locate him happiness in God and, as it were, to be willed by God. In love of other, the self is joined with the other and is broadened in such a manner as to take on a new perspective from which to order its willing. It was this loss of self-sufficiency that Lucifer rejected.

I would now like to state, in very general terms, what I take to be Anselm's ontology of the ethical situation and how his description of charity can lead to a fruitful concept of ethical practice despite the strictures of relativism. To will justly for Anselm is not to usurp and remove God from His ontological status *a se*, it is rather to accept Him as He is in that status by willing to remain *ex se*. In accepting the radical otherness of God, in respecting this ontological difference (as relativism demands of us), we become like God: as He loves those who are other than Him, so we must come to love those who are other than ourselves. We are reminded of Christ's words,

> And if you salute only your brethren, what more are you doing
> than others? Do not even the gentiles do the same? You,
> therefore, must be perfect, as your heavenly Father is perfect
> (*Matthew* 5, 47-8).

The Fall reveals that the ethical demand to be a neighbour, to practice charity, arises when there is otherness. Christ teaches that there is no charity in loving those with whom one shares ways of thought and living, your brethren. Rather, one must love those who are different and this entails loving them as different (and is this not what relativism claims ethical practice must learn to do?). This is the nature of God, who is perfect. Gilbert captures this succinctly when he describes Anselm's depiction of the structure of the Trinity:

> The opposition of paternity (the Father is not the Son, the Son
> is not the Father) is interior to the love which unifies 'singu-
> larly' the persons who give themselves the one to the other in
> their difference.[63]

Lucifer sinned when he was unable to live with the ontological difference between himself and God;[64] Lucifer wanted to be coextensive with the Good (TL, 115). This desire to be coextensive with the Good can appear in various guises: when we seek to ensure that everyone follows the practices we ourselves follow, or even when we insist that everyone must have the same concept of order and peace as ourselves and so on. Maintaining this difference, however, cannot but be painful. Saint Theresa of Avila describes her union with God in terms of pain,[65] Bonaventure in terms of death and darkness,[66] and Max Scheler has described differentiation as the ontological condition of pain.[67] Pain and suffering are found with love since love holds difference before itself. Indeed, the union of love is all the more painful for not being total despite intense hope for completion. To succumb to the abolition of difference is to forsake the other as other: to simply annul the pain but also to annul the love. Anselm instructs us then that the conditions of the ethical situation are difference, the suffering that comes of that differ-ence, and the love that must actively hold to this difference. As Bonaventure discovered through mysticism, love of the other is sensu-ous, it is the moment of an embrace (IMD, 75), a union that is never complete and must not be so.

IV

The two great theorists of relativism, with whom this paper began, believed themselves to have disclosed new and radical conditions for ethical practice. Any viable account of ethics must, they insisted, embrace an ontology of difference. Saint Anselm argues that the Devil fell because he did not love God, he did not accept an ontology of difference. Which is to say, that the theological virtue of charity is precisely an ethical practice that does acknowledge an ontology of difference. The relativism of Foucault and Heidegger does not force us to begin ethical theory anew. In an important respect, moreover, Anselm goes beyond Foucault and Heidegger. Foucault explicity ac-knowledges that he never really managed to find a theory of ethical practice that might meet the strictures of relativism. Heidegger did articulate some manner of ethical practice in the light of relativism and although sympathetic and unsympathetic commentators alike do not find it very satisfying, Heidegger's position does point in a fruitful direction. Charity does address relativism's own limitations; that it does so, is, I think, attested to by the difficulty of its practice. The practice of charity is not difficult to adopt simply because it refuses to diminish the difference that exists between the participants in the ethical situation. The difficulty of this practice resides in the call not only to live amidst the difference, but to love despite the difference and even, to suffer the difference. Only in suffering the difference do we co-operate with God's grace for only then have we helped make it possible that grace might transform.

Naturally, concerning the nature of charity and its transformative aspects, deeper questions than those addressed in this essay remain. In the complete acceptance of the other, in Christ's acceptance of the sin of humankind, for example, is there necessarily a transformation of the other? If so, would such a transformation constitute a violence done to the other? Foucault, as we saw, would insist that every ethical order is violent and he would not expect charity to be any different in this. Does the profound acceptance of otherness nevertheless lead to the violent change of the other? If so, then perhaps charity shows something fundamental about the acceptance of difference. Otherness or difference should not lead one to assume that there is an inevitable isolation of orders. *Pace* many of its proponents, the ontological conditions that relativism describes *do not also amount to ontological conditions for a peace and harmony in which all people live out their lives in whatever way they might wish.* Acts of charity show that even the most complete acceptance of difference can lead to conversion. We are reminded of the centurion at the foot of the Cross:

> When the centurion and those who were with him, keeping watch over Jesus, saw the earthquake and what took place, they were filled with awe, and said, "Truly this was the Son of God!" (*Matthew* 27, 54-55)

If in the moment of the complete acceptance of the other, and such a moment is the Cross, the other is herself transformed by witnessing charity, do we here have a violence? If so, it might be proper to call such a violence 'ontological': that charity has as one of its possible outcomes a transformative effect on the other. It could well be that an ethical practice, like charity, that fully accepts the conditions for ethical practice as prescribed by relativism contains in it an ontological movement towards a kind of unforced and gifted universalism. This requires exploration well beyond the confines of this paper. I was only concerned here to show that relativism holds to a position of particularity as essential to the ontological conditions of ethical practice, conditions which are well met by charity. This is clearly seen in Anselm's description of the ontological conditions of charity. Thus, the theological tradition has a response to relativism. In fact, this tradition leaves off where relativism arrives.

Loyola College in Maryland
Baltimore, Maryland

* * * * *

Endnotes

1. I would like to thank Jennifer DeRose, Steve Sherwood and Bob Roskos for many helpful discussions about this essay.
2. *Matthew* 5, 47-8, taken from the Revised Standard Version.
3. See Thomas McCarthy's remarks on the basic claims of poststructural ethical thought in his *Ideals and Illusions: On Reconstruction and Deconstruction in Contemporary Critical Theory*, (Cambridge, Mass.: MIT Press, 1993), 1-7. Hereafter, IAI. See Charles Taylor's comments in his *The Ethics of Authenticity*, (Cambridge, Mass.: Harvard University Press, 1991), 18-19 and 37.
4. See Cornelius Castoriadis, "Intellectuals and History," in his collection of essays, *Philosophy, Politics, Autonomy*, (Oxford: Oxford University Press, 1991). See also Hans Joas' comments on Castoriadis in his *Pragmatism and Social Theory*, (Chicago: University Press of Chicago, 1993), 155.
5. See Jürgen Habermas, *Moral Consciousness and Communicative Action*, (Cambridge, Mass.: MIT Press, 1993), 29.
6. See Jürgen Habermas, *Postmetaphysical Thinking*, (Cambridge, Mass.: MIT Press, 1992), 132-34.
7. See Christopher Norris, "Truth, Science, and the Growth of Knowledge," *New Left Review*, (March, 1995): 105-23 (= TSGK).
8. See John D. Caputo, *Radical Hermeneutics* (Bloomington and Indianapolis: Indiana University Press, 1987) (= RH).
9. K. Marx and F. Engels, *The German Ideology*, Part I, ed. C. J. Arthur, (New York: International Publishers, 1989), 91-95.
10. For an example within Christian thought, one need look no further than *Veritatis Splendor*.
11. Paul Ricoeur has also identified this peculiar character of charity in his claim that the practice of charity ultimately concerns the person as 'neighbour' and not the person as '*socius.*' The neighbour is a person who must act without recourse to any social mediation or, in the language of orders, a person who is required to act despite the radical distance between her order and that of the other. See Ricoeur's essay, "The Socius and the Neighbour," in *History and Truth*, (Evanston, IL: Northwestern University Press, 1965), 98-109.
12. I have briefly examined the fruitfulness of charity in this respect in my "Old and New: The Body, Subjectivity and Ethics," in *Philosophy Today* (Fall, 1994), 259-67.
13. This sort of problem has been nicely captured in some of Rorty's remarks about relativism. See his *Contingency, Irony and Solidarity*, (Cam-

bridge: Cambridge University Press, 1989).

14. See TSGK, 118-19.

15. Responding (in writing) to some questions put to him (in writing) on this problem, Foucault wrote in a postscript that the answer given is only the first draft which he wrote as an answer. He adds: "This was not through my faith in the virtues of spontaniety, but so as to leave the propositions put forward their problematic, intentionally uncertain character. What I have said here is not 'what I think' but often rather what I wonder whether one couldn't think." (*Power/Knowledge: Selected Interviews and Other Writings*, 1972-77, ed. Colin Gordon, [New York: Pantheon Books, 1980], 145 [= PK]).

16. See especially the second and third chapters of *The Order of Things*, (New York: Vintage Books, 1973).

17. Or the various practices that have been instituted or are being instituted by discourse.

18. "The Discourse on Language" in *The Archaeology of Knowledge*, (New York: Pantheon Books, 1972), 216 and 224 (= DL).

19. Foucault lacerates himself on the status of that which might be said to be "before" the practices instituted by discourse. In the *Discourse on Language*, Foucault criticizes historical ontologies for "eliding the reality of discourse." (DL, 227-28). But in the *History of Sexuality* Foucault insists on the priority of the unordered body. See *The History of Sexuality*, Volume I (New York: Vintage Books, 1990), 48 (= HS).

20. PK, 119.

21. "Homosexuality appeared as one of the forms of sexuality when it was transposed from the practice of sodomy into a kind of interior androgyny, a hermaphrodism of the soul. The sodomite had been a temporary aberration; *the homosexual was now a species*." (HS, 43); the emphasis is mine.

22. M. Foucault, *Madness and Civilization*, (New York: Vintage Books, 1973), 209.

23. PK, 114 and 123; See Jürgen Habermas, "The Critique of Reason as an Unmasking of the Human Sciences: Michel Foucault," *The Philosophical Discourse of Modernity*, (Cambridge, Mass.: MIT Press, 1990), 252 (=PDM).

24. Based upon power (PK, 131-33 and 141) and its scientific (PK, 112), economic, institutional and cultural forms (PK, 131).

25. Acknowledged in *Power/Knowledge*, 133.

26. Martin Heidegger, "The End of Philosophy and the Task of Thinking," *Basic Writings*, ed. D. F. Krell (New York: Harper & Row, 1977), 386 (= EP). See also "The Question Concerning Technology," *Basic Writings*, 297 and 312 (= QCT).

27. See Martin Heidegger, *Being and Time*, trans. John Macquarrie & Edward Robinson (New York: Harper & Row, 1962), 264-65).

28. See Mark C. Taylor, *Altarity*, (Chicago: University of Chicago Press, 1987), 40.

29. See Jürgen Habermas, "The Undermining of Western Rationalism through the Critique of Metaphysics: Martin Heidegger," PDM, 132-33.

30. Heidegger sets Descartes, Hegel, Husserl (EP, 380-83) and Kant ("On the Essence of Truth," *Basic Writings*, 120) alongside one another in having

made this shift.

31.   The subject, "fails to see himself as the one spoken to, and hence also fails in every way to hear in what respect he ek-sists, from out of his essence, in the realm of an exhortation or address, so that he *can never* encounter only himself." (QCT, 309).

32.   RH, 240.

33.   Speaking of the event of Being, Habermas has commented: "Dasein is no longer considered the author of world-projects in light of which entities are at once manifested and withdrawn; instead, the *productivity* of the creation of meaning that is disclosive of world passes over to *Being* itself." (PDM, 152); cf. McCarthy's IAI, 5.

34.   A revealing is, "that which has already claimed man so decisively that he can only be man at any given time as the one so claimed." (QCT, 300 and 307).

35.   Anselm is certainly influenced by Saint Augustine.   "Caritas Dei, inquit, hic dicta est virtus, quae animi nostri rectissima affectio est, quae coniungit nos Deo, qua eum diligimus." (*De moribus ecclesiae catholicae*, quoted by Saint Bonaventure, *In I Sent., Commentarius in IV Libros Sententiarum, Opera Omnia*, [Quaracchi, 1882-1902], d. 17, pars 2, c. 6, 290).

36.   As no comparative study of medieval theories of charity exists (to my knowledge) I do not know if Anselm is unique in doing so.

37.   Saint Anselm, *Cur Deus homo, Sancti Anselmi Cantuariensis archiepiscopi Opera Omnia*, 6 Vols., ed. F. S. Schmitt, (Edinburgh: Nelson, 1946-1961), Vol. II, Liber secundus, c. 1, 97 (=CDH). All translations are my own unless stated.

38.   Saint Anselm, *Monologion*, Vol. I, c. 68, 78 (=M).

39.   "Si la foi est une tension vers sa fin, une «approche», sa fin lui est accordée dans l'amour, c'est-à-dire dans l'adhésion volontaire." (P. Gilbert S.J., *dire l'Ineffable*, [Paris, Éditions Lethielleux: Namur, Culture et Vérité, 1984], 283 [= DI]). Gilbert is surely indebted to the classic of Paul Vignaux, "Note sur le chapitre LXX du *Monologion*," *Revue du moyen âge Latin*, 1947, 321-34.

40.   "La vertu théologale de charité ou d'amour constitue le point de départ de l'anthropologie d'Anselme." (DI, 269).

41.   P. Gilbert S.J., *Le Proslogion de s. Anselme: Silence de Dieu et joie de l'homme*, (Rome: Analecta Gregoriana, 1990), 210 (= PA).

42.   The French runs: "l'amour est le nom de Dieu en son unité propre, vivante entre le Père et le Fils; le nom 'Esprit' «désigne la substance du Père et pareillement du Fils» (M, 69, 10)." (PA, 211).

43.   Saint Anselm, *De concordia praescientiae et praedestinationis et gratiae dei cum libero arbitrio*, Vol. II, quaestio III [XIII], 286 (= DC); M, c. 68.

44.   M, c. 68, 79; CDH, Liber secundus, c. 1, 97-98.

45.   "Quod tamen a summe iusto summeque bono creatore rerum nulla eo bono ad quod facta est iniuste privetur, certissime est tenendum; et ad idem ipsum bonum est omni homini toto corde, tota anima, tota mente amando et desiderando nitendum." (M, c. 74, 83); Saint Anselm, *Proslogion*, Vol. I, c. 25, 120.

46.   "Denique rationali naturae non est aliud esse rationalem, quam posse

discernere iustum a non iusto, verum a non vero, bonum a non bono, magis bonum a minus bono. Hoc autem posse omnino inutile illi est et supervacuum, nisi quod discernit amet aut reprobet secundum verae discretionis iudicium." (M, c. 68, 78); CDH, Liber secundus, c. 1, 97.

47.   I have argued elsewhere that Abelard, who was much influenced by Anselm, seems to have thought that election issued from a particular kind of reason which had an affective dimension. See my "Reason, the Ethical Subject and Sin in the Thought of Peter Abelard," in *Quodlibetaria. Miscellanea studiorum in honorem Prof. J. M. da Cruz Pontes anno iubilationis suae offertae, Conimbrigae MCMXCV.* Cura Marii A. Santiago de Carvalho, praestamen iuvamen J. F. P. Meirinhos. *(Mediaevalia. Textos e Estudos,* 7-8), Porto, 1995. The role of affectivity in a variety of early scholastic treatments of charity is given in R. Wielockx's, "La sentence *De Caritate* et la discussion Scolastique sur L'Amour," *Ephemerides Theologicae Lovanienses,* (1982): 50-86, 334-356 and (1983): 26-45.

48.   CDH, Liber primus, c. 9, 61; justice is not a natural possession (DC, quaestio I [VI], 256) it is from God alone (DC, quaestio III [III], 266) and it is His grace that sustains one in justice (DC, quaestio III [IV], 267).

49.   Saint Anselm, *De casu diaboli,* Vol. I, c. 4, 240-41 (=DCD).

50.   A. Vergote, *Guilt and Desire,* trans. M. H. Wood, (New Haven: Yale University Press, 1978), 145 (= GD).

51.   DCD, c. 12, 254.

52.   E. Briancesco, *Un triptyque sur la liberté, La doctrine morale de st. Anselme: De veritate, De libertate arbitrii, De casu diaboli (L'oeuvre de saint Anselme),* (Paris: Desclée de Brouwer, 1982), 40 (= TL).

53.   Henry of Ghent, *Quodlibet X,* ed. R. Macken, *Opera Omnia Henrici de Gandavo,* vol. XIV, (Leuven: Leuven University Press, 1981), q. 15, 303.

54.   Saint Anselm, *De conceptu virginali et de originali peccato,* Vol. II, c. 4, 145 (= DCV).

55.   Anselm speaks of the will which wills to keep rectitudo as being in itself not deficient in anyway, although it can be expelled by another willing (DC, quaestio III [X], 278); DCV, c. 2, 141 and c. 28, 171.

56.   DCV, c. 13, 155; This is not to say that they experienced Paradise itself. They were, however, in a proximity to God far surpassing that of the human subject today. Adam, Anselm tells us, while not in Paradise as it would be at the end of history, was nevertheless in Paradise; the angels too, presumably.

· 57.   DCD, c. 4, 242; Christ always obeyed God as every human will ought: "Hanc igitur oboedientiam debebat homo ille deo patri, et humanitas divinitati, et hanc ab illo exigebat pater." (CDH, Liber primus, c. 9, 61).

58.   DCV, c. 5, 147.

59.   Peter Abelard, *Peter Abelard's Ethics,* ed. and trans. D. E. Luscombe, (Oxford: Oxford University Press, 1971), 4 and 6.

60.   Anselm describes this in the stunning and justly famous passage of Chapter 27 of *De casu diaboli* (DCD c. 27, 275).

61.   Briancesco speaks of the desire of the fallen angel to have "une liberté auto-créatrice." (TL, 21 and 125); cf. PA, 116.

62.   DCD, c. 5, 240 and c. 14, 258; Gilbert explicitly places Anselm within

Augustinian thought and precisely in terms of the role love plays in drawing one to God (PA, 115). Of this tradition, he writes: "Pour les augustiniens, la création n'est pas seulement le don de l'être, la position par amour dans l'existence; elle est aussi le don de l'énergie par laquelle l'étant se tourne vers Dieu. Le péché inverse l'intention de cette énergie par laquelle l'homme devrait répondre droitement au don d'amour de Dieu; par là, le libre arbitre use en faveur de soi de la force d'amour par laquelle il devrait s'orienter vers Celui qui le fait être." (PA, 114-15).

63.   "L'opposition parentale (le Père n'est pas le Fils, le Fils n'est pas le Père) est intérieure à l'amour qui unifie 'singulièrement' les personnes qui se donnent l'une à l'autre en leur difference." (PA, 209).

64.   The Fall issues from, as Briancesco puts it, "l'irrépressible desir, de la part de la créature, de faire coincider l'immanence et la transcendance." (TL, 21); see PA, 117 and 119.

65.   "The "sweet pain" St. Theresa speaks of evokes the twofold nature of the experience - the suffering of being torn out of her self-sufficient enclosure and the joy of being inhabited, as it were, by the other." (GD, 162).

66.   *Itinerarium mentis in Deum, Works of Saint Bonaventure*, Vol. II, trans. Philotheus Boehner, O.F.M., (Saint Bonaventure, N.Y.: The Franciscan Institute of Saint Bonaventure University, repr. 1990), 101 (=IMD).

67.   See his "The Meaning of Suffering," in the collection, *Max Scheler (1874-1928) Centennial Essays*, ed. Manfred S. Frings, (The Hague: Martinus Nijhoff, 1974).

St. Thomas Aquinas and the Defense of Mendicant Poverty

by John D. Jones

Like many of his contemporaries, Aquinas joined in the polemical fray over the legitimacy of the new mendicant orders: the Friars Minor and the Friars Preachers. The poverty practiced by these orders was a key flashpoint in this controversy. In particular, and especially in relation to the Franciscans, the question arose as to whether the possession or ownership of material goods could be completely renounced not only by individuals but also by a community. The relation between the use and possession of consumable goods such as food and clothing was central to this question for, as Gerard of Abbeville argued against the Franciscan position:

> To say that the use of [necessities of the body] is yours alone, and that the dominion pertains to those who have given them, until they are consumed by age, or until the food is taken into the stomach, will appear ridiculous to all, especially since among men, use is not distinguished from dominion in things that are utterly consumed by use.[1]

Bonaventure dealt with this argument explicitly in the *Apologia Pauperum* by arguing for a simple use of things which did not require the user to possess, own, or hold dominion over them.[2] The Franciscans could claim that they did not own but merely made use of all the goods to which they had access by arguing, through papal support, that the ownership of these goods was held by the Pope and not the order.[3]

Aquinas took up the defense of mendicant poverty in two works: the *Contra impugnantes Dei cultum et religionem* (CID), written about 1256, and the *Contra pestiferam doctrinam retrahentium homines a religionis ingressu* (CR), written about 1270.[4] In this paper I want to argue that there is a systematic ambiguity in Thomas's defense of mendicant poverty, especially in relation to the question of whether a community can renounce possession of all goods including consumable goods.[5] Although Thomas never explicitly adopted Bonaventure's position on the existence of a simple use of things without dominion over them, there is a distinct line of reasoning in both the CID and the CR that seems to parallel Bonaventure's position.

On the other hand, when Aquinas explicitly discusses the renunciation of possessions by a community, it isn't clear exactly whether *possessiones communes* extends to all material things that can be owned or only to some subset of them. Moreover, in other writings when Aquinas treats usury in relation to consumable goods, he explicitly adopts the conventional view that the use of and dominion over such goods cannot be separated. While Aquinas never directly utilized this reasoning in the mendicant controversy, Hervaeus Natalis, a 14th century Dominican, did so precisely in order to reject the Franciscan claim that people can completely lack common as well as personal dominion over goods.[6] I will use Hervaeus's position to spell out what I think are the implications of applying Thomas's analysis of consumables to the question of mendicant poverty.

I

In this section, I want to consider four texts in the CID and CR that seem to support Bonaventure and the Franciscan view that an order can completely renounce ownership of things both personally and in common. In the CID, Aquinas frames the discussion of mendicant poverty with the question about "whether it is lawful for a religious to relinquish everything that is his so that nothing remains possessed by him either personally or in common."[7] In the *responsio* to the question, Aquinas twice asserts that "to possess nothing"[8] on earth pertains to Christian perfection. While Aquinas does not explicitly add the phrase "personally or in common" to this assertion, Bonaventure does so in making an identical claim in *De perfectione evangelica*.[9] While Thomas makes the claim about possessing nothing in the context of showing that actual poverty pertains to Christian perfection, which he separates from the question of whether lacking common possessions pertains to such perfection, nevertheless, given the entire thrust of the Chapter 6, it would be odd if Aquinas only referred possessing nothing to something done personally and not in common.

Second, while Aquinas never directly cites Gerard of Abbeville's consumables argument, he does present what seems to be an equivalent version in the CR, "it is impossible that people possess nothing personally or in common, since it is necessary that they eat, drink and be clothed. They cannot do this if they have nothing."[10]

Aquinas regards this objection as "altogether frivolous" since, as he says, "the things which religious use to sustain their life are not theirs in terms of the ownership of one who has dominion, but these things are dispensed for the use of their necessities by those who have dominion over them, whomever they might be."[11]

Given that the objection refers to the impossibility of not possessing things personally or in common, Aquinas's response meets the objection only if it refers to religious collectively and not just personally. That is, it would mean that religious who rejected common dominion over things could only use things for their sustenance which were owned by others. This, of course, is exactly what the Franciscans claimed.

Third, although Aquinas grants that Jesus held a purse, he denies that Jesus contributed to this purse from any possessions which he had. Rather, in the CID, Aquinas writes that Jesus received goods from the women who ministered to him.[12] Unfortunately, Thomas never clarifies exactly what he means by "possessions" or whether the things which Jesus received became the common possession of him and Apostles after they were received. If possessing nothing, however, belongs to Christian perfection and Jesus and the Apostles offer the highest example of such perfection, it is certainly tempting to make the inference that Aquinas thought that Jesus and the Apostles possessed nothing.

Finally, in light of 1 Tim 6:8, that we are to be content with food and clothing, and the Gloss "that we are not to completely reject temporal things," the sixth objection to Chapter Six concludes "but the person who gives up everything and enters a religious order which lacks possessions, rejects temporal things completely."[13] Thomas responds

> quod illud quod dicitur, quod temporalia non sunt omnino abiicienda, [i] intelligendum est, quin eis utamur ad sustentationem vitae (when it is said that temporal things should not be entirely denied, it should be understood that they are to be used for the sustenance of life).... [ii] non tamen intelligit quin homo possit omnium temporalium proprietatam a se abiicere.[14]

It seems necessary to translate the last clause [ii] as: "nevertheless [the Gloss] does not mean that a person can/may[15] deny himself all ownership[16] of temporal things." If so, this text clearly counts against the idea of the possibility of "possessing nothing" and, far from supporting the Franciscan practice of poverty, it seems to contradict it. Bonaventure, for example, asserts that the same Gloss refers to "the denial of things in regard to use, but not in regard to dominion. For the use of temporal things is necessary for human life, but it can be obtained without dominion and ownership."[17]

Must Thomas's text be translated as we have? Scott Swenson certainly takes it in this way since he claims that the passage shows Thomas believes that one cannot renounce "one's property in goods used for immediate preservation of nature."[18] While the English and French translations follow the Vives edition in reading curam for proprietatem, they nevertheless provide the same syntax to the translation of the text. Proctor's translation reads: "[it] ... does not mean that a person can ignore all provision for temporal things."[19] Yet I don't think this translation is correct.

The force of objection six is that it is wrong to give up everything since this violates the injunction of the Gloss: temporal things are not to be completely renounced. Temporal things, however, can be renounced in terms of use or in terms of possession (or both). On the one hand, the Gloss could mean that one should not give up all use of things. Thomas's response clearly shows that he accepts this implication of the Gloss since he writes in [i] that religious should use things for their sustenance. On the other hand, the Gloss could mean that a person should not renounce all ownership of things. Of course, it follows from this that [iia] the Gloss does not mean that a person may renounce all ownership of things or, more simply, [iib] the Gloss means that a person should retain some ownership of things (which parallels the construction of [i]). But if Thomas accepts this implication of the Gloss, why didn't he simply construct [ii] in the same way as [i] and write [iib]? Moreover, if Thomas did hold that the Gloss does not permit people to renounce all ownership of things, then it isn't clear why Thomas uses the adversative *tamen* in [ii] for, together with the *non intelligit quin*, it clearly suggests that in [ii] he is not going to draw an implication from the Gloss in the way that he did in [i].

If Thomas agrees that the Gloss prohibits complete renunciation of things both in terms of using them and owning them, then Thomas either agrees with the force of the objection, or he has to admit that religious in orders that lack possessions do not give up everything in terms of use and ownership. Indeed, if he agrees that the Gloss carries this implication and that it should be followed, then he either contradicts himself when he writes that "to possess nothing at all in the world" pertains to Christian perfection or the meaning of "to possess nothing at all" is hopelessly obfuscated.

Despite these concerns, the syntax of the Latin seems to require translating the second clause as we have done. There are, however, two other texts in the CID where *non intelligit quin* has to imply a denial of what comes after the *quin* so that *quin* means something like *ut non*. The first text is found in the response to objection 15.[20] Commenting on the description of the primitive Jerusalem community in Acts 4:34 that "there was no need among them," Aquinas writes that *non est intelligendum quin Apostoli et primitivae Ecclesiae discipuli multas egestates et penurias sustinuerint propter Christum.* Proctor correctly, although loosely, translates this as: "[the text] does not mean that the Christians of the primitive church **did not** sustain much poverty for the sake of Christ."[21] Moreover, this is the only possible translation in light of the texts which Thomas subsequently cites about Paul's condition.[22]

The other text occurs at CID 3.5(12), ad 3. Here Thomas is discussing St. Paul's remark *etsi sum imperitus sermone,* ("although I am impaired in speech"). Aquinas remarks that *non est intelligendum quin apostolus eloquentia uteretur.* Proctor correctly, but loosely, translates this as "one cannot assume that St. Paul did not use eloquence."[23] This seems to be the only plausible translation especially since Aquinas asserts immediately above in ad 2 that Paul did indeed use eloquence.[24]

Given these texts and the other considerations I raised, it seems to me that [ii] above can properly be translated as: "[The Gloss] does not mean that one cannot completely deny himself all ownership of temporal things." In his response to objection six, then, I believe that Thomas gives essentially the same analysis of the Gloss as Bonaventure. Of course, one might claim that Thomas only refers to personal ownership in [ii] given his use of *homo*, construed as referring only to an individual. A search of the electronic *Index Thomisticus* does not show any occurrences of *proprietas commune* in the sense of "common ownership" in Thomas's writings. There are, however, texts where Aquinas uses *possidere commune* in the sense of "to possess [things] in common."[25] Further, *commune proprietas* does occur in works from this period. In particular, in the decretal *Quo elongati* (1230), Gregory IX required that the Franciscans renounce common ownership of and dominion over all the goods it used:

> We say therefore that [the Friars] ought not to have *proprietas* [ownership] of things either individually or in common. But they may have the use of the utensils, books, and movable goods which they are permitted to have ...[26]

To conclude this section: none of the Thomistic texts that we have considered explicitly or *ad litteram* defends mendicant poverty in the sense of having no common ownership at all of temporal things, yet they surely suggest that Thomas thought this sort of poverty was possible and legitimate.

II

On the other hand, while Thomas frames Chapter Six of the CID in terms of retaining no possession of things either personally or in common, he refers in the first paragraph of the chapter to religious who enter orders which have no possessions or revenues.[27] The phrase "possessions or revenues" (*possessiones aut reditus*) is found in the Dominican constitution of 1220 in which the Order renounced both possessions and revenues. However, the sense of possessions in this constitution was clearly limited in scope: it referred to those material goods, both immovable and movable, such as lands, flocks etc., which were sources of revenue, or, what we might call capital goods. The Dominicans, for example, retained ownership of their priories, books, manuscripts, the usual "tools" of scholarly life, as well as the necessities of life.[28]

Moreover, in the CID and CR, when Thomas discusses the renunciation of common possessions he seems to have the restricted sense of possessions in mind. For example, he cites the primitive community mentioned in Acts 4:32-4 as an example of those who had no common possessions since the members of the community sold all of their homes, fields, etc.[29] Indeed, in *Summa contra gentiles* III, 131, he expressly distinguishes this community from religious communities that have common possessions.

Although this is a correct way of viewing the primitive community given a narrow meaning of "common possessions," the monastic tradition generally took the primitive community, in which everything was common to all (*illi omnia erant communia*), as the model for religious life in which everything was possessed in common. Bonaventure certainly views the primitive community in this manner.[30] So too, do Sts. Augustine[31] and Benedict,[32] for although neither of them use the phrase *possessiones communes* in their rules, both cite the primitive community as the basis for a religious life in which all of the goods of the monastery were held in common. Indeed, several of the objections that Thomas cites in chapter six of the CID, take *possessiones* in a broad rather than a narrow sense.[33]

In any event, when Aquinas defends religious orders that have no common possessions he seems to have in mind the restricted sense of possessions which the Dominicans employed.[34] Nevertheless, this creates a basic dissonance in both the CID and the CR, since not having common possessions in the restricted sense is not equivalent to possessing nothing in common, for common possessions in the restricted sense are only a subset of the things which people can possess in common. Yet what is fundamentally at issue in the CID and CR is the complete renunciation of temporal things personally and in common, where temporal things are taken in the widest sense: any material things over which one can have dominion. Moreover, as we have seen, Thomas doesn't just legitimate the lack of common possessions in the narrow sense, he also asserts that possessing nothing at all pertains to Christian perfection. This makes sense only if "possessions" is taken in the broad sense of all material things which can be possessed. It would be odd to say that a community that retains the ownership of priories, books, libraries, etc. is a community which renounces all possession of temporal things.

On the other hand, a complete renunciation of possessions in the broad sense seems plausible only if there can be a simple use of consumable things which does not involve ownership of or dominion over them by their users. As we have seen, Thomas's response to objection 9 in the CR certainly suggests this. To sum up: the basic dissonance in the CID and CR is that there seems to be a more and less radical sense of rejecting the common possession of temporal goods which depends on the scope of the term "possessions." Consequently, although there is a distinct line of reasoning that lends support to the Franciscan position, it is not exactly clear that Aquinas intends to support it. Yet it seems he must support it given the tenor of his response to the objections against complete renunciation of possession and especially since he doesn't directly investigate the consumables argument in the CR which was written after Gerard of Abbeville levelled that argument against the Franciscans.

Nevertheless, in the discussion of usury in the *Summa theologiae*, he seems to reject the possibility of a simple use of consumable goods which underlies the doctrine of the absolute poverty of Christ and the Franciscans, that is, in which the dominion over the goods is not held by the one who consumes the good. So Aquinas writes:

> there are certain things the use of which consists in their consumption: thus we consume wine when we use it for drink and we consume wheat when we use it for food. Wherefore in such things the use of the thing must not be reckoned apart from the thing itself, and whoever is granted the use of the thing, is granted the thing itself and for this reason, to lend things of this kind is to transfer the dominion (*dominium*).[35]

Aquinas repeats the same argument in *De malo*, 13,4; *Quodlibet* III, 7,2; as well as III *Sententiarum*, d.37,a.6 (which was written before the CID). Given that Aquinas also writes that "something becomes one's own through the generosity of the donor"[36] it is difficult, for example, to see how the provisions which were given to Christ and the disciples did not become things over which they had dominion. The same reasoning seems to apply to all of the goods which the primitive community acquired with the funds which they had. Indeed, in his discussion of the Apostles' observation of evangelical life, he notes that "the necessities of life were possessed in common, while possessions were completely renounced by them."[37] This text exempts the necessities of life from the class of possessions, but then it certainly doesn't give an example of those who possess nothing at all on earth. Nor does it resolve the question of whether Aquinas thinks that there can be a community which does not own even the necessities of life in common.[38]

In any event, in his discussions on usury, Aquinas accepted the position that at least the legitimate use of consumable goods cannot be separated from dominion over them. He never applied this position directly, however, to the discussion of mendicant poverty. Yet, Hervaeus Natails, the master general of the Dominicans in the early 14th century, did employ it directly against the Franciscan doctrine of the absolute poverty of Christ. In his work, *de paupertate Christi et apostoloroum*, a brief submitted to Pope John XXII in connection with the mendicant controversy, Hervaeus uses an analysis of consumables that is similar to Aquinas's in order to show that Christ and the disciples had possessions in common. Hervaeus's text is worth quoting in full:

> in regard to things which are consumed in their use, one cannot separate a right to a thing from the licit use of it, nor

vice versa. For in such cases, it is not possible to separate the
licit use of a thing from the right to the use of the thing. Also,
it is not possible to separate the licit use of a thing from the
right to the thing [itself] which someone uses. However, it is
a fact that Christ and the Apostles had a licit use of such
things, for example, food, clothing, and the money with which
they bought things. Moreover, it is a fact that such things are
consumed in their use, for example, food and clothing, or
transferred, for example, money. Thus Christ and the Apostles had a right to the use of such things as well as a right to
the things themselves.[39]

Consequently, for Hervaeus, so long as a community licitly uses
goods it is impossible for it not to possess and have dominion over some
things in common, in particular, the consumable things that it uses.
Hervaeus, it seems to me, would have rejected the notion that possessing
nothing at all on earth personally and in common is a part of Christian
perfection for the simple reason that it is not possible.

III

In the final analysis, Thomas's defense of mendicant poverty remains
rather frustratingly unclear. I will suggest three considerations relevant to this matter. Unfortunately each of these considerations involve
a certain amount of speculation that is probably not resolvable in terms
of any clear textual or historical evidence. First, Thomas only formally
defines voluntary poverty in terms of the lack of possessions qua
ownership: poverty is the "privation of all property."[40] Although questions regarding the use of goods are present throughout his treatment
of poverty, he never formally (that is, definitionally) relates voluntary
poverty to the use of goods. Of course, he is aware that poverty involves
a lack of use of goods. For example, he cites the Gloss on Psalm 39:18
which characterizes the poor person as one who lacks what is sufficient
for life.[41] Nevertheless, Thomas's definition of poverty as the privation
of all property seems remarkably unidimensional when compared with
Hervaeus Natalis's definition:

poverty signifies a lack of temporal things either in regard to
a right to them and dominion over them, or in regard to the
use of them, or in regard to both of these conditions. Further,
this lack can be greater or lesser since in regard to use,
dominion and right, temporal things can be completely lacked
or lacked to such and such a degree.[42]

By failing to include both use and ownership of goods in his definition
of poverty, Thomas effectively closed off a systematic and formal treatment of the relation between use and dominion in connection with

voluntary poverty. Perhaps this explains in part why he didn't apply his analysis of consumables to the question of mendicant poverty. It certainly explains why many of his discussions about poverty in CID Chapter 6 refer in an implicit yet unclarified way to both the use and possession of goods.

Second, it is impossible to know how much attention Thomas paid to the Franciscan controversy on poverty or whether he read Bonaventure or other Franciscan authors on this issue. As was his practice, he makes no direct reference to contemporary authors. Indeed, in contrast to people like Bonaventure, Pecham and Kilwardby, he never frames his arguments concerning mendicant poverty in terms of the practices of a specific order. Thus it isn't clear that he is deliberately setting out to defend the Franciscan practice. Accordingly, the discussions of poverty in the CID and CR may have been written with two purposes in mind.

On the one hand, in the CID and CR, Thomas may have simply set out to defend the Dominican practice of poverty. His use of the phrase *possessiones aut reditus* at the beginning of the discussion of poverty in the CID suggests this. Given that these works were polemical in character, he may have simply assumed that people would read him only in this context. As such, it would not have been necessary to raise the consumables argument because it didn't apply to the Dominicans. As we have seen, however, the objections he was countering were directed to a renunciation of possessions more radical than that practiced by the Dominicans and which required a direct treatment of this question, especially in the CR.

On the other hand, Thomas may have set out to provide a general defense of mendicant poverty which would apply to both orders. As I mentioned, Thomas never explicitly restricts the CID or CR to the Dominicans alone. Certainly the two orders presented a more harmonious front against their detractors in 1250-60, than they did later. For example, even in the 1270s there were bitter disagreements between the orders as is evident in the controversy between Kilwardby and Pecham.[43]

Third, let's revisit Aquinas's response to objection nine in CR, Chapter 16. That objection explicitly rejected the possibility of an order owning nothing in common or personally. Although it doesn't explicitly state the consumables argument it seems to assume it. As I argued, Aquinas's response only meets the objection if a community which rejects ownership of all goods uses those goods which are owned by others. This condition certainly did not apply to the Dominicans; but it is possible that Aquinas was familiar with and accepted as legitimate the arrangement which was developed in Innocent IV's decretal *Ordinem vestrum* (1245) which transferred the ownership of all goods in the Franciscan order to the Pope. This arrangement technically met the consumables argument, if one felt that the dominion over consumables did not pass to the Franciscans when they obtained them. John XXII certainly did not accept the arrangement, but it is possible that Aquinas did.[44]

Admittedly, the consumables argument seems to have first been raised by Gerard of Abbeville in 1269. Neither Bonaventure or Aquinas raise or treat the issue in *De perfectione evangelica* or the CID. Given that in the early work III *Sententiarum,* however, Aquinas essentially held Gerard's view on the relation between dominion and use in regard to consumables, it is odd that someone like Aquinas would not have seen the threat of the consumables argument to the defense of mendicant poverty when he wrote the CID unless he merely set out to defend the Dominican practice of poverty or sought to defend mendicant poverty generally and accepted the papal arrangements concerning the Franciscans. Whether Aquinas would have accepted the doctrine of the absolute poverty of Christ and the disciples which supported the Franciscan arrangement is less clear. In light of his analysis of consumables in the ST, it is hard to see how he would. Nevertheless, as we have seen, his analysis of Christ's poverty in the CID suggests that he might have.

One final note. If Aquinas accepted and defended the Franciscan practice of poverty, there is a somewhat interesting and ironic implication that can be drawn from the relation between poverty and perfection that Aquinas develops in the CID and CR. In contrast to his view in the *Summa theologiae* that poverty can be said to be perfect only so far as it is adapted to the ends of an order,[45] Aquinas holds in both the CR and the CID that the lack of common possessions contributes more greatly to perfection than the possession of things in common. So, in the CR, he writes, "Among those who have followed the highest perfection, there will be no possessions ... Thus if there is a congregation in which everyone tends to greater perfection, it is useful that they do not have common possessions."[46]

Consequently, if one takes common possessions in the broad sense of any material goods which a community might possess, it would be relatively easy for a Franciscan to have claimed on the basis of this text that the Franciscans attained a greater degree of perfection than the Dominicans since they achieved a more radical renunciation of common possessions than the Dominicans.

Marquette University
Milwaukee, Wisconsin

* * * * *

Endnotes

(An earlier version of this paper was read at 29th International Congress on Medieval Studies, Western Michigan University, May 1994. I want to thank my colleague, Dr. Richard Taylor, for reading a late draft of this paper and making several helpful suggestions for improvement.)

1.   "Dicere, vero, quod usus tantum vester est, dominium eorum, qui

dederint, quosque vestustate consummantur, aut ciborum, quosque in ventrem reconditi fuerint, omnibus ridiculum videbitur, maxime cum eorum, quae per ipsum usus penitus consummuantur, ab usu dominium nullatenus inter homines distinguatur." *Contra adversarium perfectionis Christiane*, edited by S. Clasen, *Archivum Franciscanum Historicum* 32(1939):133. I use the translation provided by M.D. Lambert in *Franciscan Poverty: The Doctrine of the Absolute Poverty of Christ and the Apostles in the Franciscan Order 1210-1323*, (London, 1961), 135.

2. See *Apologia pauperum*, XI,1-6.

3. Innocent IV transferred all goods used by the Franciscans to Papal ownership in *Ordinem vestrum* (1245). (*Bullarium franciscanum*, I, 401.)

4. The *Contra impugnantes dei cultum et religionem* and *Contra pestiferam doctrinam retrahentes homines a religionis ingressu* are found in the Leonine edition of the *Opera Omnia*, vol. XLI, (Rome: St. Thomas Aquinas Foundation, 1970) and in *Opuscula theologica*, vol. 2, ed., R. M. Spiazzi, (Turin: Marietti, 1954). References to both works will cite the page and line number from the Leonine edition. The paragraph number from the Marietti edition will be placed in brackets. All references to the CID are to Chapter 6 [2.5(6) in the Marietti edition] unless otherwise noted.

5. See my "The Concept of Poverty in St. Thomas Aquinas's *Contra Impugnantes Dei Cultum et Religionem*," *The Thomist* 59, no. 3 (July 1995) 409-39. The present article continues the analysis and links Aquinas's views to those of Bonaventure and Hervaeus Natalis. It should be noted that Aquinas approaches the subject of poverty quite differently in these works than in the *Summa theologiae*, 2.2,188.7. For a discussion of this matter, see my "Poverty and Subsistence: St. Thomas and the Definition of Poverty," *Gregorianum* 75(1994): 135-49. See also, Thomas Aquinas, *La Somma Teologica*, Vol. 22, trans. Tito S. Centi (Rome, 1969), 460, n.1. Ulrich Horst provides a general treatment of Aquinas's teachings on poverty that focuses particularly on ecclesiological and theological issues: *Evangelische Armut und Kirche: Thomas von Aquin und die Armutskontroversen des 13. und beginnenden 14 Jahrhunderts*, (Berlin: Akadamie Verlag, 1992).

6. Hervaeus Natalis, *Liber de paupertate Christi et Apostolorum*, ed. by J. G. Sikes, *Archives d'Histoire Doctrinale et Littéraire du Moyen Age*, 12-13(1937-8):209-97.

7. ("Utrum liceat religioso omnia sua relinquere, ita quod nihil sibi possidendum remaneat, nec in proprio nec in communi.") (A94,1-3 [200]).

8. A98,403-5 [218], where Aquinas emphatically writes "nihil omnino possidere," and A98,410 [219].

9. Q. II, a. 1 (*Opera omnia* (Quaracchi ed.), V,129B). The article is titled *De paupertate quoad abrenuntionem*. Like the CID, this work was written in response to the attacks of William of St. Amour on the Mendicant orders.

10. "Hoc esse impossibile quod aliquis nihil in communi vel proprio possideat; quia necesse est quod comedant et bibant et induantur; quod facere non possunt, si nihil haberent." CR 14, obj. 9. (C68,92-6, [833]).

11. "Ea quibus utuntur religiosi ad sustentationem vitae, non sunt eorum quantum ad proprietatem dominii, sed dispensatur ad usum necessitatis eorum

ab his qui harum rerum dominium habent, quicumque sint illi." CR 16, ad9 (C74,158-63, [858]).

12.   Ad8 (A101-2,702-4 [242]).

13.   "Non tamen omnino abiicienda sunt haec temporalia. Sed illi qui dimissis omnibus religionem intrat quae temporalibus possessiones caret, omnino temporalia abiicit." (A94,50-55 [200]).

14.   A101,660-66 [240].

15.   It isn't clear whether *possit* has only the modal sense of being able to do something or a normative sense or having permission to do something. The Gloss and [i] seem to deal with normative claims, although the descriptive aren't far behind: we should not give up all use of things because we will not be able to go on living.

16.   "Ownership" translates *proprietatem* as found in the Leonine edition; older editions have *curam*.

17.   "De abiectione temporalium quantum ad usum, non quantum ad dominium. Nam usus temporalium necessarius est vitae humanae, qui tamen haberi potest absque dominio et proprietate, *"De perfectione evangelica"* Q.II, a, 1, ad3 (*Opera Omnia* (Quaracchi ed.), V,131B).

18.   Scott Swenson, *Emerging Concepts of Jurisdiction, Sacramental Orders and Property Rights Among Dominican Thinkers From Thomas Aquinas to Hervaeus Natalis*, [Ph.D. Diss., Cornell University, 1988]:265, n.159.

19.   The English translation is found in *An Apology for the Religious Orders*, ed., John Proctor, (London: Sands and Co., 1902), 188. The French translation is found in *Opuscules de Saint Thomas D'Aquin*, v. 5 (Paris, Librairie Philosophique J. Vrin: 1984), 22. The French reads: "Il n'entend cependant pas par là, que l'homme doive renoncer à toute espèce de soins des choses temporelles."

20.   A103,813-5 [250].

21.   Proctor, 194; my emphasis. The French concurs: "Il ne faut pas entendre par là que les apôtres, et les disciples de la primitive Eglise n'ont pas été exposés pour Jésus Christ, à toutes sorts de misères et de privations." (29).

22.   The texts are 1 Cor. 4:12 ("We go hungry, thirsty, etc.") and 2 Cor. 6:4 ("We have to show great patience in times of tribulations and need.").

23.   A136,201-3 [415]. Proctor, 285. The French translator misses the negative import of *quin:* "il ne faut pas entendre par là que l'Apôtre avoit recours à l'éloquence." (121).

24.   A136,175-6 [414]

25.   See *Summa theologiae* 2.2., 188,7,sc and CID 2,5(6),ad17 (A104,926 [252]).

26.   *Bullarium franciscanum*, I,69. Nicholas III also used the expression in that way in the encyclical *Exit qui seminat* (1279) in which he legitimated the Franciscan doctrine of absolute poverty: "we say that such renunciation of proprietas of all things, both individually and in common, for God is meritorious and holy." (*Corpus Iuris Canonicum* 2, 1112).

27.   A94,9 [200]. See also, A99,490 [226].

28.   See W. A. Hinnebusch, *The History of the Dominican Order*, (New York,

DEFENSE OF MENDICANT POVERTY

1966): I,153ff.

29.　CR 15 (C71,239 [841]).

30.　*Apologia pauperum*, VII, 4-5.

31.　*Regula* 3-4 (*La Règle de Saint Augustin*, [Paris, 1967]:I,48).

32.　*Benedicti regula*, 33. (*Corpus Ecclesiasticorum Scriptorum Latinorum*, 75,91).

33.　See, for example, objs. 13, 15,(A95,105-7; 124-7 [200]), 22 and 23 (A96, 190-6 [200]).

34.　It is instructive, for example, to read the section of the responsio to CID 6 where Aquinas explicitly discusses the relation between perfection and lacking common possessions to see if one can determine exactly what he has in mind by the scope of *possessiones communes*. (A99-100,446-521 [223-230]).

35.　*Summa theologiae*, 2.2., 78.1, res.

36.　"Fit autem aliquid alicuius ex liberalitate donantis." *Summa theologiae*, 2.2.,187,4, res.

37.　"Ut ea quae ad necessitatem vitae pertinent, possideantur communiter, possessionibus omnino abdicatis." CR 15 (C71,240-1 [841]).

38.　It should be noted that Bonaventure does not use the primitive community as such an example. In fact, in the *Apologia Pauperum*, (VII, 4-5) he distinguishes the common possession of things in the primitive community from the more radical renunciation of all possessions, personally and in common, which was practiced by the disciples on their first missionary journey. For Bonaventure, however, the common possession of things in the primitive community was exercised by the "multitude of believers" and not by the Apostles.

39.　Q. 3, A. 1 (Sikes, 283).

40.　"Privatio omnium facultatem," *Summa theologiae*, 2.2.,188,7, res.

41.　CID 2.6(7) (A114,751-2 [289]).

42.　Q.1, A.1 (Sikes, 226).

43.　See Pecham's work, *Tractatus contra Fratrem Robertum Kilwardby* in *Fratris Johannis Pecham: tractatus tres de paupertate*. Eds. C. L. Kingsford, A. G. Little, and F. Tocco. (England, 1969): 121-47.

44.　For an extensive discussion this matter see Lambert, *op. cit.*, 208-47.

45.　2.2.,188,7, res.

46.　"Unde et apud illos qui illam summam perfectionem sectabantur, possessiones not erant ... Unde si qua sit congregatio, in qua omnes ad maiorem perfectionem tendant, expedit eis communes possessiones non habere." CR 16,ad1 (C73,10-1; 25-7 [850]). See CID 2.5(6), res. (A99,474-4 [224]; A99,500-3 [227]; A100,519-21 [230]).

Beauty and Technology as Paradigms for the Moral Life

by Montague Brown

It seems to be the prevailing attitude in contemporary western society that living a morally upright life is either burdensome (but perhaps worthwhile for the sake of social harmony or to help the less fortunate or to get to heaven) or maybe even reprehensible (with those who take moral obligation seriously considered "too good" or "uptight"). Although the difficulty of being good is nothing new, and pleasure has always stood in some kind of tension with moral obligation, there seems to be a significant difference between the distinctively modern approach to ethics and that of the ancient-medieval tradition. While the modern way of thinking holds that living a morally upright life may be good for others or likely to bring one a net balance of pleasure over pain, the ancient-medieval tradition held that living a morally upright life is to be happy, that such a life is not only in accordance with duty, but is delightful, joyous. This difference in attitude is traceable, in part, to a paradigm shift, and this shift can be understood as a move from the ideal of beauty to that of technology.

It is well known that at the time of the Renaissance there occurred a shift in the meaning of science, that what had been a word for knowledge in general became specified as knowledge attainable through scientific method. With the clarification of scientific method and the realization of its success, there was a swing in the philosophical community toward a focus on the practical results of philosophy, in short, on technology. While there is much that could be, and has been, said about the shift from metaphysics to science as paradigms of theoretical reason, what I am interested in exploring here is the effect on ethical thinking of taking science and technology as ideals of human achievement. More specifically, I wish to compare the ancient and medieval paradigm for the moral life--beauty--with the modern paradigm--technology.

The degrees to which beauty and technology do or should inform ethics will become clearer as the paper progresses. I shall consider each paradigm in its historical setting and then reflect on the influence of the technological paradigm on moral thinking today. My thesis is that we have lost more than gained, morally speaking, in this shift of paradigms, for the morally good life under the technological paradigm is not regarded as good and delightful in itself, but only in its fruits.

Beauty and the Ancient-Medieval Moral Tradition: In reviewing the tradition which took beauty as the paradigm of moral excellence, we shall consider briefly the thought of Plato, Aristotle, Cicero, Augustine, and Aquinas. Let us begin at the beginning with Plato.[1] In Plato we find the identity of the good and beautiful.[2] Given this identity, if good is the paradigm of the moral life, then so is beauty. In the ladder of ascent to Beauty Itself, Plato has Diotima move from the appreciation of beautiful bodies to beauty of souls.[3] Such latter beauty is the beauty of character, the beauty of a virtuous person, one who is directed in an orderly way by the higher intelligible beauties and ultimately Beauty Itself. While Plato is not explicit about what makes bodies beautiful,[4] it is clear enough that it is the presence of order in the soul which makes the soul beautiful, and that this order is from reason.[5] This order of the soul is perfected by virtuous behavior and destroyed by vicious action. Our human happiness depends on acting in a consistently virtuous manner. The focus is clearly on intention and what our choices do to us, not on the consequences of our actions.[6]

Aristotle spends more time discussing the principles of beauty than does Plato, who is interested mostly in how beauty functions as a metaphysical principle. Thus in the *Poetics*, where he is discussing what makes a fine plot, Aristotle talks of beauty as being a matter of magnitude, order, unity, and wholeness.[7] In the *Metaphysics* Aristotle says that "the most important kinds of the beautiful are order, symmetry, and definiteness."[8] This kind of definition is applicable to the moral life in terms of describing what it means to be virtuous. In both Plato and Aristotle, virtue is a whole: that is, the virtuous person is at once, prudent, just, courageous, and temperate.[9] The virtuous person's character is ordered by reason, so that intentions always honor virtues and actions carry out these good intentions. While Aristotle does not explicitly use the term beautiful to speak of a morally good character as does Plato, it is clear that the same principles of order, proportion, and wholeness which apply to the beauty of material things and works of art are at work in his conception of the good person. So it is that a morally good life is not only demanded by reason but is also the fulfillment of human happiness.[10] Both the state and the individual are made happy by the proper ordering of internal relations.[11]

In his letter to his son, *On Moral Obligation*, Cicero explicitly mentions beauty as a fundamental human good and relates it to character.

> Another important natural capacity, to which man alone is heir, is that of discerning order, decency and a sense of proportion in words and deeds. Indeed in objects of percep-

tion no other creature can discern beauty, grace and symmetry. It is our natural reason which extends the comparison
from the eye to the mind, so that beauty, consistency and
order are thought even more worthy of observance in intentions and actions as a precaution that nothing dishonourable,
unmanly or lustful be done or even contemplated.[12]

This life of beauty, grace and symmetry involves the proper ordering of
all the virtues in one's life. It is fitting or decorous to think wisely and
to act justly and courageously, indecorous to think unwisely and to act
unjustly or in a cowardly manner. "Thus it is apparent that what I have
called 'decorum' is relevant to every good action...."[13] The beauty of
character is a joy to behold and happiness to the one who possesses it.[14]

Augustine follows Plato and Cicero on this idea of beauty being
applicable to moral character. Like Cicero he does analyze the beauty
of material things, using basically the same principles as Cicero: symmetry,[15] proportion,[16] and pattern.[17] Augustine also says that beauty
is found in brightness.[18] While this is primarily applicable to material
things, the idea of brightness or light can be applied analogously to the
intelligibility of things, to virtue, and to wisdom.[19] In *On True Religion*,
Augustine speaks of the "beauty of Justice" and then goes on to sing the
praises of the other cardinal virtues–courage, temperance, and prudence.[20] Additionally, in *The City of God*, he speaks of the proper peace
or order of the rational soul as being a kind of harmony of conduct and
intention.[21] Such a notion of happiness for the individual is also applicable to states. Order is essential: even earthly man can attain some
degree of happiness contained in the beauty of ordered relations.[22]

In Thomas Aquinas beauty is essentially related to order[23] and is
specified by three principles: integrity, proportion, and clarity.[24] While
he says, with Plato, that the good and the beautiful are ultimately
identical, he insists that they differ according to reason: the good
belongs to the nature of final causality and the beautiful to that of formal
causality.[25] As the essence of a moral act is its formal aspect (the
intention of the agent), and as character is formed by the order or
disorder (harmony or disharmony) among intentions, there is an easy
connection between beauty and the morally good person. Indeed,
Thomas speaks explicitly about beauty of character. "The beauty of the
body consists in a man having his bodily limbs well proportioned,
together with a certain clarity of color. In like manner spiritual beauty
consists in a man's conduct or actions being well-proportioned in respect
of the spiritual clarity of reason. Now this is what is meant by honesty,
which we have stated to be the same as virtue...."[26] Order within the
soul, and order between one's intentions and one's actions are essential
to virtue. Like Plato, Aristotle, and Cicero, Aquinas holds that real
virtue is a whole–the integration of prudence, justice, courage, and
temperance within the soul.[27]

Not only is the fulfillment of moral requirements in accordance with reason's demands, but it is also delightful. As Aquinas says, the beautiful is that which, when perceived, delights.[28] There is thus a contemplative aspect of morality in the tradition, a way in which moral goodness is delightful in itself, not just valuable for its fulfillment of duty or for its consequences.[29]

Technology as Moral Paradigm: That in the Renaissance there was a turn to science and technology as paradigms for all human thinking (and thus for ethics) is hardly a controversial claim.[30] Let me highlight a few of the salient features of this turn by examining briefly a few major figures: Francis Bacon, René Descartes, Thomas Hobbes, David Hume, and Jeremy Bentham. My focus will be on their doctrines of the limits on human thought and how these doctrines affect their ethical theories.

Let us begin with Bacon, self-proclaimed trumpeter of a new age. Bacon is very clear on the limitation of legitimate human thought to the study of material things by the so-called scientific method: hypothesis and verification.[31] Bacon, however, was no great scientist in search of knowledge of the material world for its own sake: behind his choice of the scientific method as the sole path open to the human mind was a desire for technological progress to make nature serve the ends of mankind.[32] Technology or "the mechanical arts"[33] was his greatest interest. Speaking of his project, Bacon says: "it is not an opinion to be held, but a work to be done;.... I am laboring to lay the foundation, not of any sect or doctrine, but of human utility and power."[34]

In his new dispensation, any discussion of forms, final causes, or first cause is to be ruled out of court.[35] This effectively does away with all metaphysics as well as traditional moral and aesthetic thought. If we are not to speak of first causes, then we are not to speak of God or providence; if we are not to speak of final causes, then we are not to speak of the final end of mankind, the purpose of our being here, or the ideal standards toward which we strive; and if we are not to speak of forms, then, in addition to the abandonment of any metaphysical or moral knowledge, we must give up any discussion of the beautiful, for beauty is found essentially in form. Hence, it is easy to see that embracing technology as the paradigm for all human thought implies the rejection of beauty as the paradigm for moral excellence. According to Bacon, our attention must be focused on the material world, on matter in motion.[36]

How does such a theory affect Bacon's moral thought? This is hard to say, since Bacon wrote very little of an ethical nature. He does, however, say that the moral and political sciences are stagnant and lack depth[37] and that these sciences, too, ought to follow the scientific method.[38] There is, of course, a moral supposition at the root of his whole system: the goodness of improving human life. Technology is to serve humankind, to make lives better. Two things can be briefly said about this position. In the first place, the human life in question is our material existence, for with the abandonment of all formal, final, and first efficient causes, Bacon has ruled out the possibility of human goods such as wisdom and virtue, as well as a human nature which transcends material influences, that is, one that is free. With no ultimate moral norms and no freedom, the prognosis for creating an ethics looks pretty bleak. Secondly (and more destructive of the consistency of his project as a whole), it is hard to see how the moral obligation to better human life could be derived from a study of matter in motion.

René Descartes, too, claims in his *Discourse* to have a new method which he thinks will be useful and provide progress in the sciences for the future.[39] While his idea of reason is not itself reductionistic like Bacon's, there is built into Descartes' as well as Bacon's conception of reason the idea of a useful tool. Studies are commended insofar as they are useful: languages and morals are useful;[40] mathematics are useful.[41]

Like Bacon, he disparages the moral theories of the ancients for being foundationless and not practically feasible.[42] Evidently, Descartes has hopes that his new method will help us know what virtues are and how to arrive at them.[43] Like Bacon, however, Descartes never gets to a discussion of morals. He adopts as provisional the morals of his society and Church in Part Three of the *Discourse* and never revisits the issue. Although he does not go as far as Bacon in explicitly equating utility and truth,[44] Descartes does judge thinking according to its effects.[45] His purpose is to replace the speculative science of the Schools with a practical one which will make us "masters and possessors of nature."[46]

As for Thomas Hobbes, his system of the social contract based on radical egoism is a bold reduction of ethics to what is useful to the individual. Reason is a technique for getting what one wants. The theme of power which is stressed by Bacon and Descartes takes on the position of first principle for Hobbes. Everything is reducible to the effects of power, from sensation and thought where the power of matter in motion is the determining factor,[47] to the political community where "justice" is whatever the sovereign power says it is.[48] The value of a human being is measured by how much the use of his power is worth.[49] All virtue is reducible to power: "*Honourable* is whatsoever possession, action, or quality, is an argument and signe of Power.... Magnanimity, Liberality, Hope, Courage, Confidence, are honourable; for they proceed from the conscience of Power."[50] Thus, the moral question becomes, for Hobbes, how best to maximize one's power so as to get what one wants. This may involve compromise with others, but only if the compromise is likely to benefit one in the long run. Again, reason as it applies to morality is a technique for getting what one wants: the technological paradigm is firmly in place.

When we turn to David Hume we find him beginning his moral philosophy by restricting the operations of the intellect to "relations of ideas" and "matters of fact."[51] Examining our moral judgments, he finds that they are not based on relations of ideas, for then we should treat logically parallel cases, such as parricide and the killing of a parent tree by its offspring, in the same way morally speaking and we do not: we disapprove of the first but not of the second. Secondly, moral judgments are not based on matters of fact (by which Hume means things that are known through scientific method), for there is no matter of fact which corresponds to virtue, or justice, nor to vice or injustice: not being material things, these moral ideas are not verifiable and hence meaningless.[52] Since reason operates only in these two ways, moral judgment must not be a matter of reason. Rather it is a passion, a feeling with which we respond to various actions either positively or negatively and which impels us to act. This is Hume's famed moral sense or benevolent impulse which he holds to be the essence of morality.[53] Reason's role, insofar as it plays one in morality, is secondary: it establishes the facts and helps us calculate means-to-ends and cause-and-effect relations. Thus its role is similar to that which it played in Hobbes's thought: it is a technique for accomplishing what the passion dictates.

Like Bacon and Descartes, Hume is motivated in his work by a concern for what benefits the public. Unlike Bacon and Descartes, who merely presuppose this principle of utility without accounting for it, Hume claims that it is a natural passion, a "public affection."[54] Reason's role is to find a way to accomplish this public utility.

The utilitarianism of Bentham is expressly based on the idea of usefulness: morality is a matter of consequences. Unlike Hobbes and Hume, however, who see reason as having but a secondary technical role in morality with the primary role played by passion (self-love in Hobbes, love of others in Hume), Bentham attempts to ground morality in reason's deliberations themselves. Thus, between establishing the facts (for Bentham the various values of pleasure and pain from any particular act)[55] and understanding relations of ideas (for Bentham the tendencies of an act to produce pleasure or pain and how many people will be affected by the act).[56] Bentham thinks that we have all the information needed to make good moral choices. Both the end (a balance of pleasurable and painful consequences) and the means or technique of accomplishing that end (the operations of reason compatible with scientific method) are oriented toward usefulness. Moral actions are to be judged by external standards, not by the standards of intention, virtue, and character. When it comes to judging moral actions, intentions do not matter, for they cannot be measured.[57]

With Bentham, the shift in paradigm is made explicit. Reason operates in morality for the sake of another end: it is a technique for getting something accomplished. The end is some future state of affairs to be accomplished by advances in technique. What has disappeared from moral consideration is the individual agent as a model of virtue or vice.

Return to the Tradition but with a Difference: Not only did modern philosophy depart from the traditional moral idea that the proper integration of moral intentions and virtues is itself good and worth appreciating for its own sake; it also affected many later philosophers who desired to restore traditional morality by removing from them any understanding of the beauty of the morally good agent. In this section I would like to limit myself to two cases: Kant's moral theory, and contemporary proportionalist natural law theory.

No one, I think, would deny that Kant had as strong a sense of duty as is found in Plato, or Cicero, or Aquinas. Duty is, in fact, the center of Kant's theory. What is missing of the tradition in Kant is a serious discussion of virtue and happiness. Happiness is relegated to the realm of the hypothetical imperative; it is a matter of natural self-interest.[58]

It is not hard, in a way, to understand why Kant missed this part of the tradition, for he follows historically the moral thinking of Hobbes and Hume. They had argued that morality must be a matter of passion since its traditional subject matter (good and evil, right and wrong, justice and injustice, virtue and vice) could not be verified by scientific method. Kant is at pains to show that moral obligation is not reducible to passion, whether that passion be self-love or natural benevolence. In making his point, he is most interested in distinguishing the heart of morality (duty) from other motivations (physical and emotional). Duty is not self-interest, nor direct inclination.[59]

Kant's departure from the moral thinking of Hobbes and Hume, however, was not based on a departure from their assumption that scientific method was the sole legitimate way to have knowledge about reality. Kant held that all real knowledge must have a component of sense experience.[60] Ideas of reason (theoretical or practical) may be inevitable and impossible to disprove (since their objects--God, the human soul, duty--are not material), but they cannot be demonstrated with certainty to the human understanding. This means that the only certain, verifiable knowledge we can have about human beings concerns their material existence, their physical drives, and passions. While he insists that there is a non-material, noumenal self, we have no access to it.[61] The elements which make up the paradigm of moral goodness of the tradition--the integration of intentions and virtues within character--are not physical things; were they to have any place, it would be with the noumenal self to which we have no access. For Kant, there is no understanding of metaphysical realities, or moral norms, or aesthetic principles. We can say no more than that we cannot avoid thinking about them and making judgments according to them.

Where Aristotle would have said that we delight (as unities of body and soul) in the order and beauty of the proper integration of intentions and virtues, Kant, who (given his theory of reality) has no way to connect the noumenal self with the phenomenal self, would say that the delight of which Aristotle speaks is merely physical or emotional experience and so ethically suspect. As physical or emotional, this delight is a consequence of heteronomous causality, completely removed from the autonomous reality that is freedom and morality. Kant is not wrong in insisting that freedom and attention to reason's commands of duty are central to morality; Aristotle would agree.[62] Kant, however, cannot see that the delight of contemplating the beauty of a good character need not be the subordination of moral obligation to pleasure or passion.[63]

The other case where the influence of the technological paradigm on traditional moral thinking is clearly apparent is in the natural law proportionalism of such thinkers as Josef Fuchs and Richard McCormick.[64] The thinking of these proportionalists is not quite the same as traditional utilitarian thinking, for it does not make consequences the only feature of correct moral decision-making. In general, one should not intentionally do something morally evil, even for the sake of a good outcome.[65] The place where attention to overall consequences is required, they think, is in the tough situations where the principle of double effect comes into play. The historical locus for this principle is in Thomas Aquinas's discussion of the permissibility of killing in self-defense.[66] In such an act, there are two effects, one good (the saving of one's life) and the other evil (the taking of another's life). The act may be done if one's intention is only for the good effect. Thomas adds to this that the means used to secure one's safety must be "proportionate" to the end, that is, that one should use the least force necessary in accomplishing that end. There is no sense here that the moral quality of an act is determined by the balance of goods and evils that it causes (that the goods must be proportionate to the evils either in number or seriousness), and yet this is what the proportionalists seem to think Aquinas meant.[67] Thus, while proportionalists do focus on intentions, these are ultimately defined by consequences; an intention is good if it intends a greater amount of good than evil in the end; the proximate object of intention of an act is viewed in proportion with the goods and evils that act is likely to bring about. For Aquinas, one should never intentionally violate a fundamental human good (say life or knowledge) even if one's overall intention is to bring about a state of affairs where good outweighs evil.

This misinterpretation of the natural law position of Aquinas is due, in part, to the prevalence of the technological model of moral thinking. Moral goodness is viewed as something external and future which we bring about by our actions. Reason is a technique for calculating the good and evil consequences of our choices and coming up with a balance which will either justify a choice or condemn it.

Such a consequentialist view has a number of serious problems. In the first place, any such calculation, however one understands it, must be extremely difficult, for it is hard to say how much and for how long the consequences of an act may last, and how many people will be affected. There is, however, a much more serious philosophical problem with such a position. Given our knowledge of what human good means (life, knowledge, friendship, beauty, etc.) a calculation of overall net good is impossible. The basic human goods are incommensurable; that is, one cannot provide a least common denominator by which one could sum up the goods of life with goods of friendship, with goods of beauty, with goods of knowledge. These goods are self-evidently good, not good by some more basic good.[68]

Besides these objections to proportionalism, there is also a theological objection. To claim the right to base moral judgments on ultimate consequences is to claim the privilege of being God, for only God knows all the consequences of any act. But we are not God; we do not and cannot understand the future with any precision or comprehensiveness.[69] Thus, proportionalism is incompatible with natural theology with its radical distinction between creatures and God. It also appears to be unsound on the basis of revealed theology. For while we believe that God will bring good out of evil, we do not understand how this can be done. Our job is to do good and avoid evil: we are not to do evil so that good may come.[70] To judge that future overall good will justify a present evil is to assume that we understand what God has in mind for the world. We put ourselves in place of God, which is at once metaphysically absurd, and theologically blasphemous. Not only is taking responsibility for overall net good not our job; it is a burden that is not ours to carry. Consequentialism implies that there is only one best choice to be made in any situation and that all we have to do is to figure out what it is. Since we cannot figure it out (given the uncertainty and complexity of the future and the incommensurability of fundamental human goods), we are bound to be immoral based on such a criterion for moral decision-making. To think that we are bound to undertake such a burden is but one more instance of our viewing moral responsibility as an onerous task.

Conclusion: The moral life is about duty, and law, and conscience, but it is also about bringing these into practice in one's life and the world. Clearly, it is important to know what one ought to do, and informing our consciences and those of our children and fellow citizens is a major task. To know, however, what one ought to do is not necessarily to do it. The morally good person is not just the one who knows what is right, but the one who puts this knowledge into practice. All sorts of things can help in providing support in this area: a loving family, a good education, good examples to imitate in the public and private sectors, religious faith. One thing that has pretty much ceased to be a factor is the notion that, besides being its own reward, being a good person is a joy, a beauty worthy of contemplative delight. With the technological paradigm of the moral life where actions are judged ultimately according to their fruits or consequences,[71] the motivation for being good rests solely on the idea of duty as a task to be fulfilled, one which, when fulfilled, will ease one's conscience so that one may get on with the things one really wants to do. According to the ancient-medieval tradition, being good is not just the bitter medicine of utility,[72] like eating spinach or zucchini so that one may be healthy, but also delightful like eating a peach or an orange--at once healthy and a delight.

In the modern tradition which assumes materialism (Bacon, Hobbes, Hume, Bentham) the reasons for ignoring this paradigm of moral goodness as beautiful are fairly clear. Since only what can be verified by sense experience is to be accounted real, the participation in the virtues of prudence, justice, courage, and moderation–which is the heart of moral goodness and its beauty in the tradition–must be rejected as unreal and the traditional moral terms deemed meaningless. While Descartes is not a materialist, he is motivated, at least in part, by the ideal of utility and progress; his metaphysics will clear the way for the success and progress in the sciences. Kant, while insisting on duty as the core of morality, also sees his project as saving the sciences, and admits that human understanding does not extend to immaterial realities. Even the natural law proportionalists, who accept the notion of the virtues and of a moral life which honors fundamental human goods irreducible to matter, tend to see the ultimate criterion for judging morality as being the overall net good or evil caused by the act.

Recovering the idea of moral goodness as beautiful and delightful not only restores one of the aids in helping us to do what we know we ought to do (which could be viewed as an argument from utility); it also brings a great joy to human beings. When morality is understood as the fulfillment of the integrated person whose morally good acts are known to be at once good for others and for that person, it ceases to be seen as something burdensome, something owed to others so that one can do what one wants, and is recognized as a beauty worthy of contemplation and delight.

Saint Anselm College
Manchester, New Hampshire

* * * * *

Endnotes

1. This, at least, is where the philosophical tradition begins. Given the example of Homer, and the great tragedians Aeschylus, Sophocles, and Euripides, it could be argued that the paradigm of beauty informed Greek culture as well as the philosophical tradition.

2. *Symposium* 201c; *Lysis* 216d.

3. *Symposium* 210a-212a.

4. At the end of the *Greater Hippias*, Socrates admits how difficult it is to define beauty (304e).

5. *Republic*, 444d-444e.

6. In a rhetorical question, Socrates asks Crito: "Above all, is the truth such as we used to say it was, whether the majority agree or not, and whether we must still suffer worse things than we do now, or will be treated more gently, that nonetheless, wrongdoing is in every way harmful and shameful to the wrongdoer"? *Crito* 49b, in *Five Dialogues*, tran. G. M. A. Grube (Indianapolis: Hackett, 1981), 51.

7. *Poetics* 7, (1450b34-51a15) in *Aristotle: Selected Works*, tran. Hippocrates G. Apostle and Lloyd P. Gerson (Grinnell, Iowa: The Peripatetic Press,

1982), 640.

8. *Metaphysics* XIII, 3 (1078b1) tran. Hippocrates G. Apostle (Grinnell, Iowa: The Peripatetic Press, 1979), 218.

9. See Plato, *Gorgias* 507b-c, and Aristotle, *Nichomachean Ethics* VI, 13 (1114b18-1145a12).

10. "Let us acknowledge that the extent to which we become happy is measured by the extent to which our ethical virtues and prudence and the actions according to these are present; and let us use as a witness God, who is happy and blessed not because of any external goods but because of Himself and His kind of nature." *Politics* VII, 1 (1323b22-26) in *Aristotle: Selected Works*, 590-91.

11. *Politics* VII, 3 (1325b19-31).

12. Cicero, *On Moral Obligation*, tran. John Higginbotham (Berkeley and Los Angeles: University of California Press, 1967), Ch. 4, 43.

13. *On Moral Obligation*, Ch. 27, 72.

14. "For just as physical beauty attracts our attention because of the perfect harmony of its component parts and is a source of great delight because of their matching charm, so this 'decorum' which shines forth in life, stirs the admiration of all around us because of its logical consistency and reasonableness in all its words and deeds." *On Moral Obligation*, Ch. 28, 73.

15. See *De Trinitate* (hereafter DT) IX, 6, *De civitate Dei* (hereafter DCD) XI, 22 & XXII, 20, and *De vera religione* (hereafter DVR) XXX, 55.

16. See *De diversis quaestionibus* LXXXIII, 78, and DCD XI, 22 & XII, 4.

17. See DCD XI, 22 & XI, 27.

18. DVR XXIX, 54, DCD XI, 23, and DT VIII, 2.

19. See, for example, *De libero arbitrio* II, 16, 52 and DVR XLIX, 97.

20. DVR XV, 29.

21. DCD XIX, 14.

22. DVR XXVI, 48.

23. Thomas Aquinas, *Summa theologiae* (hereafter ST) 1-2.49.4c; *Summa contra gentiles* 2.71+72.

24. ST 1.39.8c.

25. ST 1.5.4ad1; 1-2.27.1ad3; *Expositio in Dionysium de Divinis Nominibus* 4, 5, 356.

26. ST 2-2.145.2c, in *The Summa Theologica of St. Thomas Aquinas*, tran. Fathers of the English Dominican Province (Westminster, MD: Christian Classics, 1981), Volume III, 175-76.

27. *Disputationes, de Virtutibus Cardinalibus*, 1c.

28. ST 1-2.27.1ad.3.

29. A contemporary tradition of natural law ethics carries on this tradition of holding beauty to be the paradigm of the moral life. In his book, *The Way of the Lord Jesus: Christian Moral Principles* (Chicago: Franciscan Herald Press, 1983), Germain Grisez speaks of harmony as characterizing several of the basic goods which fulfill persons (Ch. 5-D, 121-25). John Finnis in *Natural Law and Natural Rights* (Oxford: Clarendon Press, 1980) speaks of beauty or aesthetic experience as a basic good. But he also speaks of the good of practical reasonableness as a fundamental human good, in fact, as the basis for all moral

precepts. This good of practical reasonableness he defines as involving the freedom to act, integrity (the integrations of one's intentions in accordance with the proper honoring of the basic goods) and authenticity (the proper ordering of one's intentions and one's acts). Later, in a joint article, "Practical Principles, Moral Truth, and Ultimate Ends" (*American Journal of Jurisprudence* 32 [1987]: 99-151), Grisez, Finnis, and Boyle speak of reflexive goods–each of which involves a harmony, a proper order of relation–the ordered relations of intentions within the individual, the ordered relations between people, and the ordered relations between the individual or a people and God. These are various levels of proportionate unity, of integrated wholeness which are the marks of the beautiful.

30.   The fact that the ancient-medieval moral tradition which I have traced did not choose technology as a moral paradigm is worth noting. The cause was partly cultural: the ancient Greeks were at least ambivalent about the benefits of technology, as is indicated in the religious tension between the god Prometheus, who gave humans fire and thus the means of developing technology and some independence from the gods, and the god Zeus, who demanded justice and unconditional worship (see the play *Prometheus Bound* by Aeschylus). A similar ambivalence was present in medieval culture, perhaps even stronger against technology since one was to trust in God's absolute providence.

More important for our discussion are the moral reasons for rejecting technology as a suitable paradigm for the moral life. In the Platonic tradition, where the body was often seen as in the way of the soul's achieving happiness (*Phaedo*, 65b), efforts to make the life of body more comfortable through technology would not make sense. In addition, pleasure is the great threat to virtue; hence to seek technological ends is to turn away from virtue. Augustine, while recognizing that temporal goods, the goods of the body and those which technology can provide, are really good, insisted that they are so much less good than the goods of the soul and than the ultimate good which is God. To overemphasize technology would be to make living virtuously in this life very difficult and the achievement of eternal happiness with God even more unlikely.

Although Aristotle was less suspicious of science and technology than Plato (since he insisted that the human being is the unity of body and soul), he agreed with Plato that virtue is more essential to human happiness than comfort or pleasure. "The life as a whole, too, is divided into instrumental activity and leisurely activities, into war and peace, and, of things to be acted upon, into those which are necessary or instrumental and those which are noble; and the corresponding parts and actions into which the soul is divided must be the same: war for the sake of peace, instrumental activity for the sake of leisurely activity, and the necessary and instrumental for the sake of the noble." *Politics*, VII, 14 (1333a30-37) in *Aristotle: Selected Works*, 599-600.

Like Aristotle, Aquinas recognized a hierarchy of importance among goods. Goods may be honorable, delightful, or useful: all are worthy of our attention, but the honorable more so than the delightful and the delightful more so than the useful. "Those things are called pleasing which have no other formality under which they are desirable except the pleasant, being sometimes hurtful and contrary to virtue. Whereas the useful applies to such as have nothing desirable in themselves, but are desired only as helpful to something further, as the taking of medicine; while the virtuous is predicated of such as are desirable in themselves." ST 1.5.6.ad2, Volume 1, p. 28.

31.   Bacon speaks of "entering upon the one path which is alone open to the human mind." *The Great Instauration*, Proem, tran. James Spedding, Robert L. Ellis, and Douglas D. Heath in *The New Organon and Related Writings* (New York: The Liberal Arts Press, 1960), 4.   See Aphorism XCVIII, where he bemoans the fact of things not being "duly investigated verified, nothing counted, weighed, or measured" (95).

32.   "That the state of knowledge is not prosperous nor greatly advancing, and that a way must be opened for the human understanding entirely different from any hitherto known, and other helps provided, in order that the mind may exercise over the nature of things the authority which properly belongs to it." *Instauration*, Preface, 7.

33.   *Instauration*, Preface, 8.

34.   *Instauration*, Preface, 16.

35.   *The New Organon: or True Directions Concerning the Interpretation of Nature* in *The New Organon and Related Writings*, Aphorism LXV, 62.

36.   "Matter rather than forms should be the object of our attention, its configuration and the changes of configuration, and simple action, the law of action or motion; for forms are figments of the human mind, unless you will call those laws of action forms." *New Organon*, Aphorism LI, 53.

37.   *New Organon*, Aphorism LXXX, 77.

38.   *New Organon*, Aphorism CXXVII, 115.

39.   *Discourse on Method*, Part One, tran. Donald A. Cress (Indianapolis/Cambridge: Hackett, 1980), 2. The full title of the work is *Discourse on the Method of Rightly Conducting One's Reason and of Seeking Truth in the Sciences*. It introduced treatises on optics, geometry, and meteorology.

40.   *Discourse*, Part One, 3.

41.   *Discourse*, Part One, 4.

42.   "I compared the writings of the ancient pagans who discuss morals to very proud and magnificent palaces that are built on nothing but sand and mud. They place virtues on a high plateau and make them appear to be valued more than anything else in the world, but they do not sufficiently instruct us about how to know them." *Discourse*, Part One, 4.

43.   He says (*Discourse* Part Two, 12) that he has hopes of applying his method as successfully to other sciences as he did to algebra.

44.   *New Organon*, Aphorism CXXIV, 114.

45.   "It seemed to me that I could discover much more truth in the reasonings that each person makes concerning matters that are important to him, whose outcome ought to cost himself dearly later on if he has judged incorrectly, than in those reasonings that a man of letters makes in his private room, which touch on speculations producing no effect...." *Discourse* , Part One, 5.

46.   *Discourse*, Part Six, 33. Ultimately, his practical purpose is to found a science of medicine, for "the maintenance of health, which unquestionably is the first good and the foundation of all the other goods in this life." *Discourse*, Part Six, 33. Thus, although there is a good deal of metaphysics in Descartes

work, it is subservient to practical purposes, to the sciences and the technological advances (especially in medicine) that they will bring. "I have resolved to spend my remaining lifetime only in trying to acquire a knowledge of nature which is such that one could deduce from it rules for medicine that are more certain than those in use at present." *Discourse*, Part Six, 41.

47.   Thomas Hobbes, *Leviathan*, Part I, Ch. 1 (New York/London: Penguin, 1985).

48.   *Leviathan*, Part I, Ch. 15.

49.   *Leviathan*, Part I, Ch. 10.

50.   *Leviathan*, Part I, Ch. 10, 155.

51.   *An Inquiry Concerning Human Understanding*, Section IV, Part I.

52.   *A Treatise of Human Nature*, Book III, Part I, Section I.

53.   *An Inquiry Concerning the Principles of Morals*, Section IX, Part I.

54.   David Hume, *An Inquiry Concerning the Principles of Morals*, ed. Charles W. Hendel (Indianapolis/New York: Bobbs-Merrill, 1957) Section IV, Part II, 47. "If usefulness, therefore, be a source of moral sentiment, and if this usefulness be not always considered with a reference to self; it follows, that everything, which contributes to the happiness of society, recommends itself directly to our approbation and good-will. Here is a principle, which accounts, in great part, for the origin of morality: And what need we seek for abstruse and remote systems, when there occurs one so obvious and natural?" Ibid., 47.

55.   These include the first four elements in Bentham's hedonistic calculus; see *Introduction to the Principles of Morals and Legislation*, Ch. 4 (New York: Hafner Press, 1948), 29.

56.   These include the last three elements of the hedonistic calculus; see Ibid, Ch. IV, 29.

57.   Ibid., Ch. X, 98.

58.   *Fundamental Principles of the Metaphysics of Morals*, Second Section, tran. Thomas K. Abbott (New York: Macmillan; London: Collier Macmillan, 1949), 33.

59.   Ibid., First Section, 15-17.

60.   Ibid., Third Section, 69.

61.   Ibid., Third Section, 68.

62.   "The main principle in virtue and in character lies in intention." *Nichomachean Ethics* VIII, 15 (1163a23), tran. Hippocrates G. Apostle (Grinnell, Iowa: The Peripatetic Press, 1984), 159.

63.   The position of the utilitarian Mill seems to involve such a reduction of everything to our desire for pleasure: "pleasure and freedom from pain are the only things desirable as ends" *Utilitarianism*, Ch. 2 (Indianapolis/New York: The Liberal Arts Press, 1957), 10. This is not the case with Aristotle, however, who held that human happiness is centered essentially on contemplation and virtue; see *Nichomachean Ethics* X, 7 & 8.

64.   John Finnis discusses this point in his book *Moral Absolutes: Tradition, Revision, and Truth* (Washington, D. C.: Catholic University of America Press, 1991); see especially 13-16 and 95-101. For Josef Fuchs' position see his *Christian Morality: The Word Becomes Flesh*, tran. Brian McNeil (Washington, D.C.: Georgetown University Press; Dublin: Gill & Macmillan, 1987), particu-

larly 16-17. For Richard A. McCormick's position see his *Notes on Moral Theology 1965 through 1980* (Lanham, Md.: University Press of America, 1981), 708-10 and 761-63, and "Ambiguity in Moral Choice" in *Doing Evil to Achieve Good* (Chicago: Loyola University Press, 1978) 7-53. "The basic analytic structure in conflict situations is the lesser evil, or morally avoidable/unavoidable evil. ... This means that all concrete rules and distinctions are subsidiary to this and hence valid to the extent that they actually convey to us what is factually the lesser evil. ... Thus the basic category for conflict situations is the lesser evil or unavoidable/avoidable evil, or proportionate reason" ("Ambiguity," 38). See also Bruno Schüller's *Wholly Human: Essays on the Theory of Language and Morality*, tran. Peter Heinegg (Washington, D. C.: Georgetown University Press; Dublin: Gill and Macmillan, 1986), 165.

 65. Schüller, *Wholly Human*, 165.

 66. ST 2-2.64.7c.

 67. See Finnis, *Moral Absolutes*, 97.

 68. See Germain Grisez, *Christian Moral Principles*, Ch. 6-F, 152, and John Finnis, *Natural Law and Natural Rights*, Ch. V.6, 112-15.

 69. See Finnis, *Moral Absolutes*, 12-16.

 70. Romans 6.

 71. Clearly, consequences are one of the factors to be considered in moral decision-making, for one is required to bring about good in the world. Consequences, however, are not the ultimate moral criterion; intending good consequences can never justify a direct violation of basic human goods.

 72. Aquinas, ST 1.5.6ad2.

Habermas, Fichte and the Question of Technological Determinism

by John E. Jalbert

The publication of Jacques Ellul's *La Technique ou l'enjeu du siecle* in 1954 precipitated a full-fledged debate over the alleged autonomy of technology. Stated succinctly, Ellul's twofold thesis is that technology is autonomous and human will and agency heteronomous. Although he elsewhere explicitly denies that "technology is ... a closed system,"[1] his own account suggests something quite different. Indeed, one can legitimately speak of technology as a system precisely insofar as it constitutes a reality engulfing all other realities--be they economic, political or social--such that human reality in general is completely engulfed by *technique*. That there are people caught in the grip of this phenomenon who nonetheless believe themselves to be the agents of technology is evidence of how insidious and pervasive technology's impact on human self-understanding is. Briefly, then, for Ellul and other like-minded critics of technology, the belief that human beings are in control of technology is an unfortunate illusion without warrant given the nature of *technique*.

To be certain, the autonomy of technology thesis is not without its detractors. Joseph C. Pitt, a past president of the Society for Philosophy and Technology, rejects the notion of autonomous technology, proclaiming it "a major mistake to think that there is any useful way in which we could conceive of technology as autonomous."[2] Somewhere between these extremes is Jürgen Habermas, who dismisses the notion of technological determinism when he observes that "this thesis of the autonomous character of technical development is not correct,"[3] yet, he notes that there is an implicit tendency toward autonomous technology in the positivistic conception of reason.[4] Consequently, Habermas is not particularly sanguine about the prospects of the human will and action miraculously resuming directorship over the techno-scientific order, but he nonetheless wants to keep open the possibility of an enlightened relationship to technology. Such a possibility exists for him because the problem is not technological determinism *per se*, but rather, the severely impoverished conception of rationality with which modern technology is historically connected.

By rejecting the idea of "hard" technological determinism, Habermas knows that he is not doing anything to ameliorate the genuine concern over the loss of human autonomy, a concern he shares with other critics of technology such as Ellul. On the contrary, Habermas recognizes that the stakes are actually raised by shifting the locus of concern from the alleged inertia of technological progress to the disjunction between *theoria* and *praxis*, for this disjunction amounts to the exclusion of ethics from the realm of rational discourse. The result is an existential dislocation that cannot be mitigated by simply fine-tuning our arsenal of moral precepts to deal with the torrent of new moral dilemmas arising in the wake of technological progress because what has been lost is "'ethics' as such as a category of life."[5] Although Habermas's understanding of "ethics" admittedly harks back to Hegel, to *Sittlichkeit*, rather than Fichte, his main point nonetheless holds true: it is no longer the adequacy of our moral codes that is at issue, but the restoration of life as a sphere of ethical discourse.

Once ethics has receded from the realm of rational discourse, the question regarding genuine socio-political life is transformed into a merely technical problem. One may insist, of course, that life is nonetheless rationalized in this process, at least insofar as it is remade into a product of techno-scientific reason and brought under our technical control; but, obviously, this is a one-sided and therefore misleading form of rationalization. When false dichotomies between theory and practice, knowledge and interest, fact and value claim and define the landscape of the lifeworld, variations of power and domination are all that remain of reason's work. Everything else, everything that does not arise from or is not amenable to pure theory, disinterested reason, or their close relative, technology, is immediately relegated to the sphere of the "irrational."

Of course, one can appreciate Habermas's contributions to the discussion without necessarily accepting every detail of his analysis of the separation of theory and practice as the explanation of how "a technology [that] becomes autonomous dictates a value system–namely, its own–to the domains of *praxis* it has usurped."[6] Just as unnecessary was it for Habermas to adopt Fichte's philosophical program *in toto* in order for him to appreciate what Fichte's idealism could lend to his own efforts to address the issue. The present paper will look over Habermas's shoulder, so to speak, and revisit Fichte's ethical idealism in light of the question of technological determinism. Specifically, we will return to the mainstays of Fichtean idealism, self-determination and personal responsibility, and ask whether we are not too hasty in dismissing such perspectives as too ethereal and unrealistic for the present age.[7] The first order of business, though, is to more closely examine Habermas's diagnosis of the problem.

I

It is important to remind ourselves that Habermas's exposition of techno-scientific rationality is a Kantian style critique in that it seeks to establish the limits of this conception of reason, not to undermine and discard it. Indeed, Habermas admits the legitimacy of technological rationality's separation of reason from the commitment to "adult autonomy" [*Mündigkeit*], so long as technology does not constitute the *telos* of rationality as such. In his own words: "Science as a productive force can work in a salutary way when it is suffused by science as an emancipatory force, to the same extent as it becomes disastrous as soon as it seeks to subject the domain of praxis, which is outside its technical disposition, to its *exclusive* control."[9] Unfortunately, the latter has become the rule rather than the exception. Still, to impute an imperialistic intention to positivistic science and technology, as Habermas does in this instance, gives the impression that he holds to the thesis of hard technological determinism, which of course he does not. The intoxication of expert and non-expert alike with the successes of science and technology is sufficient explanation for the transgressions of instrumental rationality. We are reminded of Edmund Husserl's uncharacteristic comment about the naiveté of "people who imagine that with sufficient progress in physics and chemistry mankind will come so far that ... it will cure not only physical but also moral syphilis."[9] How is such naiveté possible?

Looked at from a Habermasian perspective, the naiveté of the modern techno-scientific lifeworld germinated from a wedge that was driven between *theoria* and *praxis*, an event that occurred early in the tradition of Western philosophy. In the beginning, the attempts to purge pure theory of human interests did not automatically cut it off from the practical sphere of life, because, for the Greeks, theoretical contemplation accessed and assimilated an order discovered in the cosmos. In this respect, theory was imbued from start to finish with a prominent practical component. Today, however, Habermas observes that, "The conception of theory as a process of cultivation of the person has become apocryphal."[10] All that remains of the classical view, it seems, are the distorted fragments preserved in the modern conceptions of theoretical reason and dogmatic ontology, while the cosmology necessary to support the original practical intent of the ancients has long been buried and forgotten. But if this ancient cosmology is no longer available to us, perhaps there is another path to the essential connection between theoretical reason and life conceived as an ethical matter.

Positivistic science misinterprets the ideal of "disinterested reason" when it takes this to mean severing the umbilical cord that attaches reason and human interest, volitions and values. One could accept the scientific ideal of "value-freedom" if it signified nothing more than the methodological desideratum of shedding our most mundane and banal interests but not when it claims for itself the exclusion of all values, as though such a thing were not only possible, but desirable as well. As with many other critics, Habermas argues that science and technology are anything but value-free and, thus, demands the recovery of their connection and subordination to ethical life.

The project of science and technology is motivated by and, in turn, motivates, values that contribute to what the Greeks, Plato and Aristotle, would describe as "mere life," in careful contradistinction to the "good life" and the values associated with it. Still, modern science and technology do enable us to rationalize one aspect of human life--its material conditions--by constantly reconfiguring them in accordance with the values expressed in notions such as efficiency, economy and productivity. In other words, technology enables us to transport people from place to place more efficiently each day, it enables us to produce more goods at less cost, and, yes, I suppose it enables us to more adequately treat cases of physical syphilis. What impact has all of this on the "good life?" Do science and technology make us more autonomous, more responsible, and, as Aristotle would add, more honorable?

These questions raise the point that modern science and technology offer only the semblance of bridging the gap between *theoria* and *praxis*. In reality, they do nothing more than fuse the two sides of the same technological equation: technological rationality with technical control over nature and human affairs. The human dimension is treated no differently than nature, namely, it is treated as something to manipulate and control. The possibility of moving individuals and communities by persuasion rather than by control, by mutual understanding and dialogue rather than manipulation, by enlightenment and self-understanding rather than coercion is not given serious consideration. Indeed, it cannot be so considered as long as the "thesis that human beings control their destinies rationally to the degree to which social techniques are applied"[11] remains unchallenged. Here techno-scientific rationality is grafted onto a domain that, strictly speaking, lies outside of its sphere of legitimacy. Furthermore, technological rationality presupposes a commitment, an interest, that is not itself technological in nature and, thus, falls outside of its scope but not outside of the scope of reason as such. Reason is never divorced from interests and values, but forgetting this has shaped the self-understanding of positivistic science and technology and imprisoned them in the illusion of value-neutrality.

Habermas further contends, though somewhat unfairly, I should add, that Husserl's transcendental phenomenology is victim to basically the same mistake that plagues positivism.[12] Husserl, we are told, is correctly critical of the objectivism of positivistic natural and social sciences; but he then immediately turns around and ignores his own underlying assumptions regarding the claims of disinterestedness and freedom from presuppositions. First, he assumes that such things are possible, and, second, he assumes that the results of pure theory so conceived would be efficacious in the practical sphere, that is, in the lifeworld, even though the transcendental standpoint from which he launches his critique of objectivism has undermined or, at minimum, has bracketed, the necessary cosmological apparatus to make this practical efficacy conceivable. Philosophical self-reflection, including the transcendental variety introduced by Husserl, has as its correlative "cognitive interest" [*Erkenntnisinteresse*], human interest in autonomy and responsibility which it can only realize by "acknowledging its dependence on this interest and turning against its own illusion of pure theory the critique it directs at the objectivism of the sciences."[13]

At this juncture, the significance of Fichte's philosophy for Habermas's project should be evident. Fichte's philosophy does what Habermas calls for,—on the one hand by taking as its point of departure the unity of *theoria* and *praxis* and, on the other by recognizing the existential connection between knowledge and interests as the foundation of this unity and, hence, of the system as a whole. Habermas thinks, however, that Fichte pays too high a price for this insight, as evidenced by the statement that, "whoever becomes conscious of his independence and autonomy of all that is external to him ... does not require things as a support for his self, and can have no use for them, because they abolish that independence and transform it into empty semblance [*Schein*]."[14] Hence his preference for Hegel's notion of *Sittlichkeit* rather than Fichte's loftier claims on behalf of moral consciousness. What Habermas fails to sufficiently consider, however, is precisely that Fichte is expressing a moral victory over the external world and only secondarily an epistemological one. At any rate, whatever else it might be, it surely is not a flight from finitude. Additionally, because Habermas eschews what he takes to be the foundationalist element in Fichte's thought, he ignores crucial insights that have an important bearing on the question of technology and its place in a moral world order. This aspect of Fichte's thought deserves consideration.

II

Since our disproportionately long approach to Fichte's philosophy of consciousness gathers support from Habermas, we should begin by recalling the two critical points he extrapolates from Fichte's work. The first thing of import is that when Fichte asserts the primacy of practical over theoretical reason, he gives a voice to the moral command that demands the freedom and autonomy of subjectivity vis-à-vis the apparently inescapable contingency and determinism associated with the theoretical world. He may go too far in this regard for Habermas, but it is worth repeating that contingency and finitude do not simply evaporate in the Fichtean schema. It is rather a matter of achieving a perspective in which the two realms–the theoretical and practical–are integrated in such a manner that the contingency associated with the former is subordinated to and subsumed under the freedom required by the latter.

The perspective in question presupposes the free act of self-reflection whereby thought and deed are one and inseparable. The *Second Intro-duction to the Wissenschaftslehre* (Science of Knowledge) states this about as lucidly as one is likely to find anywhere in Fichte's writings: "Everybody, one hopes, will be able to think *of himself*. He will become aware, one hopes, that, in that he is summoned to think thus, he is summoned to something dependent on his self-activity, to an *inward action*, and that if he does what is asked, and really affects himself through self activity, he is, in consequence, acting."[15] Philosophizing reenacts on the level of philosophical reflection what is in fact a condition for the possibility of all reflection, particularly philosophical reflection, namely, free activity. By bending its reflective gaze back upon itself, the self gains immediate access to this freedom as its object because it acts freely. In opposition to Kant, then, Fichte thinks that there is at least one instance of an *intellectual intuition* "whereby I know something because I do it"[16] and, therefore, without the necessity of having to posit something outside myself as acting upon me. Conse-quently, he often talks about intellectual intuition as the only acceptable standpoint for philosophy because, for reasons stemming from moral consciousness, genuine philosophy cannot tolerate passivity. Or, to put it another way, intellectual intuition must serve as the philosophical point of departure because, "I *ought* in my thinking to set out from the pure self, and to think of the latter as absolutely self-active; not as determined by things, but as determining them."[17]

What really prompts someone to take up this philosophical stand-point rather than another one? Why does someone choose idealism, which makes the external world intelligible by beginning with the self, rather than dogmatism, which takes the opposite course and bases intelligibility of the self on things external to it? The latter, after all, has the weight of common sense on its side. Fichte's response to these questions is the second thing of great importance for Habermas. Accord-ing to the *First Introduction to the Wissenschaftslehre*, idealists and dogmatists (realists) have different interests, and the highest interest–the one that grounds all the others–is *self-interest*.[18] The implication is that dogmatists forfeit the self for the sake of the not-self, for the sake of things, while idealists want to preserve the active self at all costs and, therefore, make the not-self something for the self. This is where Habermas sees Fichte broaching the notion of "cognitive interests" that he himself capitalizes on in *Knowledge and Human Interests*, but Fichte is doing more than that. He is setting forth, even if only tacitly and perhaps unwittingly, a different way of looking at the question of technological determinism.

Instead of exhausting ourselves in a debate over whether or not technology really is autonomous, oscillating from one side of the antin-omy to the other, better progress could be made by adopting a Fichtean strategy and asking how we ought to comport ourselves vis-à-vis the technological world? What happens to the project of modern science and technology if it issues from the moral law and, hence, is subordinated from the outset to the task of maintaining the self as moral agent, that is, as autonomous? What are the concrete implications of this Fichtean proposal for technology and human self-understanding?

III

As a popular presentation of his thought, *The Vocation of Man*, offers a few clues. For instance, in Book III, entitled "Faith" [*Glaude*], the I/Ego that has been engaged in self-reflection from the beginning of the volume, is finally in a position to consciously grasp its freedom and dignity, and, when it does, it demands and strives for a "better world," that is to say, a world that is in accord with the moral will. So far, this is vintage Fichte, but what subsequently appears is more noteworthy given the general theme of this conference. Surprisingly, the first step in the direction of this "better world" appears, on the face of it, to take its bearings directly from Francis Bacon. Fichte writes: "Nature must gradually be resolved into a condition in which her regular action may be calculated and safely relied upon, and her power bear a fixed and definite relation to that which is destined to govern it–that of man."[19] Furthermore, reason demands the successful extortion of nature's secrets, for once nature is rendered more intelligible "human power, enlightened by human invention, shall rule over her without difficulty...."[20] Once again, Fichte sounds as though he has torn a page out of Bacon's book and appropriated Bacon's model of controlling and exploiting nature for the sake of improving our material estate.

The only obstacle to the complete conquest of nature is human vice and avarice, which makes human beings, not nature, the chief adversary of human beings, and the "mightiest inventions of the human intellect" [*mit dem Höchsten, was der menschliche Verstand ersonnen, ausgerüstet,....*] only makes them a more formidable enemy.[21] Fichte, however, does not dwell on this point because, ostensibly, such conflicts will eventually dissolve under the sway of moral consciousness, thereby leaving human beings in a position to "direct their united strength against the one common enemy which still remains unsubdued–resisting, uncultivated nature."[22] Obviously, instrumental reason remains an important part of the lifeworld even if we are looking at it from the point of view of Fichte's moral consciousness. Nevertheless, what we need to recognize is that as long as instrumental reason and technology are contained under the umbrella of moral reason, as they are in Fichte's case, they can and, indeed, they must serve the material conditions of life but not at the expense of our true destiny as moral agents.[23] By way of implication, although this is going beyond Fichte himself, one can argue that his viewpoint contains an inherent check against any unconscionable destruction and exploitation of the natural environment to the extent that it would never confuse a life of material abundance with one of spiritual fulfillment. Moreover, there are also good reasons to think that adopting this moral perspective compels us to treat nature "as if" [als ob] it too had rights, for to act otherwise would be repugnant to moral consciousness and contrary to our highest interest.

Naturally, what Fichte is thinking of when he talks about "human inventions" covers only a small portion of what is generally subsumed under the concept of technology today. Consequently, the question of technological determinism and autonomy is ours, not his, but once the question arises, as it inevitably does once the moment that the more inclusive notion of *technique* is allowed, Fichte's ethical idealism becomes immediately relevant. Of course, technological determinism is never problematic in abstraction from the conception of what the human being qua self ought to be, namely, autonomous. Moral consciousness obliges us to maintain the autonomy of the self, although in the concrete lifeworld realizing it will necessarily take the form of striving to be autonomous. Human beings, after all, are finite, but this does not relieve us of the responsibility of striving for infinitude, for freedom and for autonomy. To do anything less diminishes us. In *Computer Power and Human Reason*, Joseph Weizenbaum, following his own path, arrives at a very similar conclusion. On his account, each individual must act as if the future of the world and humanity depended on him or her. "Anything less is a shirking of responsibility and is itself a dehumanizing force, for anything less encourages the individual to look upon himself as a mere actor in a drama written by anonymous agents, as less than a whole person, and that is the beginning of passivity and aimlessness."[24] Because Fichte's moral idealism targets such passivity, indeed aggressively attacks it with a zeal that arises from a profound awareness of and interest in human autonomy and responsibility, it deserves a hearing whenever the discussion turns to the question of technological determinism. Additionally, such awareness is arguably a necessary precondition for participation in Habermas's "ideal speech situation." One thing is certain: there are far more anemic ways of responding to the problem of technological determinism.

Sacred Heart University
Fairfield, Connecticut

* * * * *

Endnotes

1.   Jacques Ellul, *The Technological System*, trans. Joachim Neugroschel (New York: Continuum, 1980), 80.

2.   Joseph C. Pitt, "The Autonomy of Technology," *Philosophy and Technology*, vol. 3, *Technology and Responsibility*, ed. Paul T. Durbin (Dordrecht: D. Reidel, 1987), 99. See also Mary Tiles and Hans Oberdiek, *Living in a Technological Culture: Human Tools and Human Values* (London: Routledge, 1995), 26-27. For a slightly more moderate view in the same vein, see Robert Heil-

broner, "Do Machines Make History?" *Controlling Technology: Contemporary Issues*, ed. William B. Thompson (Buffalo, NY: Prometheus Books, 1991), esp. 220-21.

3. Jürgen Habermas, *Toward a Rational Society: Student Protest, Science and Politics*, trans. Jeremy J. Shapiro (Boston: Beacon Press, 1970), 59. For the German see *Technik und Wissenschaft als Ideologie* (Frankfurt am Main: Suhrkamp Verlag, 1968), 116.

4. Jürgen Habermas, *Theory and Practice*, trans. John Viertel (Boston: Beacon Press, 1973), 270.

5. Jürgen Habermas, *Toward a Rational Society*, 112. Note, however, that the Hegelian influence is more evident in the original German; *Technik und Wissenschaft als Ideologie*, 90.

6. Habermas, *Theory and Practice*, 270.

7. Martin Heidegger, *Introduction to Metaphysics*, trans. Ralph Manheim (Garden City, NY: Anchor Books, 1961), 37-38, expresses his appreciation of German idealism–most likely with Fichte in mind–presumably because it has the potential of giving us a purchase on the modern technological worldview. Unfortunately, Heidegger does not pursue exactly how idealism bears on the issue, and, furthermore, there is now some doubt about the sincerity of his reference to German idealism in this context. Hans Sluga, *Heidegger's Crisis: Philosophy and Politics in Nazi Germany* (Cambridge, MA: Harvard University Press, 1993), makes a compelling case that Heidegger's brief self-alignment in 1933 with German idealism, particularly Fichte, was motivated by his political interests in National Socialism. Be that as it may, it does not vitiate the importance of idealism as a way of coming to terms the over-technologizing of human existence.

8. Habermas, *Theory and Practice*, 281,

9. Edmund Husserl, *Phenomenology and the Foundation of Science. Book Three: Ideas Pertaining to a Pure Phenomenology and to a Phenomenological Philosophy. Collected Works, Vol. 1*, trans. Ted E. Klein and William E. Pohl (The Hague: Martinus Nijhoff Publishers, 1980), 12.

10. Jürgen Habermas, "Knowledge and Human Interests: A General Perspective," appendix to the English translation of *Knowledge and Human Interests*, trans. Jeremy J. Shapiro (Boston: Beacon Press, 1971), 304. The German text *"Erkenntnis und Interesse"* in *Technik und Wissenschaft als gies*, 150.

11. Habermas, Theory and Practice, 275.

12. Habermas does not mention for instance that in "Appendix IV: Philosophy as Mankind's Self-Reflection: The Self-Realization of Reason" in *The Crisis of European Sciences and Transcendental Phenomenology: An Introduction to Phenomenological Philosophy*, trans. David Carr (Evanston: Northwestern University Press, 1970), Husserl emphasizes, possibly with intentional Fichtean overtones, that reason is motivated by the will to reason, that a rational existence is an "infinity of living and striving toward reason," that what "becomes, in this life, is the person himself" and, finally, that "individual-personal and communal-personal being" are correlative and mutually constitutive. Even without citing any additional texts, there are already enough points of contact here to suggest that Husserl is already looking at the terrain that

Habermas will cross. In this vein, one should see James G. Hart, "Husserl and Fichte: With Special Regard to Husserl's Lectures on Fichte's Ideal of Humanity," *Husserl Studies*, vol. 11, no. 3, 1994, 135-52.

13.   Habermas, *Theory and Practice*, 311. Or, *Technik und Wissenschaft als Ideologie*, 159.

14.   As cited in Habermas, *Theory and Practice*, 260-61. See also J. G. Fichte, *Science of Knowledge (Wissenschaftslehre), with the First and Second Introductions*, eds. and trans. Peter Heath and John Lachs (New York: Appleton-Century-Crofts, 1970), 15 and J. G. Fichte, *Gesamtausgabe der Bayerischen Akademie der Wissenschaften*, Vol. 4, eds. Reinhard Lauth and Hans Gliwitzky (Stuttgart-Bad Cannstatt: Freidrich Frommann Verlag, 1970), 194.

15.   Fichte, *Science of Knowledge*, 37; *Gesamtausgabe*, v. 4, 215.

16.   *Science*, 38; *Gesamtausgabe*, 217.

17.   *Science*, 41; *Gesamtausgabe*, 219-20.

18.   *Science*, 15; *Gesamtausgabe*, 194.

19.   J. G. Fichte, *The Vocation of Man*, trans. William Smith, L.L.D. (La Salle, Ill: Open Court Publishing Co., 1965), 116. *Gesamtausgabe der Beyerischen Akademie der Wissenschaften*, Vol. 6, 1981, 268.

20.   *Vocation*, 117; *Gesamtausgabe*, 269. It is interesting to note too that Fichte has a citation from Bacon's *The Great Instauration* as "*Vorerringerung*" to the *First Introduction to the Wissenschaftslehre*.

21.   *Vocation*, 118; *Gesamtausgabe*, 269.

22.   *Vocation*, 127; *Gesamtausgabe*, 275.

23.   This is the implication of Fichte's critique of the "present age" in "Characteristics of the Present Age" where he observes: "With respect to the influence which it exerts upon nature and its employment of her powers and products, such an Age looks everywhere only to the immediately and materially *useful*–to that, namely which is serviceable for dwelling, clothing, and food–to cheapness, convenience, and, ... to fashion; but that higher dominion over nature whereby the majestic image of man as a race is stamped upon its opposing forces–I mean the dominion of ideas.... –this is wholly unknown to such an age; ... and thus art itself, reduced to its most mechanical forms, is degraded into a new vehicle of fashion, the instrument of a capricious luxury, alien to the eternities of the ideal world." See Johann Gottlieb Fichte, *Characteristics of the Present Age and the Way Towards the Blessed Life or The Doctrine of Religion*, ed. Daniel N. Robinson (Washington, DC: University Publications of America, 1927), 29. Interestingly, Husserl's Fichte lectures accurately and perceptively discern the shift of meaning that occurs within Fichte's conception of "dominion over nature:" "God's essence becomes manifest in each pure beauty. Again, it is manifest in perfect domination of nature.... In other words, God reveals himself wherever *Ideas* luminously beckon us in the empirical realm. They are divine Ideas." In the same vein, Husserl speaks of "the noble technologist whose love aims at creating for humans domination of nature (and not for lower sensuous goals)." See Edmund Husserl, "Fichte's Ideal of Humanity [Three Lectures]," trans. J. G. Hart in *Husserl Studies*, vol. 12, no. 2, 1995, 127-28.

24.   Joseph Weizenbaum, *Computer Power and Human Reason: From Judgment to Calculation* (New York: W. H. Freeman and Company, 1976), 267.

Medieval Technology and the Husserlian Critique of Galilean Science

by Timothy Casey

Edmund Husserl's critique of modern mathematical physics (or what he simply calls Galilean science) is notable for its attempt to trace the highly formalized idealizations of this science back to the lifeworld. He does this, in part, to reestablish the sense of modern science and the scope and validity of its truth claims. Left unclear in Husserl's account is the historical character of the lifeworld out of which modern science arose.[1] In this paper I will examine that lifeworld in some historical detail as a way of filling in the technological origins of the formalization of knowledge in general and of science in particular that Husserl subjects to phenomenological analysis and critique.

According to the eminent historian of medieval technology, Lynn White Jr., the Industrial Revolution of the eighteenth and nineteenth centuries was in fact no revolution at all but rather a managerial innovation which ushered in the emergence of large factories.[2] The "mechanical stage" of Western technology begins, he claims, prior to modernity in the great agricultural revolution of the early Middle Ages. Some scholars have even come to identify the signal achievement of this allegedly backward time not with any one invention, such as the Gothic cathedral or scholasticism, but with what Whitehead has called "the invention of invention," that is, the impulse to invent for invention's sake. Of course, there have always been inventors and inventions. The ancient Greeks, for example, had their fascination with machines and mechanical gadgets. Even the great Archimedes, however, lacked what Robert Brumbaugh describes as the inventor's mentality and its thoroughly technological approach to problem solving. Clearly, then, something radically new began to emerge in Western Europe around 600 A.D., something more than simply a change in intellectual habits, so that by the late Middle Ages Western civilization reached a turning point, a central vantage point, as Lewis Mumford puts it, from which "one has at hand most of the important elements derived from the past, and the germ of most of the growth that is to take place in the future."[3] Medieval technology not only provided an agricultural, industrial, military and navigational base for the great geographical discoveries and colonizations of the modern era, but it also instigated the programatic approach to technological innovation as well as the more aggressive attitude toward nature necessary to effect such an approach.

Of all the changes occuring in the Middle Ages, perhaps the most striking was the dramatic increase in population largely attributable to the agricultural revolution in Northern Europe between 500 and 800 A.D. At the heart of this change was the introduction of the heavy plough, the three-field crop rotation system, and, later, the shoulder harness, the tandem harness and the nailed horseshoe. With improvements in the water mill and the invention of the wind mill, the benefits of nonhuman power became increasingly apparent, as did a new, more technological relationship with nature, as depicted in ninth century calendar scenes of plowing, harvesting, pig-slaughtering and so forth.[4]

This new milieu encouraged a change from empirical groping to a widespread program of mechanically exploiting the forces of nature for human purposes.[5] Beneath the Aristotelian-inspired philosophy and science of the day was stirring a more coercive attitude toward natural resources and their manipulation by mechanical means. The invention of the more efficient windmill, for instance, joined forces with the extensive application of water-power in the eleventh century, hastening the replacement of human labor in basic industries for the first time in history. Wind and water power were extended beyond their traditional function of grinding grain to such processes as sawing, fulling, polishing, rolling, tanning, laundering, crushing, treating hemp, making pulp for paper, powering bellows and driving hammers of the forge. The implementation of nonhuman power sources provided an impetus for the improvement of machinery and machine design. According to White, the single most important mechanical invention was the crank, "since it is the chief means of transforming continuous rotary motion into reciprocating motion and the reverse."[6] Other mechanical devices indicative of this new approach to labor included the flywheel, treadle and spinning wheel; the cross-bow, horizontal frame-loom and kick wheel; not to mention the spring, pipe organ and various aids to mining and warfare.[7] In short, the desirability of mechanical power was prevalent by the thirteenth century, and by the fourteenth, the common person was for the most part comfortable in the presence of power machinery.

Mumford contends, however, that the creation of the mechanical clock—and not the crank, as White argues—was the primary medieval technological achievement. He traces its origins to the call for more precise time-pieces needed to regulate the ritualized activity of the Benedictine monks. In the monastery, order and regularity constituted the external manifestation of a devotion to God: routine had been raised to the level of a virtue. The sun-dial, hour glass and eventually the clock were thus indispensable for the new mode of religious life; but their effects reached far beyond the monastery walls. The clock, argues Mumford, synchronizes the actions of people and submits organic functions to a mechanical measure both more efficient and precise than the sun—a portent of the coming social regimentation implemented by the army, bureaucracy and factory. The automation of time became the model for all subsequent systems of automation inasmuch as the clock was the first automatic power-machine to produce a standardized product with stunning accuracy via a regular mode of production. What is more, the "production" of seconds and minutes "helped create the belief in an independent world of mathematically measurable sequences"[8] and hence the modern, quantitative approach to nature where space and time eventually exist in an objective sphere indifferent to human concerns and designs. White adds credence to this interpretation when he observes that the astronomical clocks of the fourteenth century were more than mere toys or curiosities. Rather, they captured the popular imagination and were seen as "symbols related to the inmost, and often universalized, tendencies of that age."[9] Nicholas Oresmus was in fact the first to employ the clockwork metaphor for the universe, a metaphor that was to become an explicit metaphysics, if it had not already become an implicit one.

The employment of this metaphor expressed nothing less than the growing moral enthusiasm for machines in the West. Just as God, the divine clockmaker, regulates the celestial mechanics, so too should man regulate his own body, a machine composed of various parts, by temperance. In his provocative essay, "The Iconography of *Temperantia* and the Virtuousness of Technology,"[10] White examines the increasing presence of mechanical devices in the iconography of the late Middle Ages and concludes that mechanization not only affected everyday life but had actually taken on an aura of virtuousness and religiosity. Unlike the Byzantine Church of the East, Latin Christianity eventually allowed the placement of astronomical clocks and pipe organs inside its churches, the Roman rite apparently being more inclined to accept the clock as a reflection of God's orderly cosmos. Thus the Italian depiction of temperance as a clock worked its way north and became the "icon of the Christian life." In one version, temperance is shown wearing a clock for a hat, while standing on a wind mill with bit and bridle in mouth and rowel spurs at her feet. In brief, the acceptance of the new technology had gone beyond the simple awareness of meeting life's necessities and providing for a few of its luxuries. A new concept of *measure* had arisen, symbolized by the clock and associated with the new technology in general. No longer were the routine and mechanical considered a denial of the human spirit and a threat to the moral order. The enhancement of technique as embodied in the machine and fostered by the monastic communities was now an essential part of the good life. As White summarizes the matter, "Engineering is so creative in Europe partly because it came to be more closely integrated with the ideology and ethical patterns of Latin Christianity than was the case with the technology and the dominant faith of any other major culture."[11]

Out of the medieval mechanical tradition grew the great technological innovations of the fifteenth and sixteenth centuries. Guttenberg and Leonardo were heirs to an empirical program of technological development long set in motion. Already Western humanity was in thrall to the machine and was adapting its needs to a new technical order. By the end of the Middle Ages, then, we have two profound ways of looking at things existing side by side, one looking back to the past and the other to a future still dimly perceived. The stage had been set for the quarrel between the ancients and the moderns. It awaited only a realization on science's part of the vast implications of the new technology, metaphysical and utilitarian implications that in the process of being taken up by science were to transform the very conception of knowledge itself.

Galileo is certainly the most illustrious and instructive example of the intimate relation between modern science and the new technology. Both the separation of natural science from theology and the utilization of the principles of mechanics developed by direct analogy with machines came to fruition in his work. Indeed, the impact of technology on Galileo was profound. His early career, during which practical and mechanical interests predominated, demonstrates that prior to 1600 "European technology had been both more sophisticated than European science and little related to that science."[12] He therefore made more explicit the new "feel" for matter and motion that was "the result not of any book but of the new texture of human experience in a new age."[13] Galileo embodied a new type of scientist, one who combined philosophical and mechanical interests in order to overcome the perceived sterility of the ancient, speculative approach. This new combination, in alliance with astronomy and mathematics, effectively nullified the traditional conception of knowledge. No longer would science seek out the formal and final causes of things for the sake of knowledge itself. Instead, it would focus on efficient and material causes mechanically construed; and, although religious aspirations initially lay behind the need to demonstrate the mathematical orderliness of the universe (for example, Kepler), such concerns were soon to fall by the wayside in favor of more utilitarian motives.

As early as the thirteenth century, the new, mechanical technology, as Nicholas Lobkowicz has pointed out, "tended to result in a new interest in methods for solving practical problems better than the age-old method of trial and error. Intellectuals, not only artisans, began to be interested in 'practical problems' and techniques."[14] Most important in this respect was the revival of mathematical study in the late Middle Ages. The improvement of mathematical techniques, especially in the field of algebra, paved the way for a less representational and more analytic approach to geometry. According to E. A. Burtt, "various mechanical schemes were invented to aid the poor mathematician's endeavors."[15] What these schemes resulted in was nothing less than the mathematization of nature via the rescuscitated Platonism of Galileo. That the new interest in method accompanied the flourishing of the mechanic arts cannot be emphasized enough. Simply recall that for the

Greeks mechanization was limited to the sphere of gadgetry and curiosity, most likely because of its intrinsic lack of teleology and overstress on technique, for machines are the physical embodiment of technique, efficient in their productivity and seemingly neutral with regard to ends and purposes.

It would seem, then, that the early modern emphasis on method reflects and, more importantly, is rooted in the earlier emergence of practical mechanics in the Middle Ages and their subsequent growth in the Renaissance. This is to say that "apart from the transference of ideas and techniques, there must have been an appreciable effect of a subtle kind upon the way in which problems were tackled"[16] and indeed upon what constituted a problem worth solving. While the desire for and inquiry into a new scientific method became explicit in the sixteenth century, it virtually dominated scientific and philosophical thinking in the seventeenth. Of course, it was only natural and right that the fledgling science demand serious reflection on methodology in order to secure its position vis a vis the established Aristotelian conception of knowledge. What is most striking, however, about the early moderns was the novel prominence they attributed to the efficacy of technique. R. M. Blake sums it up best when he writes that "it is altogether characteristic of the attitude of the champions of the new science to manifest an almost boundless confidence in man's power, if only the right method be employed, of attaining the absolute and final truth about nature."[17] From the outset scientific progress was conceived as essentially tied to the improvement and refinement of method, regardless of whether the emphasis fell on mathematical or experimental techniques, on deduction or induction.

What concerns us here, though, are not theories of methodology as such, that is, their correspondence to and justification of actual scientific practice, but rather the significance of the "grand preoccupation" with method for our understanding of modern science in its relation to and dependency on the mechanic arts. Husserl's analysis of the origins of mathematical physics in *The Crisis of European Sciences and Transcendental Phenomenology* sheds a good deal of light on this nexus.

There Husserl is concerned with an uncovering of the original meaning of modern science, a meaning which has been "sedimented" and hence forgotten through an increasing formalization of that science. Employing Galileo as emblematic of the early scientific movement, Husserl attributes to him a desire to revive the ancient Greek idea of *episteme* through the universal applicability of mathematics to the sphere of nature. By means of the mathematical method, which consisted in quantifying natural phenomena in order to turn a scientific problem into a mathematical one, Galileo believed he had found a way of conceiving of the universe as completely rational and hence of overcoming the relativity and fundamental indeterminacy of perceptual experience. It is precisely in this mathematical reduction, however, that Husserl detects a radical transformation of the traditional conception of *episteme* and, with it, a consequent loss of meaning.

By "loss of meaning" Husserl means here the inability of modern natural science to give an account of itself, especially as it relates to the most important problems of human life, namely, problems of values, ethics, freedom, religion, etc.[18] At the inception of modern science there occurred a fateful loss of the lifeworld, a loss traceable to Galileo's naive utilization of geometry as an established method capable of yielding "being as it really is in itself." The revitalization of the Pythagorean element in Platonism was thus perceived as possessing a power for scientific interpretation hitherto unrealized. The fascination with this new intellectual tool quickly swept aside any serious consideration of what it meant or how it fit into human life as a whole. Consequently, it underwent a transformation of sense and was seen as an autochthonous and self-sufficient kind of *episteme* rather than as the mental and cultural accomplishment that it is, rooted in both the empirical art of measuring and the horizon of everyday, practical life. Liberated from the imprecision and doxic character of the lifeworld, the art of measuring became in modern science a mere technique shorn of any teleological direction or orientation. It ceased to be an art in the fullest sense of the word, for, as Aron Gurvitsch comments, "the perfection of techniques of measurement is now emancipated from the concern with practical goals and needs arising in concrete situations."[19] Instead, the Galilean mathematization of nature rested on the identification of science with method. The demands of the method in turn determined the shape and direction of the new science. As Husserl observes, "There must be measuring methods for everything encompassed by geometry, the mathematics of shapes with its a priori ideality. And the whole concrete world must turn out to be mathematizable ... that is, if we work out the appropriate method of measuring."[20] Since the method always has a "general sense," that is, is related to the objective world by subsuming particular cases under algebraically expressed laws of functional dependency, it can develop independently of the world, remarkably unshackled by any concerns with content. Its applicability to nature is assumed a priori—and herein rests its awesome power. For "if one has the formulae, one already possesses the practically desired prediction of what is to be expected with empirical certainty in the intuitively given world of concretely actual life, in which mathematics is merely a special form of praxis."[21] Thus, the task bequeathed by the Galilean idea of a universal physics to its scientific successors consists in the constant improvement and development of those methods necessary to the continuing formation of mathematical formulae and hypotheses.

For Husserl, then, the crucial factor in the modern transformation of the meaning of *episteme* is the dominance of method. It is evident, he believes, even before Galileo in those mathematical studies that initiated the arithmetization of geometry and its liberation from any reference to spatio-temporal realities. What is more, "this process of method-transformation" eventually surpassed in Galileo's time the stage of arithmetization and completely formalized the algebraization of numbers and magnitudes. As a sheer technique, it became "a mere art of achieving" according to a set of technical rules that a priori excluded the type of thinking needed to situate the method in its proper context of meaning. Husserl, it should be emphasized, has no intention of denying the legitimacy of technique, even in this highly formalized sense. Rather, he is interested in probing what he perceives as the loss of meaning through which the method "has the sense of achieving knowledge about the world."[22] The recovery of this meaning takes on the character, not of eliminating the technicization of mathematical natural science, but of reestablishing through phenomenology the horizon within which technique and method can be meaningful at all.

What has historically prevented this reestablishment of sense is the false equation of the method with "being as it really is in itself." The essence of this technique as a technique, that is, its remarkable capacity to improve upon the predictions and inductions of everyday life, is covered over with what Husserl calls a garb of mathematical ideas, and "it is through the garb of ideas that we take for true being what is actually a method."[23] The formalization (or what I am calling the mechanization) of technique therefore carries with it metaphysical implications far beyond the practice of modern science. These include the replacement of the lifeworld with a self-contained sphere of mathematically determined bodies and spatio-temporal events, as well as the subjectivization of secondary qualities and a portentous obfuscation of human subjectivity. The Hobbesian and Lockean treatments of the soul merely round out the triumph of method in the new conception of knowledge.[24]

Husserl's analysis of the origin and meaning of modern science allows us to focus more clearly on the underlying sense of *techne* that took root in the Middle Ages and came to expression in the seventeenth century. For what Husserl is in essence battling is a philosophical conception of *episteme* modelled on scientific method and its character as "a machine, reliable in accomplishing very useful things, a machine everyone can learn to operate correctly without in the least understanding the inner possibility and necessity of this sort of accomplishment."[25] Husserl's opposition to a positivistic construal of science is thus rooted in a more fundamental rejection of a similarly positivistic conception of *techne* that omits reference to the larger context of human existence. Such in fact was the new sense of *techne* and of technology in particular that shaped Galileo in the Venice arsenal and his experience there of the mechanic arts.

At the core of the new technology was a wholly novel way of looking at and living in the world, a mechanistic attitude that science took over and turned to profit, and nowhere is it more evident than in the Baconian and Cartesian obsession with method as a mechanical technique. Indeed, since the beginning of the seventeenth century, the formalization of procedure has increasingly come to be seen as the key to the "one best means," as Jacques Ellul puts it,[26] necessary to insure those results whose sum will somehow add up to the good life. It should therefore come as no surprise that we now equate technological progress itself with the constant retooling of method, and that machinery has become the focus—though not the exclusive recipient—of that retooling.

One can trace the outstanding success of modern technology to a "systematic perfecting of procedures" and hence to the importance given to methodological analyses of all sorts and to mechanical ways of thinking in general. The machine now serves as inspiration, model, and—in the case of the computer—the device for introducing the mechanical perspective into art, politics, literature, philosophy, architecture, the law, etc. Thomas Carlyle recognized this drift in the nineteenth century, noting that "mechanical genius ... has diffused itself into quite other provinces. Not the external and physical alone is now managed by machinery, but the internal and spiritual also."[27] There is, in short, a shift in the modern understanding of techne toward a rationalization of technique which transforms it into an instrumental or calculative form that is more and more associated with rationality itself and our knowledge of the world as a whole.

Husserl's analysis of Galilean science in effect foreshadows Ellul's description of our attempt "to bring mechanics to bear on all that is spontaneous or irrational"[28] in the hope of establishing mathematical models that, in their universality, will guarantee the most efficient courses of action deemed possible. The importance of medieval technology and its turn toward the machine, though overlooked by Husserl, initiates the modern transformation of techne into sheer technique and prepares the way for the nexus between science and technology in the modern age. But more than this, it makes clear the need to bring science and, strangely enough, technology itself back to the lifeworld where the so-called universal solutions proffered by mechanical technique are yielding results that increasingly are destructive of the unique, the particular, and the local—destructive, that is, of what makes the lifeworld livable in the first place.

University of Scranton
Scranton, Pennsylvania

* * * * *

Endnotes

1.   While Husserl does trace geometry to its roots in the ancient art of land surveying, his philosophical bias towards knowledge rather than practice probably accounts for his lack of interest in real, historical lifeworlds.

2.   Lynn White, Jr., *Medieval Religion and Technology: Collected Essays* (Berkley: University of California Press, 1978), 80.

3.   Lewis Mumford, *Technics and Civilization* (New York: Harcourt Brace

& World, Inc., 1963), 65.

4.   Previously, medieval as well as Roman calendars depicted the months of the year as static personifications holding symbolic attributes. See White, *Medieval Religion and Technology*, 251.

5.   Ibid., 79.

6.   Lynn White, Jr., *Medieval Technology and Social Change* (London: Oxford University Press, 1981), 114-15.

7.   For warfare, see Mumford, 266.

8.   Ibid., 15.

9.   White, *Modern Technology and Social Change*, 125.

10.   White, *Medieval Religion and Technology*, 181-204.

11.   Ibid., ix.

12.   Ibid., 132.

13.   Herbert Butterfield, *The Origins of Modern Science* (New York: The Free Press, 1957), 130.

14.   Nicholas Lobkowicz, *Theory and Practice: History of a Concept from Aristotle to Marx* (Notre Dame: University of Notre Dame Press, 1967), 96.

15.   E.A. Burtt, *The Metaphysical Foundations of Modern Science* (Garden City, N.J.: Anchor Press, 1954), 44.

16.   Butterfield, 104.

17.   R. M. Blake, "Natural Science in the Renaissance," in R. M. Blake, C. J. Ducasse and E. H. Madden, eds., *Theories of Scientific Method: The Renaissance Through the Nineteenth Century* (Seattle: University of Washington Press, 1966), 19.

18.   "But the mathematician, the natural scientist, at best a highly brilliant technician of the method — to which he owes the discoveries which are his only aim — is normally not at all able to carry out such reflections. In his actual sphere of inquiry and discovery he does not know at all that everything these reflections must clarify is even in need of clarification." Edmund Husserl, *The Crisis of European Sciences and Transcendental Philosophy* (Evanston: University of Northwestern Press, 1970), 56-57.

19.   Aron Gurvitsch, *Studies in Phenomenology and Psychology* (Evanston: University of Northwestern Press, 1966), 409.

20.   Husserl, 38.

21.   Ibid., 43.

22.   Ibid., 47.

23.   Ibid., 51.

24.   Here Husserl refers to "Galileo's famous doctrine of the merely subjective character of the specific sense-qualities, which soon afterward was consistently formulated by Hobbes as the doctrine of the subjectivity of all concrete phenomena of sensibly intuitive nature and world in general." Ibid., 54.

25.   Ibid., 52.

26.   Jacques Ellul, *The Technological Society*, trans. John Wilkinson (New York: Vintage Books, 1964), 21.

27.   Quoted in Leo Marx, *The Machine in the Garden: Technology and the Pastoral Ideal in America* (Oxford: Oxford University Press, 1980), 171.

28.   Ibid., 78-79.

Donagan on Cases of Necessity

by Michael J. Malone

Judeo-Christian moralists traditionally maintain that certain acts are impermissible regardless of their consequences. Augustine, for example, commenting on the suicides of Christian women during the horrific sack of Rome, claimed that suicide is always wrong.[1] Traditional morality also absolutely prohibits rape and the killing of innocent human beings. This moral absolutism is a pivotal element that distinguishes it from consequentialism. Philosophers who formulate rational accounts of traditional morality must contend with difficult "cases of necessity."[2] Through these cases, consequentialists profess to show the intuitive implausibility of always prohibiting an action when that action is *necessary* to prevent disasterous consequences.

Alan Donagan in *The Theory of Morality* constructs a deontological explanation of the Hebrew-Christian moral tradition.[3] He also responds to difficult cases of necessity. In his account, Donagan permits the intentional killing of innocent persons in a certain kind of situation. In this essay I argue that Donagan has made a mistake in granting this permission. Among other things, I argue that (i) there is no morally relevant difference between the kind of case Donagan permits and others he considers impermissible and (ii) intentionally killing the innocent in this kind of situation contraverts his fundamental principle and its moral foundation.

I

Cases of Necessity: The fundamental principle of Donagan's theory of morality is the following: "It is impermissible not to respect every human being, oneself or any other, as a rational creature."[4] This principle is grounded in the recognition that rational nature is an end in itself. Murder, as a description of a whole class of actions, always fails to respect another as a rational being, since it annihilates the rational nature of another. Donagan concludes, "It is absolutely impermissible to commit murder."[5] In *The Theory of Morality* he defines murder as "killing the innocent." "Innocent" is understood in the mate-

229

rial sense as "referring to those who are neither attacking other human beings nor have been condemned to death for a crime."[6] In his account, therefore, it is absolutely impermissible to kill materially innocent persons. He writes,

> Common morality as traditionally conceived absolutely forbids, as we shall see, either torturing or murdering the innocent, no matter what may be gained thereby in any possible system of nature.[7]

Donagan relates two cases in which consequences seem to necessitate taking the life of an innocent person. The first is the celebrated case of *R. v. Dudley and Stephens.*[8] During a storm a group of English sailors was forced to abandon their yacht. They were stranded in an open life boat and adrift at sea. After twelve days they had consumed all their food. After passing eight more days without food and six without water, two of the sailors killed a cabin boy who was seventeen or eighteen years old. The two sailors, together with another, then fed upon the boy's body and blood until they were rescued four days later. The boy when killed was lying on the bottom of the boat and was very weak and unable to resist. The two argued that if they had not killed and eaten him, everyone would have expired. They believed the boy, being in the weakest condition, would not have survived the ordeal anyway. Although the killing and subsequent cannibalism were ghastly, they claimed their actions were *necessary* in order to save the greater number of lives.

According to Donagan, the sailors' action was "clearly murder." It was not justified by necessity. The only necessity for killing recognized by the Judeo-Christian moral tradition is the necessity of defending oneself from violent assault. The "unoffending and unresisting" boy obviously was not attacking anybody and so was materially innocent. Since the sailors' action violated the duty never to kill the innocent, it was objectively immoral.

In *The Theory of Morality*, however, Donagan also discusses second-order precepts that deal with personal culpability. In some instances, fear, ignorance, or other factors may reduce the blame or responsibility of agents who perform morally impermissible deeds. Although the sailors' action was impermissible, they may not have been fully culpable. Donagan praises the decision of the judge who convicted the two but who commuted their sentences to just six months imprisonment without hard labor. Donagan claims, "the danger of imminent death, and the extreme fear arising from it, may mitigate blame but cannot exculpate..."[9]

The second case is a fictitious one discussed by Kai Nielson, the "Case of the Innocent Fat Man."[10] In this scenario, a group of unfortunate spelunkers is exploring a cave near the ocean. When high tide occurs, it fills with water. As the tide approaches, the group makes its way to the surface, but a large man who is leading them becomes stuck in the cave's mouth. Since his head is above the water line, he is safe. The others behind him, however, will be drowned. As it so happens, a bystander has with him a stick of dynamite. When it becomes clear that the spelunker cannot be unstuck, a decision must be made: either he is blown up and the others saved, or he is allowed to live and the others perish. It is one life or many.

Nielson states,

> Our Christian absolutist presumably would take the attitude that it is all in God's hands and say that we ought never to blast the fat man out, for it is always wrong to kill the innocent. Must or should a moral man come to that conclusion? I shall argue that he should not.[11]

Nielson argues that the moral status of any action depends upon its consequences, and in this situation, blasting the spelunker from the entrance of the cave produces much better consequences than refraining from blasting. Therefore, someone ought to blast him.

Surprisingly, Donagan agrees with Nielson's conclusion, but not because of its beneficial results. The action is permissible because it is reasonable to assume that prior to risky joint enterprises, such as the exploration of caves near the ocean, each member of the group agreed to a policy in which, if *necessary*, a smaller number of their party would be sacrificed to save a larger number.[12] Each would agree because it is in each's "own probable interest." Donagan writes,

> An agreement of the kind described would have force even if it was tacit; that is, even without saying so, all members of the group take it as accepted by all that they should conduct themselves in accordance with it.... And with most such groups, such agreements will be in force.[13]

The tacit agreement of participants in risky ventures justifies killing some to save more. Because the victims consented to their own deaths before the calamity, killing them to save the others does not fail to respect them as rational creatures.

Although the stuck spelunker is innocent, that is, he isn't attacking anyone but only struggling to free himself, nevertheless, killing him in this situation should not be described as an act of murder. In a subsequent paper, Donagan alters the definition of murder to exclude this kind of situation:

> It follows that the definition of murder should be supplanted: murder is killing another rational being who is materially innocent *and* who has ... not entered into a permissible agreement for a risky joint enterprise by which members may

be killed under circumstances that have turned out to be his.[14]

To demonstrate that it is the victim's prior, tacit consent that justifies his death and not its beneficial consequences, Donagan constructs a variation of the cave scenario. Spelunkers are again trapped in a cave with rising water, and one must be blown up to save the rest. This time, however, two picnickers who are lunching near the cave inadvertantly will be killed by the explosion. Since the deaths of three people are still preferable to the deaths of the many more trapped inside, consequentialists will argue for the explosion. Since, however, the picnickers are not part of the risky cave-exploration enterprise, they could not have participated in any tacit agreement. Therefore, in this case, Donagan concludes it is impermissible to blast the stuck spelunker.

II

Analysis of the Argument: What should one make of the argument from tacit consent in risky enterprises? It allows Donagan to save the greater number of persons without conceding that this beneficial result is its justification. The argument, however, generates inconsistencies and problems for his moral theory.

In comparing the case of the sailors to that of the spelunkers, for instance, why couldn't the sailors also employ this argument to justify their action? Yachting is a risky, joint enterprise. It is just as reasonable to assume before their catastrophe they also tacitly agreed to sacrifice a smaller number for a larger one. The contract is in each's probable self-interest. When the disaster struck, they acted on their agreement. They rationally chose the one with the slimmest chance of surviving. If killing the spelunker is permissible, then killing the sailor should be also.

The introduction of *risk* as a relevant factor in cases of necessity is also dubious.[15] The prior perception of an activity's risk level seems unrelated to the ethics of killing some to save more. Riskiness pertains only to the *likelihood* of a tragedy occuring, not to what is permissible once it has occured. Suppose, for instance, the spelunkers had merely been tourists in an amusement park, exploring the "Mine Shaft of Menace." If a menacing event really did occur and the mine filled with water, would the less risky nature of their activity alter the permissibility of blasting one to save the rest? Pretend the sailors had been cruising on a massive, luxury liner rather than a small yacht. Should the less risky character of this activity affect the moral status of desperate actions performed in a lifeboat? Intuitively, risk seems morally irrelevant.

It is also not clear why the participants of an activity should sanction the principle of killing some to save more only when the activity is especially risky. One could argue with as much plausibility (however much that might be) that rational people, *qua* rational, would tacitly consent to it out of probable self-interest in all their activities. Since life itself is a risky joint venture, it might be in everyone's probable interest to accept such a principle. The enlarged scope of this argument, however, would cover virtually any case of necessity. This, of course, would lead Donagan to a position equivalent to straightforward consequentialism.

The reason one may kill the spelunker to save the others is because the victim *consented* to his own death. This justification, however, becomes intuitively more questionable in other cases that are identical in all morally relevant respects. Consider, for example, a case in which three young people desperately need life-saving organ transplants. There are no transplant organs available, but a woman who is donor-compatible with each volunteers to sacrifice her life so they may receive her heart, liver, and lungs. She asks her physician to painlessly kill her to save the others. She is not ill, depressed, or mentally deranged but acts from benevolent motives. No other compatible organs will be available in time. In this case, her death is necessary to save the three and she *explicitly* consents. Most people, Donagan included, however, probably would not permit it. If killing the spelunker is permissible, however, killing the donor should be also.

Donagan, himself, seems to back away from consent as the permitting condition in another case of necessity. This time a captain threatens to horribly kill twenty innocent Indians unless a stranger agrees to kill one of them.[16] The Indians "beg" the stranger to comply. Donagan concludes, however, that it is impermissible to kill one Indian to save nineteen. He argues that the stranger should refuse because he cannot know his refusal will result in the deaths of the twenty. The captain, after all, might be bluffing. Using the same reasoning, however, one also cannot know that blasting the spelunker will save the others. Perhaps the explosion will only seal the cave entrance. As Terrance McConnell notes, it is difficult to understand why the prior, tacit consent of the spelunkers is sufficient to allow killing an innocent person when the present, explicit consent of the Indians is not.[17]

The tacit consent argument is also not consistent with Donagan's criticism of the rational choice theory of John Rawls.[18] Rawls's two principles of justice possess intuitive justificatory force because everyone in an original position of fairness would agree to them. Everyone agrees because they are in everyone's probable interest. Donagan concurs with Ronald Dworkin[19] that Rawls does not appeal to consent in the original position as the foundation of his moral system but as a "contrivance by which his 'deeper' theory of morality can be applied."[20] Donagan, however, seems overly critical of the contrivance given his own tacit consent argument. He asserts that the consent of rational agents in the original position falls "ludicrously short" of other theories "in not even appearing to justify that system."[21] He asks, "Why should anyone accept a moral system merely because a rational contractor in the original position would accept it?"[22] But then why should anyone accept the morality of killing some to save more during risky ventures because rational contractors prior to those ventures would accept it? The position of the members of the cave exploration party prior to their disaster is analogous to the original position of Rawls's agents in this relevant respect: the spelunkers also agree to their principle in an original position of ignorance. The veil of ignorance is not lifted until the disaster occurs, and they learn their unique status in the cave as either victim or redeemed. Like Rawls's agents, they agree because it is in their probable self-interest. The moral intuition (consent in a position of fairness) that supports both is the same.

By adopting the tacit consent argument, Donagan seems to identify the fundamental principle with respect for the autonomy of others. The spelunker may be blasted because he has (tacitly) consented to his own death. The explosion is not permitted in the modified case because two other victims have not consented. Respect for another as rational seems to mean respecting another's right to determine *the good* for oneself. This might be a viable interpretation of the fundamental principle except that Donagan forcefully rejects it. He argues elsewhere against strong autonomy-based ethics.[23] In *The Theory of Morality*, the duty never to kill the innocent (to extinguish their rational natures) is independent of the wishes of the victim. Donagan's subsequent redefinition of murder, however, allows him to retain the absolute prohibition of murder while permitting the deaths of consenting innocents in cases of necessity. As a result, the new definition appears rather contrived. It also serves to increase confusion about the role of consent and autonomy in his moral theory.

Donagan's fundamental principle is based on a recognition of the incomparable moral worth of beings with rational capacities. By virtue of such capacities, all persons are "ends in themselves." He writes,

> It is as ends in themselves that rational beings find in their own natures a ground for the law they lay down to themselves.... And that status furnishes the content of the fundamental principle of morality.[24]

Donagan considers each rational agent an "ultimate end" for human action.[25] As Thomas Hill notes, for Kant and for Donagan, an "end in itself" is incommensurably valuable.[26] One is not permitted to "trade off the value of one person's thriving as a rational being for more value of the same kind in several persons."[27]

This trade off, however, takes place if the hapless spelunker is intentionally killed to save the others. The value of the spelunker's rational existence is exchanged for the "greater" value of a sum of rational agents. In the exchange, the spelunker is not treated as an end in himself. Instead, he is regarded as an impediment to a better end. Even if the spelunker consents to be treated as such, blasting him is not compatible with Donagan's moral foundation, since the unfortunate spelunker clearly is not the "ultimate end" of the other's action.

Cases of necessity differ from cases of voluntary, active euthanasia in this important regard: the one who is euthanized is still the "ultimate end" of the other's action. The relevant question is whether death really is in her best interest. In cases of necessity, however, where a smaller number of persons must be sacrificed to save a larger one, the victims are never the "final end" of the action. There is no question, for instance, that killing the stuck spelunker is not in his best interest.

In order to keep Donagan faithful to his traditional foundation and to avoid the inconsistencies and problems noted above, the tacit consent argument should be abandoned. The spelunker should be treated in the same manner as the sailor. Blasting the spelunker is morally impermissible. In such a desperate situation, however, the culpability of someone who kills may be greatly diminished.

Intuitively, one of the strengths of traditional morality is its central conviction that the moral worth and dignity of an innocent person should not be sacrificed for some perceived greater goal. Killing the innocent to produce beneficial results constitutes such a sacrifice. Even in war, that most desperate and risky of all human endeavors, traditional morality forbids the intentional killing of non-combatants (innocents). Traditional just war theories regard the atomic bombings of Hiroshima and Nagasaki, for example, as objectively immoral, since they represent the intentional killing of non-combatants on an unprecedented scale. Even if the bombings were necessary to save a greater number of lives, traditional morality cannot condone them without becoming radically inconsistent and unfaithful to its central conviction. Neither can Donagan condone the killing of innocent people engaged in risky joint enterprises without inconsistency and without contraverting his own fundamental principle.

Saint Joseph's College
Rensselaer, Indiana

* * * * *

Endnotes

1. *De Civitate Dei*, I.16-20.
2. For a list of such cases, see Kai Nielson, *Ethics Without God* (Great Britain: Pemberton Books, 1973), 69-75.
3. Alan Donagan, *The Theory of Morality* (Chicago: University of Chicago

Press, 1977).

4.   Donagan (1977), 66.
5.   Donagan (1977), 88.
6.   Donagan (1977), 87.
7.   Donagan (1977), 36.
8.   Donagan (1977), 175-76.
9.   Donagan (1977), 175.
10. Nielson, 69-70.
11. Nielson, 70.
12. Donagan (1977), 178-79.
13. Donagan (1977), 178-79.
14. Alan Donagan, "Comments on Dan Brock and Terence Reynolds" *Ethics* 95 (July 1985): 875.
15. These criticisms regarding risk were suggested by Professor Richard Eggerman of Oklahoma State University in his commentary on an earlier version of this paper which I read at the annual "Mountains-Plains Philosophy Conference" in 1994.
16. Donagan (1977), 207-08. This case was first presented by Bernard Williams in *Utilitarianism: For and Against* (Cambridge: Cambridge University Press, 1973), 98-99. See Terrance C. McConnell's criticism of Donagan's treatment of this case in "Moral Absolutism And The Problem of Hard Cases," *The Journal of Religious Ethics* 9 (Fall 1981): 286-97.
17. Terrance McConnell, 291-92.
18. John Rawls, *A Theory of Justice* (Cambridge: Harvard University Press, 1971).
19. Ronald Dworkin, "The Original Position," *University of Chicago Law Review* 40 (1973): 25-26.
20. Donagan (1977), 224.
21. Donagan (1977), 223.
22. Donagan (1977), 223.
23. See Alan Donagan, "Comments on Dan Brock and Terrence Reynolds," *Ethics* 95 (July 1985): 874-77. Here Donagan argues that his fundamental principle is teleological and that autonomy in the Kantian sense is the "right (which is more fully conceived as the inescapable duty) of each to determine what the practical law requires" (876). In my opinion, this is equivalent to saying that autonomy is "the obligation of each to discover what is right." This is a weak sense of autonomy. In the case of the spelunkers, it is precisely what is right that is at issue. So claiming, as Donagan does, that the spelunker may be killed in a case of necessity because he has consented to his death (it respects his autonomy) begs the question, given his understanding of autonomy.
24. Donagan (1977), 232-33.
25. Donagan (1985), 874.
26. Thomas E. Hill, Jr., "Donagan's Kant," *Ethics* 104 (Oct. 1993): 41-42.
27. Hill, 41-42.

Modal Voluntarism in Descartes's Jesuit Predecessors

by Jeffrey Coombs

Modal voluntarism holds that modalities such as logical necessity, possibility, contingency, and impossibility have their ontological ground in God's power. Since Modal Voluntarism has been ascribed to René Descartes, the question I wish to address is to what extent was his espousal of modal voluntarism new?[1] Was his support for the view unprecedented in the Early Modern period, or was he able to borrow the view from other sources? I will show that although modal voluntarism was by no means a popular view during the sixteenth and early seventeenth centuries, there was at least a fledgling tradition among some Jesuit philosophers to champion the view. Before introducing the views of these Jesuits, I will first introduce the "Second Scholastic" movement to which they belonged. I will then define modal voluntarism and briefly describe the philosophical problem it was meant to resolve. Next, I will present the views of the modal voluntarist Jesuits and discuss the likelihood of their influence on Descartes.

Second Scholasticism was the continuation of Medieval Scholasticism into the modern period. It extends roughly from 1450 to the nineteenth century when it becomes Third Scholasticism. The main characteristic of Scholasticism narrowly construed is that it approaches philosophical problems using Aristotelianism tamed by Christian dogma. Other names for roughly the same school are Early Modern "Latin philosophy," "post-medieval" philosophy, and "late-scholastic philosophy."[2] There exists now a Third Scholasticism which should not automatically be equated with "neo-scholasticism," often associated with the neo-Thomist movement of the late nineteenth and early twentieth centuries. Third scholasticism is thriving because it has rediscovered what Medieval and Second Scholastics knew: Scholastic Philosophy is at its best when it allows debate among its many, many schools, of which the Thomist, Scotist, and Nominalist (Ockhamist) are just the main ones. I would take all of the "philosophical theology" going on these days as evidence of Third Scholasticism's health.

One major philosophical problem which bothered a great many

237

Second Scholastics was: what aspect or element of the world do logically possible entities possess which the impossibles lack? In other words, what is it about me which makes it true to say of me that I am logically possible, but of the round-square or the chimera that they are impossible? Although Second Scholastic philosophers offered many different answers to this question, I will focus only on the modal voluntarist perspective since we are looking for sources of Descartes's views.[3]

Modal voluntarism is the view that modalities such as logical necessity, possibility, contingency, and impossibility have their ontological ground in God's power. The main alternatives to this view in Second Scholasticism were what Simo Knuuttila has called *modal transcendentalism*, which claims that modalities are independent of God, and the view that God's intellect or His Ideas are the primary source of modalities, which, for lack of a better expression, we may call *modal intellectualism*.[4]

Descartes expressed the core idea of modal voluntarism in this way:

> The mathematical truths which you [Mersenne] call eternal have been laid down by God and depend on him no less than the rest of his creatures. (...) Please do not hesitate to assert and proclaim everywhere that it is God who has laid down these laws in nature just as a king lays down the laws in his kingdom.[5]

Later Descartes added that mathematical truths are "therefore something less than, and subject to, the incomprehensible power of God."[6]

Modal voluntarism was often, but probably incorrectly, attributed to Gabriel Biel (c. 1425-1495), and thus to late-medieval Nominalism, by Second Scholastics.[7] In fact, Biel, and Ockham before him, held that God's power only extends to what God can do without contradiction and therefore God's power is restricted by the logical relation of consistency.[8] Allan Wolter pointed out long ago that the "neophyte in philosophy can scarcely avoid receiving a distorted view of Ockham's doctrines if he reads the neo-scholastic textbooks." He supported this claim by pointing out that Scholastic textbooks of the nineteenth and twentieth centuries attributed modal voluntarism to Ockham even though Ockham did not hold the view.[9] I would add that this distortion already appears in Second Scholastic textbooks of the sixteenth and seventeenth centuries.[10]

Nonetheless, even if Medieval Nominalism did not espouse modal voluntarism, the Jesuits Pedro Fonseca (1548-1599) and Hurtado de Mendoza (1592-1651) did. Pedro Fonseca, for example, holds that logical connections between terms are based on God's power. Fonseca attacks none other than Thomas de Vio Cajetan (1468-1534) on this point. Cajetan had said that even if everything in the universe including God were annihilated except for one person, that person's knowledge of a rose would still remain.[11] It would also be true under such a hypothesis that a rose is still rose, and that a rose is odoriferous.[12] Fonseca, however, disagrees, and believes that if the causes, including the ultimate cause, of an entity do not exist, then one could not even claim that the entity was a *possible* creature. If God, the creator of all entities did not exist, neither would the rose be a real being, nor would it even be a rose. In fact, if there were nothing else besides *me*, which means my ultimate cause would not exist either, then I would not even be who I am, nor would the rose be a rose, nor would it be odoriferous.[13]

Fonseca concludes that the possibility of creatures requires the existence of their ultimate efficient cause because he believes that created possibles (as opposed to God, the uncreated possible) are only properly understood as creatures. In other words, a possible human necessarily contains the property of being created and thus being depedendent on God.[14] In this way, Fonseca undercuts Cajetan's original thought experiment. If *per impossibile* God did not exist, would creatures be possible, entities which by definition require a God to exist? Fonseca's answer is "no."

Fonseca also rejects the view that possibles should be reduced to God's intellect, specifically to the divine ideas. For Fonseca, the divine ideas are neither prior to God's creative power, nor to His practical knowledge. The ideas, as exemplars, pertain to God's creative activity, and thus they cannot "be" except insofar as they can be imitated by the divine power. Fonseca goes so far as to claim that if God did not have the power to produce creatures, then God would not be their exemplar.[15]

Another Jesuit, Pedro Hurtado de Mendoza, borrows many arguments from Late Medieval Nominalism to reduce modality to divine power.[16] Hurtado was a theology professor at the University of Salamanca. His notorious discussion of the ontological source of possibility appears in disputation 8 of his *Metaphysica* which comprised part of his *Universa Philosophia*.[17] Disputation 8 is "on essence and existence" and section I asks the question "what does it mean for creatures to be possible?"

Hurtado considers two examples of possible creatures, Peter and an angel, whom he later names "Gabriel." Peter is meant to be a generic human, and because Peter is subject to death, it is possible for him not to exist. Also, because he was born, before he was born he was possible, and his parents had the potency to engender him before he was born. Peter thus serves as an example of an existing possible. Hurtado's angel is a "pure" or "mere" possible which never exists. God now has the potency to create an angel which He does not create, never did create, and never will create. Since God, however, has the potency to create the angel, the angel is possible.[18]

This preamble leads to Hurtado's main conclusion, which would make him the object of attacks by Second Scholastics throughout the seventeenth century. He states: "the term 'possible' is formally derived from God's omnipotence."[19] His argument for this conclusion runs thus. Because I am capable of seeing color, color is visible. This holds because the seeing "in" me is the same as the being seen "in" the color and are, Hurtado claims, derived from the same *"forma."* Thus, something is called possible because it can be made, and one says that it can be made because God has the power to make it.[20]

Hurtado next tells us that the terms 'possible' and 'can be made' belong to entities which are really (*realiter*) distinct from God.[21] This statement may seem strange after Hurtado has told us that the meaning of 'possible' is formally and by definition derived from omnipotence. Hurtado's point is that the expression "a possible entity" refers to some entity distinct from God's omnipotence. It is Hurtado's *angel* which is said to be possible, not God's omnipotence, even though what it means for the angel to be possible is that it can be made by God. Hurtado says

> A potency which is physically able to produce something has some other object towards which it acts and does not act towards itself. ... Therefore, it is capable of making something else distinct from itself which can be made because that which can be made resides in the notion of 'something which can make'. Now, the omnipotence of God is something which can physically make. So, it follows that there is something which can be made which is really distinct [from God's omnipotence].[22]

That which is really distinct from God's omnipotence and which is said to be 'possible' is the essence and existence of the created entity.[23] Hurtado is making a common move among Second Scholastics not only to include the existence of entities among the "make-ables" (*factibiles*) but also the essence.[24] His argument for including the essences in the category of possibles and "make-ables" proceeds in this way. When one conceives of the existing Peter, our generic created entity, one conceives of Peter as an entity which depends on another. That is, God is both the efficient and final cause of Peter and thus Peter depends on God not only for his existence but for his purpose as well.[25] So, as in Fonseca, being a creature and being dependent on God are part of the very nature of a created being.

Hurtado also offers this second argument: consider that there are only two types of things, those whose existence depends on another and those whose existence depends only on themselves. Now the only entity whose existence depends on itself is God. When we think about Peter, we discover that there is no part of him which is self-created or independent of everything else. Therefore, every part of Peter is created, including his essence. So, Peter's essence, as a possible, is something which can be created by God's omnipotence.[26]

Hurtado does respond to what he calls the "common adage" that essences are immortal and cannot possibly be produced. The argument supporting this "common adage" is as follows. Essences belong to entities necessarily, not freely. God makes everything distinct from

Himself freely, and thus contingently and not necessarily. Therefore, God does not create essences.[27]

Hurtado answers this argument by claiming that essences in "objective being" and in potency cannot be produced and are immortal because they are "necessarily possibles" and the objects of God omnipotence. When the essences exist, however, they are produced and "in relation to existence, they are capable of being produced." Furthermore, it is correct *both* to say that God freely creates essences *and* that they necessarily belong to entities. Essences freely belong to entities when they exist because they come about freely from God. On the other hand, one may also say they belong necessarily to entities because if God wants to create a human, He cannot do so by giving the human the essence of a stone. He would in that case create a stone and not a human.[28]

At this point one would think that Hurtado would be committed to the contingency of all statements about creatures. Since possibility is formally derived from God's power, it should follow that what is possible is under God's control and that God has the power to determine what is possible and what is not. Thus, Hurtado should be making the jump into a theory of modality in which the modal properties of entities are not fixed and necessary, but are contingent. Possibles would *not* be necessary, that is, whether an entity is possible or impossible is a contingent matter. In other words, it seems that he should support Descartes's idea that God could change eternal truths.[29]

Hurtado, however, does not accept the contingency of all creaturely truths. For Hurtado, the potential Peter is not a true being. No mere possible is a being *simpliciter*.[30] Peter is only a being *secundum quid*.[31] Nevertheless, although possibles are only beings *secundum quid*, Hurtado makes the rather surprising claim that the *unio* or union of essential predicates with their subjects is eternally a union *simpliciter*. Thus, the essence of human is eternally united with the essence of animal since unions are unions *simpliciter* even between mere possibles. Perplexingly, the essential predicates are not themselves eternal because essences are not *simpliciter* essences if considered only as potential essences.[32] God's power is thus restricted to the extent that "God cannot remove the rational from a human because it belongs to his concept."[33]

Also, Hurtado says that even in potential existence an entity still retains the concept of its possible existence and the concepts of its quiddities. Thus, even in potency an angel is not a horse, but neither is the possible existence of a horse the same as the possible existence of an angel.[34] Of course, Hurtado insists that potential existence is not existence *simpliciter* but only *secundum quid*, but what he seems to embrace is that there is type of conceptual, pseudo-being in which possible existences and essences reside. This conceptual being includes details of all possible beings down to their individuality and ultimate differences.[35] For him, this realm of possible beings is generated by God's power, but this unbounded, infinite power, for some unexplained reason, only supports a consistent structure of possibilities and does not extend to the power possibly to make a round-square. It would not take other Second Scholastics very long to point out that Hurtado had not provided an explanation as to why the infinite power of God could not make impossibles.[36]

Hurtado's position is not without difficulties. He claims that "the essence and existence are not prior in reason (*ratione*) to the denomination from omnipotence, nor is [the denomination from omnipotence] prior to the object denominated."[37] If the phrase 'in reason (*ratione*)' means that the denomination of an essence as possible by omnipotence is not definitionally or logically prior to the essence so denominated, this would conflict with Hurtado's claim that possibility is formally, and thus, definitionally derived from God's power. For what is the status of the existence or essence of a human prior to God's power? The best answer that can be given to this difficulty seems to be that to call a human's essence or existence possible is the same as saying that that human's essence and existence has been denominated possible from omnipotence and vice versa. I suspect he is most interested in making a point about the logical priority of the two. He is echoing Ochkam and Gabriel Biel's claim that they are logically equivalent, or in the language of the day: convertible.[38] He would still have to grant a definitional priority to the denomination from God's power since possibility is formally derived from there.

Hurtado's *Metaphysica* first appeared in 1615 about the same time Descartes was leaving the Jesuit college La Flèche for the Université de Poitiers.[39] Thus, it is not very likely that Descartes was directly influenced by Hurtado's work (unless he read it later, but there is no evidence of this). It is very likely, however, that Fonseca's *Commentariorum in libros metaphysicorum Aristotelis* was the text for Descartes's last year at La Flèche.[40] The first two volumes of Fonseca's *Commentary* were published in Rome in 1577 and elsewhere throughout the next three decades.[41] So, that two very influential Jesuit teachers had argued for modal voluntarism makes it highly probable that these ideas were well discussed at Jesuit Colleges, and it would not be considered so very strange for a Jesuit-trained young man of the period to espouse the view.

Community College of Southern Nevada
  Las Vegas, Nevada

* * * * *

Endnotes

1.   For example, see Jonathan Bennett, "Descartes's Theory of Modality," *The Philosophical Review* 103 (1994): 639. Simo Knuuttila calls the view "constructivistic" in *Modalities in Medieval Philosophy* (London and New York: Routledge, 1993), 148. Margaret J. Osler in *Divine Will and the Mechanical*

Philosophy: Gassendi and Descartes on Contingency and Necessity in the Created World (Cambridge: Cambridge University Press, 1994), 130, raises some doubts about whether Descartes's was truly a voluntarist, but her conclusions on 134-35 seem to indicate that she ultimately thinks that he is.

2. In order: John Deely, *New Beginnings: Early Modern Philosophy and Postmodern Thought.* (Toronto: University of Toronto Press, 1994). E.J. Ashworth, *Language and Logic in the Post-Medieval Period.* (Dordrecht: D. Reidel, 1974). E.J. Ashworth, "Late Scholastic Philosophy: Introduction" in *Vivarium* 33 (May, 1995): 1-8. In Jorge J. E. Gracia, "Hispanic Philosophy: Its Beginning and Golden Age." *Review of Metaphysics* 46 (March, 1993): 475-502, one finds a good introduction to the problems of naming the period as well as further bibliographical information.

3. For an overview of Second Scholastic perspectives on this problem see Jeffrey Coombs, "The Ontological Source of Logical Possibility in Second Scholasticism" in *Modality in the Schools: Studies in the Theory of Modality in Second Scholasticism (1500-1750)*, eds. Jeffrey Coombs and Gino Roncaglia. (Dordrecht: Kluwer, forthcoming).

4. Thomas Morris calls a similar view "modal activism" in Thomas V. Morris, *Anselmian Explorations: Essays in Philosophical Theology.* (Notre Dame, Indiana: University of Notre Dame, 1987), 168. Osler uses the term "intellectualism" in *Divine Will*, 18.

5. Descartes's letter to Mersenne, 15 April 1630 (AT I 145-146; CSM III 23). "CSM III" abbreviates: René Descartes, *The Philosophical Writings of Descartes: Volume III: The Correspondence*, trans. John Cottingham, Robert Stoothoff, Dugald Murdoch, and Anthony Kenny (Cambridge: Cambridge University Press, 1991). "AT" refers to the Adam-Tannery edition of Descartes's *Oeuvres* as discussed in CSM III, ix. Volume number and page numbers follow the AT and CSM.

6. Descartes's letter to Mersenne, 6 May 1630 (AT I 149-50; CSM III 24-5)

7. For example, see Bartholomew Mastrius, *Philosophiae ad mentem Scoti Cursus integer, Vol. 5, Disputationes ad mentem Scoti in duodecim Aristotelis Stagiritae libros Metaphysicorum*, 4th ed. (Venetiis, apud Nicolaum Pezzana, 1727), 25a. Reprinted in the microfilm series "Rare and Out-of-Print Books of the Vatican Film Library" (St. Louis, Missouri) List 84, roll 15. See also, John Poinsot, *Joannis a Sancto Thoma, O.P., Cursus Theologici* (Paris-Mardrid: typis Societatis S. Joannis Evangelistae, 1931-), vol. III, 578b. Scholars still insist that the Nominalists were voluntarists, as in Osler, *Divine Will*, 30. Osler overlooks the passages quoted in Knuuttila, *Modalities*, 146-47, which support Knuuttila's claim that Ockham accepted transcendental modalities.

8. Both Ockham and Biel hold that God's "absolute" power is limited to what God can do without contradiction. For Ockham, see William Ockham, *Quodlibet septem*, ed. Joseph Wey, *Opera Theologica*, vol. IX (St. Bonaventure, NY: Franciscan Institute, 1980), 586. For Biel, see I *sent.* d 43, quoted in Heiko A Oberman, *The Harvest of Medieval Theology: Gabriel Biel and Late Medieval Nominalism* (Durham, NC: The Labyrinth Press, 1983) 37, n. 25. Osler, *Divine Will*, 30-31, concludes that such logical and "analytic" necessities "in no way limit God's absolute power. They do not presuppose an ontology of necessary

connections in the world as Aquinas had." There are many confusions here. The use of the term 'analytic' is anachronistic since logicians of the medieval and early modern period often considered 'humans are animals' to be a *logical* truth whereas now most logicians would not. Ockham, at least, considers 'human are animals' to be necessary and requires an "analysis" as a conditional or *de possibili* proposition to guarantee that necessity in *Summa Logicae*, ed. Philotheus Boehner, O.F.M., Gedeon Gál, and Stephanus Brown (St. Bonaventure, NY: Franciscan Institute, 1974), 512-13). Thus, Ockham must believe in some type of "connections" on which to base the necessity of logical truths. Knuuttila (*Modalities*, 146-47) has made a good case for these connections being "transcendental" and thus not subject to God's will. Surely this would count as a "limit" on God's absolute power since his absolute power cannot create the logically inconsistent. Biel's views are discussed in greater detail in Coombs, "Ontological Source" (forthcoming).

9.   Wolter, Allan B. "Ockham and the Textbooks: On the Origin of Possibility," *Franziskanische Studien* 32 (1950): 70.

10.   Fonseca, however, uses a correct interpretation of the Ockham-Biel argument to support his view that modality is based on divine power in Pedro Fonseca, *Commentariorum in Libros Metaphysicorum Aristotelis Stagirita, Tomi Quatuor* (Coloniae: sumptibus Lazari Zetzneri, 1615; Reprinted in Hildesheim: Georg Olms, 1964), 326F-27B. Given the popularity of Fonseca's commentary on the *Metaphysics* it is likely that Second Scholastics (and thus neo-scholastics) are interpreting Ockham through Fonseca's eyes.

11.   Thomas de Vio Cajetan, *In libros Posteriorum analyticorum Aristotelicos* (Venetiis: 1505), fol. 14(ra). (Reprinted in the microfilm series "Italian Books before 1601," Roll 438). "Nihil enim minus remaneret scientia mea de trianguli passionibus et rosae, etc., si omnia annihilarentur me solo remanente, quam si remanet prima causa aut corpus celeste, etc. Extranea enim haec sunt a scientia iam habita. Nulla enim mutatio ex hiis sequitur in ea." For others who were fond of this argument, see Norman J. Wells, "Javelli and Suárez on the Eternal Truths," *The Modern Schoolman* 72 (1994): 25-26 and 30, n. 77. The argument about the rose comes from Robert Grossteste probably through Capreolus.

12.   Fonseca (*Metaphys*, II, 324E-25B) adds the odoriferousness to the rose as an example of a *passio* of the rose. Cajetan has said (see previous note) that the rose and its *passiones* would exist even in a solitary mind.

13.   Fonseca, *Metaphys.*, II, 325A-B. "si Deus, qui est rerum omnium effector, non existeret in rerum natura, nec rosa utique esse ens reale, ac proinde, nec rosa. (...) [S]ublatis enim coeteris rebus omnibus, neque ego essem, qui sum, nec rosa esset rosa, ut ostensum est, quare nec de illa ostendi posset esse odoriferam, attributumve aliud reale."

14.   Fonseca, *Metaphys.*, II, 326C. "[E]ntia omnia infra Deum ... perfecte autem ac distincte nullo modo concipi possunt, nisi quatenus pendent a Deo, ut a prima causa ef[fi]ciente."

15.   Foncesca, *Metaphys.*, II, 326E-F. "[N]egandum est, ideas divinas praecedere omnino ratione scientiam Dei practicam, aut eius potentiam executivam. Nam, cum illa sunt exemplaria divina ac proinde obiecta ad actionem pertinentia, intelligi utique non possunt nisi quatenus imitabilia a scientia practica,

potentiaque divina exequutiva. (...) [S]i Deus non haberet vim productivam creaturarum, fore etiam ut non esset earum exemplar..."

16. Note the irony of this move. Nominalists as committed logicians tended to support, or at least be committed to, a transcendental view of modality. By the early seventeenth century, however, the prejudice that Nominalists are all modal voluntarists is so engrained that Hurtado uses Nominalist arguments to support a view they did not hold.

17. Pedro Hurtado de Mendoza, *Universa Philosophia* (Ludovici: Prost, Haeredis Roville, 1624). Nova Editio. I have used copy in the microfilm collection "Rare and Out of Print Books in the Vatican Film Library," list 86, no. 27, roll 3.1.

18. Hurtadus, *Metaphys.*, 817a. "Ratio monstrat, Petrum, exempli gratia; quia est morti obnoxius, ita nunc existere, ut potuerit non existere, et quia ortu gaudet, eum fuisse possibilem, et eius parentes antequam illum genuere, habuisse potentiam illum gignendi. Item Deus nunc habet potentiam creandi unum Angelum, quem nec creat, nec creavit, nec creabit, et eiusmodi Angelum posse a Deum creari. Itaque Angelus est creatura possibilis."

19. Hurtado, *Metaphys.*, 827a. "Denominatio possibilis sumitur formaliter ab omnipotentia Dei."

20. Hurtado, *Metaphys.*, 827a. "Per eam rationem, qua ego sum factivus, aliquid est factibile, tum quia per eam rationem per quam sum visivus coloris, color est visibilis, quia videre in me est idem ac videri in colore, est enim ab eadem forma."

21. Hurtado, *Metaphys.*, 827b. "Dico secundo. Haec denominatio possibilis, factibilis cadit in rem distinctam realiter a Deo."

22. Hurtado, *Metaphys.*, 827b. "Et quidem potentia productiva physice est ad aliud, et no ad se ... ergo est factiva alicuius factibilis ab ipsa distincti quia in ratione factivae respicit id quod potest fieri, sed omnipotentia Dei est factiva physice, ergo refertur ad factibile, ab ipsa distinctum."

23. Hurtado, *Metaphys.*, 828a. "Dico tertio. Id quod denominatur possibile et factibile est essentia et existentia verum creatarum."

24. See Norman Wells' Introduction to his translation of Francis Suárez' *On the Essence of Finite Being as Such, On the Existence of That Essence and Their Distinction* (Milwaukee, WI: Marquette University Press, 1983), 8-13, for a discussion of Second Scholastic views concerning created essences.

25. Hurtado, *Metaphys.*, 828a. "In Petro existente nullus est conceptus, quin sit ens ab alio, ergo nullus est conceptus non productus a Deo, et consequenter non producibilis. Consequentia est certa quia ens ab alio est ab illo per veram dependentiam in aliquo genere causae physicae. A Deo autem pendemus ut a fine cuius gratia simus, et ab efficiente."

26. Hurtado, *Metaphys.*, 828a. "Omnes conceptus existentes in Petro sunt vera entia realia. Ens autem reale dividitur in ens a se et ens ab alio, sed in Petro nullus est conceptus entis a se, quia esset Deus, ergo omnis conceptus Petri est entis ab alio."

27. Hurtado, *Metaphys.*, 828a. "Obiicis primo, essentias rerum esse improducibiles et immortales. Existentias vero fluxas, atque caducas, ergo essentiae non sunt factibiles. Antecedens est commune adagium. Item essentiae

rebus conveniunt necessario, non libere. Quidquid autem Deus ad extra producit, producit libere, ergo non producit essentias."

28.   Hurtado, *Metaphys.*, 828a. "Respondeo, essentias in esse obiectivo, et ut sunt in potentia, esse improducibiles et immortales quia necessario sunt possibiles, obiicuntur omnipotentiae Dei. At quando existunt, produci, et in ordine ad existentiam esse producibiles. Ad probationem respondeo, adagium sic intelligi, ad alterum dico, essentias libere fiunt a Deo. Dicuntur autem necessario convenire, quia Deus volens producere hominem, non potest illi tribuere essentiam lapidis, quia non faceret hominem, sed lapidem."

29.   As in the Letter to Mesland, 2 May 1644 (AT IV 118-19; CSM III 235) and to Arnauld, 29 July 1648 (AT V 224, CSM III 358), which indicate that God's power is not restricted to our notions of what is consistent. He is of course talking about possibility from God's perspective and not ours, which is restricted to our notion of consistency.

30.   Hurtado, *Metaphys.*, 829b. "Dico, ens in potentia, non esse simpliciter ens, sed simpliciter non ens; ens autem secundum quid, id est in potentia, sive posse esse."

31.   Hurtado, *Metaphys.*, 830b. "[R]em esse possibilem non facit illam simpliciter talem, ergo eodem modo essentia in potentia, ens, homo, Petrus, non sunt simpliciter essentia, ens, homo, et Petrus, sed tantum secundum quid."

32.   Hurtado, *Metaphys.*, 832a. "[I]llae ab aeterno sunt simpliciter partes, et earum unio ab aeterno est simpliciter unio, cuius essentia est unire, sicut essentia hominis esse animal, ergo partes ab aeterno sunt unitae simpliciter, quia ab aeterno est unire simpliciter, quia unio est unire, et ab aeterno est unio simpliciter. Ergo haec praedicata non sunt simpliciter ab aeterno, quia essentiae non sunt simpliciter essentiae, nec entia dum sunt in potentia, sed entia secundum quid." Hurtado is moving towards a view which I have called the "modal string view" in Coombs, "Ontological Source" (forthcoming). He is also moving closer to modal transcendentalism or intellectualism and away from voluntarism.

33.   Hurtado, *Metaphys.*, 831a. "Deus non potest auferre ab homine rationale, quia est de eius conceptu."

34.   Hurtado, *Metaphys.*, 830b. "Existentia in potentia retinet conceptum existentiae possibilis, ac retinet conceptus quidditativos, quibus in suo esse constituitur, et distinguitur tum ab essentiis aliarum rerum, tum ab aliis existentiis, ut existentia Angeli ab existentia equi."

35.   Hurtado, *Metaphys.*, 831a. "[E]xistentiae hae individuae, et singulares, secundum ultimas differentias, et individuas sunt possibiles."

36.   Some of these are documented in Coombs, "Ontological Source" (forthcoming).

37.   Hurtado, *Metaphys.*, 828b. "Haec essentiae et existentiae non sunt priores ratione, quam denominatio ab omnipotentia, nec haec est prior obiecto denominatio."

38.   See Coombs, "Ontological Source" (forthcoming) for more details.

39.   Daniel Garber, *Descartes' Metaphysical Physics* (Chicago: University of Chicago Press, 1992). 5. For a description of the curruiculum and what is

known about Descartes's education at La Flèche, see also 6-9.

40.   I say "likely" instead of "definitely true" because no documentation as to the texts used exists.   However, Fonseca, however, was a recommended author for the *Ratio Studiorum* for Jesuit education, according to William Shea, *The Magic of Numbers and Motion: The Scientific Career of René Descartes* (Science History Publications, 1991), 4. Of course, this may refer to Fonseca's logic text, which was very popular as well.   Descartes remembers reading the commentaries on Aristotle composed at the University of Coimbra (Descartes, Letter to Mersenne, 30 September 1640, AT III 185 and CSM III 154, and Shea, 5-6). Fonseca's was the commentary on the *Metaphysics*. (I would like to thank Dr. Clarence Bonnen for pointing out Shea's book to me).

41.   Max Wundt, *Die deutsche Schulmeaphysik de 17, Jahrhundert* (Tübingen: J.C.B. Mohr, 1939), xiv.

Christian Identity and Augustine's *Confessions*

by Christopher J. Thompson

Questions concerning the nature of personal identity have plagued Christian philosophers and theologians alike. At issue is how precisely one's "self" endures through the history of our fallenness and, by the grace of God, the conversion from fallenness to redemption. Despite the variety of positions adopted, one thing seems clear throughout the tradition: an identity endures through the vagaries of sin as well as the conversion from sin to redemption. This paper explores Augustine's defense of that enduring identity as it emerges in the Confessions.[1]

An analysis of this nature is particularly challenging since his defense encompasses many forms of inquiry which have, over time, become discreet disciplines of analysis. Yet for Augustine, these aspects comprise an organic whole, one that is animated by the question of human identity: a philosophical consideration of the nature of existence, a theological consideration of the nature of God, a psychological consideration of the nature of human action and thought, and finally, a Christological consideration of the nature of redemption. The *Confessions* forces us to consider, then, how our notions of ourselves are situated within the broader framework of the Christian tradition. Specifically, it forces us to consider how our conceptions of identity are shaped by our revelatory tradition. In other words, the *Confessions* suggests that there may be good grounds for considering a truly Christian moral psychology, for rejecting conceptions of personal identity precisely for their failure to reflect adequately our doctrinal claims. It demands an analysis of personal identity, not for reasons pertaining immediately to us, but to God.[2]

The Manichaean Context of the Confessions: The relationship between Christian revelation and Christian identity is brought to light when one considers the anti-Manichaean tenor of the *Confessions*.

From a reading of the *Confessions* one learns of a vast theological drama, though it is important to note that the religious doctrines of the Manichees are often difficult to clearly determine.[3] Briefly stated, the Manichees hold that at the origin of the cosmos two principles existed—one good (Light), one evil (Darkness). From an attack initiated by the

Evil Principle,[4] good and evil became engaged in a cosmic struggle.[5] As a result of this struggle, the Evil Principle fashions animated bodies, especially human beings, as a means of keeping a part of the Good Principle trapped.[6] Through procreation within species, the good trapped inside living individuals remains in physical bodies. Thus through successive procreation and generation the Evil Principle reigns.

According to the Manichees, the human being is not a substantial unity; rather, it is the amalgam of rival forces. It is liberated or saved only to the extent that it comes to accept the *"gnosis"* of its condition through the revelation of Mani.[7] Salvation is achieved through a rigorous asceticism which serves to release the truer, though passive, part of one's self from the tyranny of the human body.

The world, moreover, did not originate, as orthodox Christians maintained, in the creative act of a Supremely Good Creator, in which all that is created is good. The Good Principle is not in any way the supreme creator of all that exists, nor is it the creator of the human being. It is, rather, an unwilling contender with the Evil Principle and thus is a passive participant in an unfortunate evolution.[8] The order of the world, the coming to be and passing away of creatures, and especially the human being, are not the effects of one, supremely good creator.[9] They are instead the results of a primordial violence between the two original principles.

Finally, and most significantly, the human being does not comprise a responsible agent, nor does human experience reveal an integrated self who is the subject of both good and evil actions. Man is, instead, merely the battleground between cosmic forces.[10] In Book V of the *Confessions* Augustine recalls, for polemical reasons, this Manichaean doctrine. "I still thought [while in Rome] it was not ourselves who sin," Augustine reports, "but that some sort of different nature within us commits the sin."[11] That alien nature within us refers to that particle of primordial evil which in the process of creation becomes mixed with the particle of the divine during the formation of the human being. When we engage in evil actions, the Manichees contend, it is due to the causality of the Evil Principle, and thus such actions are not rightly attributed to what they would identify as one's true self. "I liked to excuse myself," he says, "and to accuse some unidentifiable power which was with me but was not 'I'."[12] The term, "Augustine," then, identifies the conjunction of opposing primordial forces of good and evil. It names the locus of confrontation and does not always name, according to Manichaean beliefs, his true self, his true I.

He eventually comes to see, however, that "I was one whole [totum ego eram]."[13] It is precisely his insistence on this identity that drives the argument of the *Confessions*.

Before one simply dismisses the Manichaean world view as a "comedy of innocence," it should be noted that Augustine's initial interest in the Manichaean cult stemmed from a desire to pursue what was true and was not motivated by a specific interest in abandoning Christianity.

As Richard Lim notes, "The searching Augustine discovered that the Manichaeans offered him what he and many others regarded as a more rigorously rational form of Christianity."[14] Manichaeism persisted as a threat *from within* the Christian community, for they professed to be "true Christians," engaged in evangelizing the world through the newer revelation of Mani, a revelation which made much of Christian doctrine.[15] If Augustine is going to defeat the Manichaean position, it will not be sufficient to wage an external critique. For those convinced of its veracity—and many who did, considered themselves true Christians—Augustine must defeat them on their own so-called "Christian" terms. It is in light of these more specified requirements that Augustine offers his variety of arguments: (a) sometimes illustrating the internal incoherence of their cosmology, (b) sometimes illustrating their inadequate account of what is true of Christian conversion, (c) sometimes illustrating how they fail to be consistent with revelation itself.[16] Yet the nexus of arguments is especially driven by a concern to highlight the notion of an enduring identity throughout the life of sin, as well as from sin to salvation, and it is this emphasis on enduring identity which serves as the focal point of his energies.

a.) The Cosmological Dimensions: A significant facet of Augustine's challenge against the Manichees comes from the neo-Platonic tradition of Plotinus. The notions of the convertibility of being and goodness, of emanation and participation, of the hierarchical nature of being, and of the conception of evil as "privation" all played an essential role in further developing a notion of a Christian identity. The encounter with the neo-Platonic texts sets in motion the beginning of a positive alternative to the Manichaean schema.

Once being and goodness are understood to be convertible, evil can be seen as a privation. The Manichaean cosmology, then, with its claims of a being which is said to be entirely evil, becomes incoherent. Anything which was said to be wholly evil would, on Augustine's newly discovered neo-Platonic terms, not exist. In terms of personal identity, then, the Manichaean claims of an Evil entity which causes undesired action in the human being is demonstrated to be false. Just how evil functions within the human psyche will remain a complex question, but Augustine is certain at least of this much: the descent into wickedness on the part of a free creature does not signal the presence of an actually existing, alien principle within the human being as such.

In addition to the notion of the convertibility of being and goodness, there is the additional doctrine of creation *ex nihilo*.[17] Together these serve as the metaphysical context in which Augustine was to construct further the outlines of his counter-argument. In contrast to the Manichaean creation story, the creature is created not from the substance of a malleable god, but out of nothing by a supremely good, immutable Creator. This notion in conjunction with the doctrine of convertibility forces the problem of sin to be recast as an instance of "woundedness," of "sickness," of the loss of some initial integrity, rather than as a sign of a division within the human psyche itself. In terms of personal identity, the sinful self will comprise a single entity (though wounded), not the composite structure which the Manichees had presented. Progress in one's salvation, then, will entail the further integration of a disordered psyche by means of healing, not simply the final acquiescence of one source of power over and against another.

b.) *The experience of conversion*: The details of his own conversion are given in Book VIII, and with this newer conception of identity on the line, much is at stake in casting the conditions properly. Augustine must account for the struggles within himself in such a way as to convey the seriousness of the situation, yet he must do so in a way that does not allow it to fall prey to Manichaean terms. The experience of indecisiveness cannot betray the psychological dualism of the Manichees.

In the center of his stormy conversion recorded in Book VIII, Augustine insists that, "In my own case, as I deliberated about serving my Lord God which I had long been disposed to do, the self which willed to serve was identical with the self which was unwilling. It was I."[18] As if to clarify any misinterpretations on the part of his readers, Augustine is distracted from the flow of events and turns to his former sect and addresses, once again, their questions. "Let them perish from before your face, O God, ... who detect in the act of deliberation two wills at work, and then assert that in us there are two natures of two minds, one good, the other evil."[19] As indicated earlier, Manichees hold that the competing interests within the self signal the presence of competing substances, competing principles of human action. His account of intense deliberations, they might argue, fully confirms their dualism.

By maintaining such a position, however, Augustine argues that the Manichees, "tear apart the soul *[animus]*."[20] For if there are as many competing causes as there are inclinations, it would appear that the self is the amalgam of much more than the two opposing forces the Manichees identify.[21] This claim that the Manichaean position "tears asunder the soul," introduces another dimension of his counter-attack against the Manichaean schema. The more cosmological, neo-Platonic language of Books V and VII is set aside. Now the reader is asked to consider the effects of Manichaean analysis on his own psychological and moral experience. By way of a *reductio*, Augustine leads the reader to conclude that something is woefully lacking in an account which yields such counter-intuitive results, namely, that my complex motives prove that I am impelled to go where I go by forces beyond control or recognition. By positing as many distinct causes for actions as there are diverse and competing interests, the Manichees destroy the very fabric of moral integrity, the very notion of an "I". To be a human being, Augustine argues, is to be the consistent, enduring subject of one's deliberations and actions. One and the same soul wavers between conflicting interests. The controversy in Augustine's heart is a war within the self, not among "selves" or rival forces.[22]

Yet there is more to Augustine's challenge against the Manichees than simply charging them with an inadequate moral psychology. Rather, Augustine extends the argument to include considerations of the nature of creation and redemption itself. It is in this context especially that one begins to appreciate how our theological presuppositions have served as the foundation upon which we construct notions of identity.

In the midst of his account of his conversion, almost in passing and in a frustrated tone, Augustine insists that the Manichees would become renewed if only they would accept the true doctrines of the Church, "so that your Apostle may say to them, 'For you were heretofore darkness, but now light in the Lord.'[Ephesians 5.8]"[23] Here, as in several other places, Augustine muses upon the implications of this passage from Ephesians.[24] The same claim and attending text of Paul is echoed in *Contra Secundinum*, a piece described by Augustine as one of his best works against his former sect.[25]

This possibility of being at one time darkness but now "light in the Lord" strikes to the very heart of Augustine's defense of identity and highlights the final dimension of his effort. In this passage from Ephesians, Augustine takes the "you" as an enduring subject over time, a single "I" that is progressively brought by the grace of Christ from darkness to light. His appeal to Paul's description of people as at one point "darkness" and now "light in the Lord" suggests that something of one's identity endures through the conversion from sin to salvation. Unlike the "Heaven of Heaven" mentioned in Book XIII, which cleaves to God from the moment of its creation with "no interval of time," human beings undergo "distinct moments of time," in their experience of redemption Augustine says, "since at one stage we were darkness and then were made light."[26] If the shift from sin to redemption entails distinct moments of time, then, something of us must endure over time through the conversion from sin to salvation. That something is the created human soul—good from the point of view of its existence, yet wounded by sin.

This continuity of identity through distinct moments of sin and grace marks an important aspect of Augustinian anthropology. It also marks precisely where he departs from the Manichaeans, for the enduring quality of the subject through the change from being darkness to "light in the Lord," is precisely what the Manichaean schema had eliminated for two reasons. First, Darkness and Light, one may recall, are distinct and diametrically opposing entities. Hence, nothing, according to the Manichees, can change from darkness to "light in the Lord." Second, the Manichees hold that the good portion of the self is a part of the divine, and hence as part of the Good Principle, can never be in a state of darkness, even if changing from darkness to light. Augustine's appeal to Paul, his appeal to conversion, to seek healing by God's grace, is grounded in an altogether different conception of the human being as a creature created and redeemed by a good God.

The Manichees err in their anthropology, Augustine argues, in positing the thesis that the human person is itself a part of that Divine substance: "But they wish to be light not in the Lord but in themselves because they hold that the nature of the soul is what God is."[27] Augustine suggests that if that were to be the case, no entity could be accurately described as having endured from being at one time "darkness but now light in the Lord." Yet this is precisely what Paul states. Hence Paul supplies the critical element of Augustine's attack against the Manichaean schema: an identity which is sustained by participation "in the Lord," who endures from "the beginning" to the eschaton, from darkness to light. The Christians endure from sin to salvation because they are "in the Lord."

c.) Christology as counter-argument: The Lord, of course, is God the creator and redeemer and it is finally in this Christological context that Augustine's account of Christian identity is brought to fruition. It is in the Lord that they participate, in God who "makes" and will "remake" them and give them consolation.[28] For, "How can salvation be obtained except through your hand remaking what you once made?"[29]

The fact that God is one, that the Logos is both source of creation and redemption helps fill in the Augustinian counter-attack in important ways. Consistent with the Church, Augustine insists that it is one and the same God who is both Creator of the human being and Redeemer.[30] It is the doctrine of the creative and redeeming Word which supplies the broadest context in which human beings understand themselves as enduring characters within the journey from sin to salvation.[31] The monotheistic doctrine of God as the Creator and Redeemer facilitates the possibility of a conversion-as-development and not conversion-as-rupture; it is an invitation to become more complete, to be made whole by Christ. Augustine's constant insistence on the unity between the cause of one's existence and the source of one's redemption is one of the most important facets of his remedy to the Manichaean self, as it marks a paradigm shift from the dualistic principles of Manichaean theology.

As a result of this new theological paradigm, Augustine presents Christian conversion as a healing, as being made whole, as an alteration in quality that suggests continuity, and it is precisely this notion which the Manichaean schema had eliminated. In their schema, part of the human being is either of the same substance as the Good Principle, in which case there would be no need for conversion, or of the Evil Principle, in which case there would be no possibility of change.[32] In any event, all the Manichee could hope for was either to return to earth in still another mortal existence (as an Elect); or to be extinguished.[33] The Manichees suggest that to be redeemed from the conditions of one's sin will entail a kind of disintegration of the human being, a substantial change from the state of fallenness to redemption. Strictly speaking, no one undergoes a conversion, only an "illumination," a "gnosis" of one's condition. For Augustine, on the other hand, conversion entails the process of becoming more complete, more whole.

When Christians construct personal identity in the manner that they do, they do so because they believe their "selves" are sustained by one and the same Creator and Redeemer. Their journey from sin to grace, from sickness to healing, is framed in terms of continuity and growth—notwithstanding the psychological adequacy of that a cosmological dualism might offer. The creature participates in the one created order of the one creator as he or she journeys from sin to salvation. The progress from sin to salvation entails a qualitative improvement of one's "disposition" or "weight" in relation to the supreme good, not a substantial rupture of one's identity.[34] Supreme Goodness grounds our Christian existence: it sustains us in our historical, creaturely existence; it reconciles us to Himself through the redemptive action of His Son. This insistence on the unity between the source of being and the source redemption provides the context for his frequent appeals to seek healing from the Word made flesh who is also the Word through whom all things are made good: Christ, the Physician.[35]

From the point of view of moral theology, Augustine offers to those struggling with the life of sin the real hope that Christ can heal *them*, that growth in holiness is a real possibility for them despite the appearance of being "ruled" by an alien principle. No such alien principle exists, Augustine insists, and by dismissing cosmological dualism he clears the way for the pastoral promise of genuine conversion and growth. For the God who made us, can re-make us—into that life for which we are originally created.

In closing, I have no intention of suggesting that the arguments of the *Confessions* solve all of Augustine's problems. The logic of Augustine's insistence on the integrity of the individual in the light of the dynamics of sin and grace continued to plague him throughout his career. In one of his last works of his life, an unfinished piece directed against Julian of Eclanum, he explicitly continues along similar lines.

> The Manichees assert an evil nature ... from which all evil flows; whereas Catholics, which you refuse to be, assert a nature created good, but vitiated by sin, in need of healing by Christ.... Wherefore Manichees think that evil has to be separated from good in such a way that the evil be removed outside; but we, though we separate evil from good intellectually, we do not believe that which is called evil to be any nature (*substantiam*). It is to be removed from those who are liberated from it not in a way such that it will be outside them, but it is to be healed within them so as to cease to be.[36]

Augustine's appropriation of himself as "totum ego" was mediated through the Church's doctrine. I suggest that this doctrine may supply us as well a normative foundation for appropriating a distinctly Christian identity.

University of St. Thomas
St. Paul, Minnesota

* * * * *

Endnotes

1. Special thanks to Joe Koterski, S. J., Dr. John F. Boyle and Amanda Osheim for their helpful comments.
2. See also, Charles Norris Cochrane, *Christianity and Classical Culture: A Study of Thought and Action from Augustus to Augustine*. (Oxford: Oxford University Press, 1940). "For in the Trinity, [Augustine] discovered a principle capable of saving the reason as well as the will, and thus redeeming human personality as a whole." (1974 rpt., 384).
3. See John P. Maher, "Saint Augustine and Manichaean Cosmology," *Augustinian Studies* 10 (1979) 91-104. An extensive discussion of Gnostic religions, including Manichaeism, can be found in Kurt Rudolph's, *Gnosis: The Nature and History of Gnosticism* 2ed, trans. P. W. Coxon, J. J. Kuhn, and R. McLachlan Wilson, (San Francisco: Harper and Row, Publishers, 1984). Also Hans Jonas, *Gnostic Religion: The Message of the Alien God and the Beginnings of Christianity* 2ed. (London: Routledge, 1992); Henri-Charles Puech, *Le manichéisme: son fondateur, sa doctrine* (Paris: Civilisations du Sud [S.A.EP.] 1949); *Sur le manichéisme et autre essais* (Paris: Flammarion, 1979); Prosper Alfaric, *Les écritures manichéenes* (Paris, E. Nourry: 1918); *L'évolution intellectuelle de saint Augustin* (Paris: E. Nourry, 1918); François Decret, *Aspects du manichéisme dans L'Afrique Romaine* (Paris, 1970); and especially, Samuel N. C. Lieu *Manichaeism in the Later Roman Empire and Medieval China* 2 ed. (Tubingen: Mohr, 1992); "Some Themes in Later Roman Anti-manichaean Polemics: 1," *Bulletin of the John Rylands University Library of Manchester* 68 (Spring,1986) 434-72; Part 2, *Bulletin of the John Rylands University of Manchester* 69 (Autumn, 1986) 235-75.
4. The fact that the events were initiated by an attack on the part of the Evil Principle becomes a point of criticism in Augustine's refutation of the Manichees. See *Confessiones* VII.ii (3).
5. See *De natura boni* 40.2.
6. Saklas is the evil demiurge who is employed by the Evil one for this purpose. See *Kephalia* LVI.137-138. Carl Schmidt (ed.), *Kephalia: Part 1 of Manichäische Handschriften der Staatlichen Museen Berlin* (Stuttgart: W. Kohlhammer Verlag, 1940). See also Samuel N. C. Lieu, "Some Themes in Later Roman Anti-Manichaean Polemics: 1," *Bulletin of the John Rylands University Library of Manchester* 68 (1980) 460; also, Peter Brown, *Augustine of Hippo* (Berkeley: University of California Press, 1969), 52.
7. In the Manichaean creation myth Jesus the Splendor, in the form of the serpent, is said to have revealed humanity's true condition to Adam. See Lieu,

Manichaeism 2 ed., 123.

8.   *Confessiones* VII.ii (3).

9.   *Confessiones* XIII.xxx (45). A brief description is provided in the *De natura boni* 46.

10.   In the *Kephalia* LXXXVI, for example, a disciple asks Mani why he sometimes is at peace and at others is disturbed and confused. The answer, Mani argues, lies in his impure diet and astrological causes.

11.   *Confessiones* V.x (18) "Adhuc enim mihi videbatur non esse nos, qui peccamus, sed nescio quam aliam in nobis peccare naturam...." *Corpus Christianorum Series Latina* Vol. 27. (Turnholti: Brepols, 1981) 67. Hereafter *CCL*. Unless otherwise indicated, all English translations are taken from Henry Chadwick (trans.) *St. Augustine: Confessions.* (Oxford: Oxford University Press, 1992).

12.   *Ibid,* "sed excusare me amabam et accusare nescio quid aliud, quod mecum esset et ego non essem." It is not clear how Augustine may have reconciled this position with the apparent confessional formulas articulated at the Bema Festival, of which he was more than likely aware. See Jes Peter Asmussen. *Xuastvanift: Studies in Manichaeism.* Trans. Niels Haislund. (Copenhagen: Prostant apud Munksgaard, 1965). It should also be noted that in casting the issue in the manner he did, Augustine has already begun to tip the scales against the Manichees. For in naming the evil principle as "alien" and "not I" he shifts the paradigm from that of an otherwise evil self with some particle of good trapped inside to that of an otherwise good self with an evil dimension.

13.   *Ibid.,* "Verum autem totum ego eram...."

14.   Richard Lim, "Manichaeans and Public Disputation," *Recherches Augustiniennes* 26 (1992): 251-72.

15.   *Confessiones* III.iv (8); see note in Chadwick translation.

16.   Manichees accepted some "selected" texts of St. Paul and the gospel of Luke.

17.   *Confessiones* XII.xxviii (38); XIII.xxxiii (48); XII.viii (8); XII.vii (7); XII.xxii (31); XII.xvii (25); also, *De Genesi ad litteram* VII.xxviii (43).

18.   *Confessiones* VIII.x (22). "Ego cum deliberabam, ut iam seruirem domino deo meo, sicut diu disposueram, ego eram, qui volabam, ego, qui nolebam; ego eram." *CCL* 27, 127.

19.   *Ibid.,* "Pereant a facie tua, deus, secuti perunt, uaniloqui et mentis seductores, qui cum duas uolantates in deliberando animaduerterint, duas naturas duarum mentium esse adseuerant, unam bonam, alteram malam."

20.   *Ibid.* " ... discerpunt enim animum ..."

21.   *Confessiones* VIII.x (24).

22.   *Confessiones* VIII.xi (27); *CCL* 27, 130: "Ista controuersia in corde meo non nisi de me ipso adversus me ipsum."

23.   *Confessiones* VIII.x (22); *CCL* 27, 127: "Ipsi uere mali sunt, cum ista mala sentiunt, et idem ipsi boni erunt, si uera senserint uerisque consenserint, ut dicat eis apostolus tuus: *fuistis aliquando tenebrae, nunc autem lux in domino.*"

24.   Cf. *Confessiones* VIII.x (22); IX.iv (10); XIII.ii (3); XIII.viii (9); XIII.x

(11); XIII.xii (13); XIII.xiv (15); also indirectly, IV.xv (25).

25.   *Contra Secundinum* XXVI.602: Quod ideo tibi loquor, quia mens tua nec natura mali est, quae omnino nulla est; nec natura Dei, alioquin incommutabilis frustra loquerer: sed quoniam mutata est deserendo Deum, et ipsa mutatio eius malum est; mutetur conversa ad incommutabile bonum, in adiutorio ipsius incommutabilis boni; et talis eius mutatio erit a malo liberatio.... Si autem hanc admonitionem prudenter accipis, conversus ad incommutabilem Deum, mutatione laudabili fies in eis de quibus dicit Apostolus: *fuistis aliquando tenebrae, nunc autem lux in Domino.* Quod nec Dei naturae dici posset, quia nunquam fuit mala et digna isto nomine tenebrarum; nec naturae mali, quae si esset, nunquam posset mutari, nec lux fieri: sed recte ac veraciter hoc dictum est ei naturae quae incommutabilis non est, sed deserto incommutabili lumine, a quo facta est, obtenebratur in se; ad illud autem conversa fit lux, non in se, sed in Domino. Non enim a seipsa, quoniam non est lumen verum, sed illuminata lucet, ab illo de quo dictum est: Erat lumen verum quod illuminat omnem hominem venientem in hunc mundum. *Bibliothèque augustinienne* 17 (Paris: Desclée de Brouwer, 1961), 632.

26.   *Confessiones* XIII.x (11). "Beata creatura, quae non nouit aliud, cum esset ipsa aliud, nisi dono tuo, quod superfertur super omne mutabile, mox ut facta est attolleretur nullo interuallo temporis in ea uocatione, qua dixisti: Fiat lux, et fieret lux. In nobis enim distinguitur tempore, quod tenebrae fuimus et lux efficimur...." In the latter books especially, Augustine will play with the analogy between the "light" of Genesis 1 and the "light" into which Christians have been brought.

27.   *Confessiones* VIII.x (22). *CCL* 27, 127: "Illi enim dum uolunt esse lux non in domino, sed in se ipsis, putando animae naturam hoc esse, quod deus est...."

28.   *Confessiones* V.iii (3): *CCL* 27, 58: "sed tu, domine, qui fecisti, reficis et consolaris eos."

29.   *Confessiones* V.vii (13): *CCL* 27, 64: "Aut quae procuratio salutis praeter manum tuam reficientem quae fecisti?"

30.   See *Confessiones* XI.viii (10).

31.   For a discussion of the relationship between the Word and time see my, "Theological Dimensions of Time in *Confessiones* XI," in Joseph T. Lienhard, S. J., Earl C. Muller, S. J., and Roland J. Teske, S. J. (eds.) *Collectanea Augustiniana: Augustine Presbyter Factus Sum* (New York: Peter Lang, 1993), 187-93.

32.   See *De Moribus Manichaeorum,* XI.22; *Contra Faustum* XIX.24; *De civitate dei* XI.15 .

33.   See *Contra Faustum* II.5.

34.   *Confessiones* XIII.x (11).

35.   *Confessiones* VII.viii (12).

36.   *Contra secundam Juliani responsionem opus imperfectum* V.60. Cited in R. A. Markus, "Augustine's *Confessions* and the Controversy with Julian of Eclanum: Manicheism Revisited," *Augustiniana* 41 (1991): 918.

Clearing A 'Way' for Aquinas: How the Proof from Motion Concludes to God

by David B. Twetten

I

The Problem: Why has a thinker of Aquinas's evident weight left behind him a path to God's existence that is so hard to trace? Even among those who accept the *Summa theologiae*'s five ways as sound, many deny that these ways, in fact, conclude to God. A case in point is the proof from motion, which Aquinas introduces as the first, 'most manifest' way, but which has been dismissed even by many sympathetic readers as the first, most nebulous way. In this century preeminent Thomists who have written on this proof unanimously concede its soundness. Nevertheless, Gilson, Van Steenberghen, Owens, and Pegis each denies that Aristotle's original proof of a prime mover satisfactorily concludes to the creative God of monotheism. Hence, they concur that insofar as Aquinas's first way relies on Aristotle, it fails to prove God's existence. As Gilson says, to prove the existence of the Christian God it is necessary to prove a creator.[1] Van Steenberghen's subsequent critique of the first way is the fullest, developed over the course of some three books and many articles since 1946. According to Van Steenberghen, the first way, like the other four ways, should no longer be blithely repeated by serious defendants of God's existence. For, the proof from motion fails to arrive at a unique, provident, and creative cause of the universe, three attributes essential to the very notion of God.[2]

Now, this critique of the first way is not unique to this century but prevailed even in Thomas's own century because of Henry of Ghent, the most important theologian in Paris during the condemnations of 1277. To paraphrase Henry, there lies a great gap between what religion and what Aristotle teaches on the issue of God.[3] Aristotle's prime mover, says Henry, is a cause of motion only, a first among equals, that is not free but needs the heavens for its perfection. Aristotle's failure, according to him, lay in attempting to prove God *a posteriori*, from created effects. Henry himself, of course, admits Aristotle's physical proof as perfectly sound and introduces it, in the very words of Aquinas, as 'the

259

first and most manifest way from motion'.[4] Because of the proof's failings, however, Henry prefers an *a priori* metaphysical proof from our first concepts, a proof that concludes not from creatures but from God's very essence or quiddity to God's existence.[5] For Henry, then, the only conclusive proof knows not only that God is but also what God is: an utterly simple being whose very existence is identical to his essence.[6]

We may appreciate the importance of Henry by observing that from him John Duns Scotus borrows a similar attitude to the proof from motion. For Scotus, the proof from motion is sound, at least for some motions,[7] but not conclusive, beginning as it does from a contingent, existential premise.[8] Instead, Scotus's preferred proof, like Henry's, is quidditative, concluding from the nature of God as a possible first cause to the actual existence of God.[9] Ultimately, of course, Scotus is not satisfied with the relational knowledge of God as a cause, and he proceeds to prove that God exists through such absolute properties as 'intellectual,' 'volitional,' and 'infinite'—this last being the highest concept of the divine essence attainable in this life.[10]

The Renaissance scholastic Francisco Suárez is influenced by this same tradition. In his view, a physical proof, even if sound, fails to reach an immaterial mover distinct from the heavens.[11] Nevertheless, even proof of an uncreated being would not justify the conclusion 'God exists'.[12] This conclusion, for Suárez, is inferred from the divine quiddity when it is proved that the uncreated being is unique.[13] Only by arriving at a subsistent being that is also unique and therefore the cause of all other beings do we arrive at the God of worship.[14]

My point in linking the Thomist critique of the first way to the Avicennian tradition found variously in Henry and Scotus is not to show that this critique is false but to suggest that its inspiration is other than Aquinas himself. The question for this paper is merely whether Aquinas's proof from motion represents at least a viable approach to God. Even if it is sound, how can it conclude to God if Aristotle's proof fails?

II

Critique of Two Solutions: The 'Physical' and 'Existential' Readings: The two answers to this question much discussed in recent literature on Aquinas rely on either an 'existential' reading of the proof or on a 'physical' reading. In order to point us in another direction, I shall indicate why neither reading, as I see it, provides an adequate answer.

A. First, according to the 'existential' reading proposed by Gilson, Aquinas elevates the arguments of Aristotle's physics to the superior plane of Christian metaphysics.[15] Thus, the proof concludes to a first cause of motion that is the creator of motion, a self-subsistent being.[16] Joseph Owens has explained how Aquinas has thoroughly transformed Aristotle's reasoning, replacing the finite, essentialist concept of motion with the existential judgment of existence.[17] For Owens, the first step of the proof, "something *is* moved,"[18] establishes the only possible middle term of a Thomistic proof: existence.[19] The proof's subsequent reasoning is drawn not from Aristotle but, as for the other four ways, from Aquinas's early treatise *On Being and Essence*:[20] any new existence of motion, since it is (conceptually) distinct from motion's essence, must be caused by something already in the act of existence, and ultimately by an existent identical to its essence, the "I am who am" of Genesis.[21]

The existential reading explains well, then, contrary to Van Steenberghen, how the proof from motion concludes to God. The problem with this reading stems entirely from its lack of textual basis. Owens himself identifies six points absent from the proof that must be read in from Aquinas's existential metaphysics.[22] The first way contains (i) no explicit mention of existence or (ii) of the mental acts that for Owens allow existence to be known as logically distinct from essence; (iii) no identification of motion's 'being reduced into act' with 'being reduced into the act of existence';[23] (iv) no consideration of the priority and accidentality of *existence* necessary to establish that motion's existence requires an efficient cause; (v) no explicit tracing of the actuality of received existence to something that is existence itself; (vi) finally, no express identification of the unmoved prime mover with subsistent existence. Why, then, does Aquinas not dispense altogether with the skeletal structure that is all that is left of Aristotle's proof from motion? Owens agrees that a strictly philosophical proof would exclude motion. The *Summa*, however, is a theological treatise for students well-versed in the traditional philosophy of Aristotle, whereas the existential proof is historically unprecedented and would require a separate monograph for its full development as a proof of God.[24] Owens himself must admit, however, that independent of this monograph, the causal structure of the first way alone does not justify the conclusion to God.

B. The problems with the existential reading led Gilson himself to reject it in the 1960's, a little known fact clear only in the sixth and last edition of *Le Thomisme*, an edition never translated into English.[25] Anton Pegis, adopting Gilson's revised view, defended alternatively a 'physical reading' of the first way. According to a physical reading, the two central steps of the proof from motion should be taken just as found in Aristotle's *Physics*. First, everything in motion has a cause other than itself. Second, that cause *may* be in motion and therefore caused by another, but if every prior cause in an essentially ordered series is in motion, the entire series is left unexplained. There must, therefore, be a first unmoved mover, namely, the immaterial, non-physical prime mover of Aristotle.[26]

Now, as Pegis[27] and Gilson[28] admit, Aristotle's own proof from motion concludes to a prime mover that turns out, in fact, to be an exclusively final cause. Does Aquinas, then, mistake a pagan deity for God? Pegis and Gilson answer that the theologian strategically reports the arguments of philosophical authorities[29] and does not thereby affirm Aristotle's prime mover except in the general sense as an originator of motion.[30] To prevent any doubt about the first way's conclusion, says

Pegis echoing Scotus, Aquinas subsequently introduces in *Summa theologiae* qq. 3-44 his own superior way of knowing God, of knowing him not relationally but absolutely, as he is in himself,[31] the subsistent existent and creator.[32] According to the physical reading, then, the proof from motion concludes to a prime mover understood indistinctly or generically enough to be accepted as God.[33] On one explanation of this reading, the express identification of the prime mover with God is not a philosophical but a theological conclusion, made only on the basis of faith.[34] Thus, the physical reading is unconcerned by Van Steenberghen's or Henry's objection that the God of the monotheistic religions is unrecognizable in the conclusion of the first way.

Why should we not adopt the physical reading of Pegis and others? I have discovered major difficulties for this reading in Aquinas's own interpretation of Aristotle's *Physics*, as Aquinas understands it. In essence, the *Physics* for Aquinas does not simply conclude to an indistinct prime mover. It also affirms, concretely, what turns out to be a 'soul' of the heavenly spheres, a 'soul' that may be the first unmoved mover proved in physics. Surprisingly, this doctrine of celestial animation, attacked in the condemnations of 1277 by Henry and others,[35] Aquinas himself either adopts or at least considers completely consistent with and only verbally different from his own personal view.[36]

How does Aquinas discover this doctrine in the *Physics*? I shall list the four main elements of Aquinas's interpretation without defending them here. First, Aquinas takes seriously an argument in the middle of *Physics* 8 that the first of all moved movers must be self-moved.[37] Consequently, for him, the *Physics* reaches two distinct conclusions: that there is a first unmoved mover, and also that there is a first self-mover.[38]

Second, the *Physics* in Thomas's eyes proves that this first self-mover is a composite whole and is moved by one part that is itself not moved even accidentally.[39] This moving part, since it is absolutely unmoved, cannot inhere in its body as an immanent form that actualizes matter, unlike the soul of a self-moving animal. Consequently, although this celestial unmoved mover is part of a self-moved whole, it nevertheless is separate from its body and immaterial: a 'soul' not informing but transcending its body, the heavenly sphere. Aquinas borrows this interpretation from Averroes,[40] yet it is consonant with his personal understanding of astrophysics according to which the heavens are moved by angels 'intrinsically' united to the spheres as their mover in one 'self-moving' whole.[41]

Third, for Aquinas, unlike for his teacher Albert, the *Physics*' prime mover does not have to be identified with this celestial soul. For Thomas, the *Physics*' ultimate conclusion is drawn disjunctively: there must be a first mover that either is part of a self-moved whole or is completely separate.[42] Only for this reason can the epilogue of Aquinas's *Physics* commentary end with a prayer that quotes Paul: thus ends Aristotle's *Physics*, says Thomas, with the first principle of all nature, who is God beyond all things, forever blesed, Amen.[43]

Nevertheless, fourth, only Aristotle's *Metaphysics* properly concludes to God. This element indicates why this epilogue should not be regarded as drawing a formal conclusion from the *Physics*. In the *Contra gentiles'* account of Aristotle's arguments circa 1259, Aquinas states this element as follows: "But because God is not a part of some self-mover, Aristotle in his *Metaphysics*, starting from this mover that is part of a self-mover, searches out further another, *entirely* separate mover, who is God."[44] In fact, I maintain, precisely because Aquinas sees the *Physics* as failing to conclude definitively to God, he develops an interpretation of Aristotle, without precedent in the commentary tradition, on which book 12 of the *Metaphysics* continues and completes the account of the prime mover begun in *Physics* 8. Thus, according to a neglected passage of Aquinas's *Metaphysics* commentary, Aristotle reasons in book 12, chapter 7, to a celestial cause that is beyond a celestial soul, and that can only be the first being that is pure act.[45]

As a result of these four elements, I argue, Aristotle's *Physics* does not expressly arrive at God for Aquinas, but at best only indicates the manner of reasoning by which the *Metaphysics* alone properly and expressly concludes to a first mover that can only be the first being or God. Aristotle's *Physics* proves a first mover that *may* merely be the celestial soul also proved there—or that may be a completely separate intellect or even God. But at the point at which we come to know that the immediate first cause of physical motion that is discussed in the *Physics* is a subordinate intellect that is itself caused by God, as Aquinas himself will hold, would we not look back on the *Physics'* proof and admit that the proof *by itself* did not conclude to God? In any case, such an indistinct first mover as the *Physics'* can scarcely be identified by Aquinas at the outset of his *Summa* with the very subject matter of the science of theology, namely, with God. Nor can Aquinas justifiably take the *Physics'* proof as the 'first and most manifest' way of concluding to a being that, as he says, 'all understand to be God' (ST 1.2.3). Can Aquinas have forgotten that Albert himself, to say nothing of Avicenna, has denied that the *Physics'* prime mover can be God? or, that Aristotle and Averroes themselves affirm separate intellects superior to the celestial souls proved in physics?

For these reasons, I argue that the physical reading of the first way, like the existential, is unsatisfactory. Must we, then, agree with Van Steenberghen that the proof from motion fails? Here I shall defend a third solution, a 'metaphysical' reading of the first way.[46] First, however, it is necessary to see how Aquinas judges what a successful proof of God's existence will be. In my view, Aquinas's criteria are less stringent than those of his critics going back to Henry. In formulating his proofs, Aquinas simply follows the scientific method of Aristotle as he interprets it. We shall first examine this method's logic of existential arguments; then we will be able to see how Aquinas develops his nominal definition of 'God.'

III

Aquinas on What Constitutes a Proof of 'God': A. The Logic of Existential Arguments. According to Aquinas, God is not something that presents himself to our senses in the way that the whiteness on which these words are printed presents itself. Nor is God a form in some way immediately recognized in nature by the intellect, like dog. Thus, God, whatever he is, is not immediately evident to us but must be proved to exist if he is to be known at all apart from faith. In order to address this situation, Aquinas applies two rules that he derives from *Posterior Analytics* 2.1 based on the distinction there between the questions whether x is and what x is. First, no one genuinely knows the whatness or essence of some x without first knowing whether x exists.[47] Second, no one genuinely knows that x exists without first knowing what x is.[48] How does Aquinas avoid the circularity apparent in following both of these rules?

In answer, Aquinas's Aristotle distinguishes between two kinds of existential argument. The first proves the simple fact that x exists (*quia*); the second subsequently demonstrates why x exists (*propter quid*).[49] Thus, each of Aquinas's rules is satisfied. Without yet knowing x's essence, one may nevertheless prove the simple fact of x's existence by using as middle term some nominal definition of x. Later, having discovered the essence of x, one may demonstrate *why* x exists using the essence itself as a middle term. The seismologist, for example, proves the fact that a mild earthquake exists from the fact that a meter is fluttering. The geologist then demonstrates why it exists from the fact that tectonic plates are actually shifted.[50]

Aquinas makes one further distinction relevant here regarding the two kinds of existential argument: in each the term 'exists' in the conclusion has a different meaning.[51] In the *quia* proof, the 'exists' refers to the third sense of being distinguished by Aristotle in *Metaphysics* 5.7: the 'is' that signifies the truth of a proposition. Thus, 'it is true that' there 'is' a nominal earthquake merely if it is true that a meter is fluttering. The *propter quid* demonstration, by contrast, predicates 'exists' in Aristotle's fourth sense: 'is' as a substantial predicate signifying an essence's actually existing. Thus, an earthquake actually is or exists because the earth's plates are actually shifting.

Now, when Aquinas applies this logical doctrine on existential arguments to the case of God, he insists that the second kind of argument is impossible: a *propter quid* demonstration why God is. Humans, he argues, cannot conceive the essence or true definition of God required for the middle term of this strict demonstration. For, the divine essence surpasses the grasp of our intellect, because all our concepts are drawn from sensible, material things.[52] Nevertheless, for Thomas, since an effect reveals the proper cause on which it depends, by knowing effects we can know the existence of their first cause through a *quia* proof.[53]

Accordingly, Aquinas draws two consequences regarding the nature of a proof for God's existence. First, the *quia* proof that God exists proceeds from a middle term based on effects.[54] As Aquinas spells out this doctrine in the *Summa theologiae*, the proof of God's existence takes as its middle term a nominal definition of God based on effects.[55] Second, the existence of God known in this proof is only that of the truth of a proposition and is not God's own act of existing.[56] For, God's own act of existing must be as unknowable to us as God's essence, since, as Aquinas subsequently proves, the two must be identical in God.

Observe what a different path Aquinas takes through his logic of existential arguments from the path of Henry and Scotus. Recall that for them, God's existence is arrived at through our knowledge of God's essence or quiddity. Otherwise, the proof of God's existence will not reach God as he is in himself. Thus, for Henry and Scotus, the question whether God exists is answered only through the question what God is.[57] Existence is deduced from God's nature. Scotus even goes so far as to dismiss as irrelevant to the proof of God Aquinas's distinction between propositional existence and God's actual existence. Scotus will settle for nothing less than a demonstration of existence as it is found in the infinite God himself.[58]

B. The Nominal Definition of 'God'. We have now seen why for Aquinas a proof of God's existence is only possible through a nominal definition of God based on effects. The question now becomes: which nominal definition will satisfy Aquinas that he has concluded to the true God? The answer to this question will help us to evaluate his first and most manifest way. For, Aquinas answers this question differently from contemporary Thomists. He does not presuppose that his proof must reach any particular notion such as the creator, the unique, infinite, or self-subsistent being.[59] He does not begin with an extended discussion of which concept of God would be satisfactory given the use of the word 'God' in current culture.[60] He does not base his concept of God on revelation, although, admittedly, he has accepted the existence of God on faith from the outset of the *Summa* of theology. What, then, is Aquinas's nominal definition of God?

In question thirteen Aquinas explains what he means by the word 'God'. Unlike other names predicated of the same being, this name, he says, signifies this being's nature or essence.[61] 'God,' in other words, means the kind of thing that this being is, as distinct from all other kinds of things. But, again, we do not know God's kind or essence 'as it is in him'.[62] We only know it through effects. Hence, concludes Aquinas, the word 'God' means 'the nature of something that is above all things, the principle of all things, and removed from all things'. For, these are the three ways of knowing God from effects, which Thomas adopts from pseudo-Dionysius: the ways of eminence, causality, and negation.

A proof of God's existence, then, for Aquinas amounts to a search for something that is above all or the principle of all or removed from all.

In fact, in his earliest catalogue of proofs for God's existence, Aquinas classifies the four arguments of Peter Lombard precisely through the three dionysian ways.[63] Once we have arrived at what is the cause of all, what is perfect beyond all, or unlike all, we have arrived at the kind of thing that all name 'God.' Notice four points. First, this nominal definition of God is properly philosophical and does not rely on faith. Second, the definition does not depend on our having already proved God, so that the approach is circular.[64] We can formulate the concept of what is the cause of all or beyond all without having proved the existence of such a thing. Furthermore, we can argue that such a thing *can* be proved before it has been proved, if we know that things are effected, that they require a cause. Third, Aquinas's nominal definition is not prescriptive but rather allows for many different conceptions of God. Thus, when treating the question of God's existence *ex professo*, Aquinas offers a plurality of proofs of God. Even if some effects are more indicative of their cause than others,[65] many suffice, and Aquinas's *corpus* contains at least ten different proofs that proceed to God under various names.[66]

At the same time, observe in the fourth place that not any divine name for Aquinas will suffice as a conception of the divine nature and therefore as a terminus of a proof. God is goodness, but we do not conclude to God by proving that goodness exists. God is a mover or a cause; but not every moving cause is God. In this regard, Cajetan rightly points out that the five ways conclude to proper predicates of God.[67] A proof's nominal definition, for Thomas, must be proper to God; that is, it must belong to God and only to God—it must be convertible with 'God'. How do we know this fact? Otherwise, the definition will not signify the divine essence or nature. Consequently, the resulting proof will not conclude to what can only be God.

Take an example. In order to investigate whether any Martian exists, an Aristotelian starts with a nominal definition like 'intelligent life on Mars.' Suppose that traces of intelligent life, namely, primitive shelters, are discovered. Yet, further research discloses records of secret Russian visits during the cold war. One is not entitled to conclude that a Martian existed because there was intelligent life on Mars. Instead, one must discard this nominal definition on the grounds that it is not proper to Martians. A proof that concludes to a mover that need not be God is not an adequate proof of God.

Only names proper to God, then, can serve as a proof's nominal definition.[68] Nevertheless, not any name proper to God will suffice. The nominal definition cannot be a merely absolute term but must include some relation to effects. For, again, Thomas argues that God is only known through effects. In this regard, an objection of Suárez is on target. The names 'most perfect being' or 'most noble being,' he argues, do not sufficiently reveal the concept 'God,' even though these are proper to God.[69] For, a materialist may acknowledge humans as the most noble beings and may still deny God's existence. Instead, a potential nominal definition must relate a property to effects: for example, 'most perfect being by which all others are perfect'. Precisely insofar as the materialist does not see this relation in humans, he or she does not call them 'God'.

In summary, Aquinas's account of what counts as a proof of God's existence will differ greatly from what I call the 'prescriptive approach' found alike in the Latin Avicennian tradition and in contemporary scholarship on Aquinas. I refer to an approach that first determines a single name that must be the quidditative target of a proof of the unique God of monotheistic religion. Thus, even Suárez admits that the questions what God is and whether God exists cannot be separated.[70] For, he argues, it is not enough to conclude to existence unless this existence is known in the way that it constitutes the divine quiddity: self-subsistent, unique, creative existence.

By contrast, according to Aquinas's account of existential arguments based on the *Analytics*, a proof of God begins only with a nominal definition of God, which cannot contain the divine quiddity and need not contain even what for us will ultimately stand for the divine quiddity, namely, subsistent existence. Rather, the proof begins with any name proper to God and only God by which God is related to effects as their cause. One cannot adequately criticize Aquinas's ways to God without criticizing the methodology on which they entirely depend. Yet, to define natures only after establishing existence is, arguably, proper procedure within a realist methodology.

IV

Defense of a Third Solution: the 'Metaphysical' Reading: Aquinas's proof from motion concludes to a first unmoved mover. The question now is, can this concept 'first unmoved mover' serve as an adequate nominal definition of 'God'? Clearly, the name 'mover' expresses a relationship to effects, namely, to be the cause of motion. Yet, does *first* mover' or *'first* unmoved mover' further name what is proper to God and only God? Here Cajetan failed to see that on the physical reading of the proof this name is not proper to God. For him it does not matter whether or not the prime mover is a celestial soul in order that the proof conclude to God.[71] Nonetheless, if the first unmoved mover that is proved can be a being that is itself caused by God, then 'first unmoved mover' is not a name that signifies the divine nature as the cause of all; that is, 'prime mover' cannot be a nominal definition of the term 'God,' since it is not convertible with 'God'. Thus, our examination of the logic of a proof of God points up the weakness, again, of the physical reading of the first way.

As an alternative to Cajetan, Báñez proposed to take 'motion' in the first way not only in a physical but also in a 'metaphysical' sense, a sense referring to spiritual and 'final' motions.[72] Here I shall first present the subsequent 'metaphysical' reading of the first way. Then I shall offer new evidence for this reading and defend it against a major objection.

According to the 'metaphysical' reading of the first way, the terms that ground the proof, namely, 'motion,' 'act,' and 'potency,' are to be understood in the broadest possible sense, a sense proper to metaphysics, the first and most universal science. For, act and potency divide being in general, the subject of metaphysics.[73] Furthermore, I argue, wherever that which is in potency is said to be actualized, the metaphysician can speak analogously of a 'motion' or 'change'. 'Motion' in this sense includes all physical motions but extends, as well, to non-physical changes, such as to new acts of intellect and will, which Aquinas frequently calls 'motions'.[74] When 'motion' is taken in this universal sense, then, the subsequent proof will conclude to a first mover that is unmoved in the sense of having no potency for further actuality. This mover, I maintain, can only be God.

What evidence is there for this reading in the text of Summa theologiae 1.2.3? First, in response to the second objection there, Aquinas himself speaks of changes in the mutable human will as requiring a prior immobile cause because of the very reasoning of the first way. Second, the first way's reasoning itself reveals why it covers even such non-physical motions. For, the reasoning grounds motion's need for a cause not in the properties of physical bodies but in the universal notions of act and potency. The reasoning's basic causal premise is that all potency is brought into act only by something in act. Given this premise, the first way proceeds to prove that everything in motion needs something other than itself for its cause (the second step previously mentioned).[75]

For Aquinas this second step is proved simply by disproving the sole alternative, namely, the case of strict self-caused motion. Again, the proof works through potency and act. Aquinas reasons that strict self-motion entails an impossibility: that something is simultaneously in potency and in act. First, he argues, what undergoes motion is in potency toward some goal. Second, what causes motion is in act. For, what causes motion is defined as what brings potency into act, which, as we have seen in the causal premise, must be something in act. Therefore, whatever both causes and undergoes motion simultaneously and in the same respect must be simultaneously both in act and in potency—which is impossible.

My point, then, is that the first way is founded, both in the causal premise and in the refutation of strict self-motion, on the unrestricted terms 'act' and 'potency'. Accordingly, the entire proof extends to 'motion' in the general sense of any 'reduction' from potency into act. It follows that the proof concludes to a mover 'unmoved' in the sense of 'not further reduced or reducible from potency into act'. Otherwise, the mover arrived at will not be first but must have some cause prior to itself. The first way, in other words, concludes to a first 'irreducible reducer'. Such a mover could not even be an immaterial angel, which undergoes successive operations of intellect and will.

Strong confirmation that the first way's reasoning operates not merely on the physical plane stems from the unappreciated fact that Aristotle's *Physics*, in proving the same second step, uses neither the same causal premise nor the same refutation of self-motion as Aquinas. In fact, the best argument for this step in Aristotle depends on properties unique to corporeal things (in *Physics* 7.1).[76] Aquinas, admittedly, does borrow his refutation of self-motion from a proof used in *Physics* 8.5 for a different purpose. Nevertheless, Aquinas alters Aristotle's reasoning so that it turns not on the simultaneous presence and absence of a specific form, like heat, but precisely on the general opposition of potency and act.[77] Accordingly, Aquinas alone appeals to the axiom that nothing is simultaneously and in the same respect both in potency and in act. Aquinas has tailored Aristotle's proof, I argue, precisely so that it will establish the causality of motion in an absolutely general sense.

Still, one may object, how do we know that this result is not merely an unintended consequence of Aquinas's reasoning? The earlier parallel passage in *Contra gentiles* I.13 reports the many steps of Aristotle's *Physics* beyond the first three of the *Summa theologiae*. At the end, however, it admits, as we have seen, that all these conclude merely to a celestial soul and not to God.[78] Does it not follow, as Scott MacDonald has recently argued, that the *Summa theologiae*'s first way presents merely an abbreviated version of Aquinas's complete argument from motion, one that is therefore 'parasitic' on the subsequent third way in order to conclude all the way to God?[79]

Aquinas's own revisions to chapter 13 in his autograph of the *Contra gentiles* provide previously unnoticed evidence to the contrary. As the autograph reveals, when Aquinas originally summarized twelve arguments from Aristotle's *Physics*, he saw them as constituting five different proofs of God's existence.[80] While composing the last of these proofs, however, he realized not only that it concludes to nothing more than a sphere-soul, as we have seen, but also that the same fact applies to all but the first proof.[81] Hence, he combined the last four into one continuous proof, as in Aristotle's original, and added an argument from the *Metaphysics* that concludes to a separate mover that is God. Surprisingly, however, Thomas continued to see the first proof as arriving at God even without the supplemental detour through Aristotle's *Metaphysics*.

I submit the following explanation why. Although the *Contra gentiles*' first proof contained only the same steps familiar to us from the *Summa theologiae*'s proof, at least one of the three arguments given for each step used reasoning not limited to physical bodies. Later, when Thomas looked back over these same arguments in composing his simpler *Summa theologiae*, it is no accident that he chose for his first way there merely the arguments that could be taken universally, as applying to all 'reductions' from potency to act, and that therefore could dispense with the metaphysical supplement that is required for merely physical proofs.

In short, then, the revisions of the *Contra gentiles* give decisive confirmation that 'motion' in Aquinas's first way is to be understood metaphysically. Still, the metaphysical reading must meet a major objection: does the first way when read thus conclude *exclusively* to God? Why cannot the reasoning there lead to an everlasting first mover that is absolutely immobile even with respect to spiritual changes of intellect and will, whose existence is nevertheless perpetually emanated from God? Avicenna envisions precisely such a subordinate intelligence as the prime mover and consequently rejects the Aristotelian proof of God.[82] Furthermore, even Aquinas will concede the possibility of an everlasting intelligence as a secondary cause.

Although nothing in Báñez accounts for this objection, I argue that it can be resolved on the metaphysical reading. Aquinas reasons in the *Summa theologiae* that anything whose existence is caused cannot have actions that are identical to its very substance.[83] For, in such a thing substance must be in potency to the existence it receives from its cause. But nothing that is partly in potency can be identical to its action, because to act is the very opposite of to be in potency. Thus, only what is pure act can be its own action. By contrast, something in potency acts only insofar as it is in act.[84] Now, if a thing's action is other than its substance, the action must be an accident, and therefore the act of that which is in the same genus, an accidental potency or power.[85] Thus, Aquinas argues that even angels have powers of intellect and will really distinct from their substance.[86] These powers are 'reduced' from potency to act when an angel acts. Such a 'motion,' like the first natural motions of the human intellect and will, must be caused ultimately by the same being that caused an angel to be: the first unmoved mover.[87]

Let us suppose, then, that there exists as a mover an everlasting Avicennian intelligence below the first cause. Nevertheless, for Aquinas, it cannot be an absolutely unmoved mover but must have actions that are 'reduced' from potency to act by a prior mover. Aquinas proves that every creature must be moved.88 My point is not that this proof is part of the first way or an essential supplement, but rather that in arriving at a first absolutely unmoved mover, Aquinas arrives at what in his philosophy can only be the first being beyond all others.

According to the metaphysical reading of the first way here defended, then, Aquinas's proof begins not on the 'existential plane' but rather with sensible change understood most generally as a 'reduction' from potency to act. The proof is most evident because change most reveals the character of an effect. For, change reveals a potency's going into act. But no thing as in potency goes into act or acts on its own, since what is in potency *as such* does not even exist.[89] For Thomas, only what is in act can act. Therefore, potency's 'reduction' into act requires something else in act as its reducing cause. If that 'reducer' is itself reduced from potency to act, however, the series must nevertheless end in some first cause unreduced and irreducible into further act.

Aquinas's account of the nominal definition explains why he can now identify such a first mover as God. A first physically unmoved mover, I have argued, need not be identified with the divine nature beyond all other natures. A 'metaphysically unmoved' mover, however, for Thomas, cannot be a caused being, even an everlasting intelligence that moves the spheres, but must refer to the supreme nature alone. The first way's explicit metaphysics of potency and act suffices to conclude to what can only be the first being without appealing to the *De ente*'s existential reasoning. At the same time, nothing in the proof from motion expressly excludes there being many such first beings. Nor is it expressly known that the first mover is self-subsistent, creative, or even purely in act. Such attributes—even that there is only one God—need not be proved in order to conclude to 'what all name God'. For Aquinas, any name that exclusively designates the divine nature from effects will satisfactorily serve in answering the question whether God exists. All other properties beyond that name or names belong equally to the question what God is.

Aquinas, then, does not follow a 'prescriptive approach,' determining in advance some special attribute of the deity as the target of a proof. This approach in contemporary Thomism, I maintain, is inspired more by Suárez and the spirit of 1277 than by Aquinas's peculiar appropriation of Aristotelianism. For Aquinas at least, Aristotle's proof from motion truly concludes to God, not in the *Physics*, but in the *Metaphysics*. Aquinas's first way, as a comparison with Aristotle's original and with the *Contra gentiles*' revisions reveals, is a novel reformulation of the same metaphysical reasoning that Aquinas discovers in Aristotle.[90] The proof is founded on motion that is most universally conceived as the 'reduction' of any potency into act. As a result, it concludes to a first cause that lacks all potency for further act, a name that, for Aquinas, adequately defines God alone. If we grant Aquinas's conception of motion and his reading of aristotelian methodology, this proof appears to be a possible way, although not the only way, of arriving at God.[91]

Marquette University
Milwaukee, Wisconsin

* * * * *

Endnotes

1. *The Spirit of Mediaeval Philosophy*, trans. A.H.C. Downes (New York, 1936), 72-73.

2. *Le Problème de l'existence de Dieu dans les écrits de s. Thomas d'Aquin* (Louvain-la-Neuve, 1980), 287-96. See also "Le Problème philosophique de l'existence de Dieu," *Revue philosophique de Louvain* 45 (1947): 5-20, 141-68,

301-13; *Ontologie*, 4th ed. (Louvain, 1966); *Dieu caché* (Louvain-Paris, 1961).

3. *Summa quaestionum ordinariarum theologiae* 25.3 ad 1 (Paris, 1520), f. 166V, quoted on p. 158 in Anton Pegis, "Henry of Ghent and the New Way to God," *Mediaeval Studies* 30 (1968): 226-47; 33 (1971): 158-79.

4. *Summa* 22.4, f. 132-33M.

5. Ibid., 22.5, f. 134CD. Cf. Pegis, "Henry," 241-42.

6. *Summa* 22.5, f. 135E.

7. See Roy Effler, "Duns Scotus and the Physical Approach to the Existence of God," in *John Duns Scotus, 1265-1965*, eds. J. K. Ryan and B. Bonasea (Washington, D.C., 1965), 171-90.

8. See *Tractatus de primo principio* 3.1, in Allan B. Wolter, *John Duns Scotus: A Treatise on God as First Principle*, 2nd ed. (Chicago, 1983).

9. *Ordinatio* 1.2.1, in *Opera omnia* (Vatican City, 1950-), vol. 2; cf. I.3.1 (vol. 3), p. 6.

10. Ibid., 214-15.

11. Francisco Suárez, *Disputationes metaphysicae* 29.1, n. 7, in *Opera omnia* (Paris, 1856-1877), vol. 26.

12. Ibid., 29.2, n. 1-2.

13. Ibid., n. 4-5; 29.3, n. 2.

14. Ibid., 29.2, n. 2, 5.

15. *Le Thomisme: Introduction à la philosophie de saint Thomas d'Aquin*, 5th ed. (Paris, 1944), 114, 119. Cf. *Spirit*, 442 n. 13. For other existential readings, see John Knasas, *The Preface to Thomistic Metaphysics: A Contribution to the Neo-Thomist Debate on the Start of Metaphysics* (New York, 1990), 126, 141-46, 156-58; "*Ad Mentem Thomae* Does Natural Philosophy Prove God?" *Divus Thomas* (Piac.) 91 (1988): 408-25; Gerard Smith, *Natural Theology: Metaphysics II* (New York, 1951), 88, 108-12.

16. *Le Thomisme*, 5th ed., 119-20. See *Spirit*, 76, 80, 81 n. 13.

17. *St. Thomas Aquinas on the Existence of God: Collected Papers of Joseph Owens, C.Ss.R.*, ed. J. Catan (Albany, 1980), 163; cf. p. 200.

18. Ibid., 162-63, 173-77, 190-91.

19. According to *An Elementary Christian Metaphysics* (Milwaukee, 1963), 349-51, a strict demonstration can be founded on only the one, proper cause; and, the proper effect of God is *esse*. Every other starting point, for Owens, would be essentialist, and hence could conclude only to one being among many others. All true proofs for God's existence are simply variants of the one metaphysical demonstration, of that through existence. Cf. *Aquinas on God*, 141, 168, 205.

20. *Christian Metaphysics*, 341-51. Cf. *Aquinas on God*, 135, 179-86.

21. *Christian Metaphysics*, 343-45, 71-80. Cf. *Aquinas on God*, 135, 158-60, 182, 190, 205-06.

22. Ibid., 183-86; cf. 227.

23. Despite Owens's early suggestions, at 158-59, 163, there seem to be good reasons against making this identification: (1) for existence to be already known as the act of an essence in potency would entail a real distinction; (2) being caused to exist cannot always be identified with being moved, since no

potential essence preexists so as to be reduced into act.

24.   Ibid., 186-89.

25.   *Le Thomisme*, 6th ed., (Paris, 1965), 7, 92-97. Because Gilson denies that *De ente* 4 offers a proof of God's existence, he can hold that Aquinas never offers a proof of God's existence through the properties of being as such; cf. "La Preuve du 'De ente et essentia,'" *Doctor Communis* 3 (1950): 257-60; "Trois leçons sur le problème de l'existence de Dieu," *Divinitas* 5 (1961): 23-87, at 27-28. He traces his reversal to his work on the Avicennian distinction between agent and moving causes. After having taught and interpreted the *prima via* for some fifty years, he came to observe that Thomas nowhere there uses the word *causa*. Gilson sees in this fact, together with his understanding of Christian philosophy, a solution to the 'labyrinth' of Thomistic interpretations of the five ways. In particular, he sees a resolution of the inevitable difficulty entailed by his 'existential' interpretation: why does Thomas offer both a first and a second way if each arrives identically at a first efficient cause? Gilson's answer is that the first way does not employ efficiency but arrives at a first moving cause, which need be no more than a final cause. See *Elements of Christian Philosophy* (Garden City, N.Y., 1960), 81-86; "Trois leçons," 31-32, 39-63; "Notes pour l'histoire de la cause efficiente," *Archives d'histoire doctrinale et littéraire du moyen âge* 29 (1962): 7-31; "Prolégomènes à la 'prima via'," ibid. 30 (1963) 53-70.

26.   Other defendants of a physical reading include S. Bersani, "De mente Cardinalis Caietani circa vim conclusionum quinque viarum," *Divus Thomas* (Piac.) 36 (1933): 429-34, at 430-32; Vincent Edward Smith, *Philosophical Physics* (New York, 1950): 31, 91-92, 260; Eric A. Reitan, "Aquinas and Weisheipl: Aristotle's *Physics* and the Existence of God," in *Philosophy and the God of Abraham: In Memory of James A. Weisheipl, OP*, ed. James Long (Toronto, 1991), 179-90, at 189-90; Kevin D. Kolbeck, "The *Prima Via*: Natural Philosophy's Approach to God," Ph.D. diss. (Notre Dame, 1989), 191-239.

27.   "St. Thomas and the Coherence of the Aristotelian Theology," *Mediaeval Studies* 35 (1973): 67-117, at 112-13; cf. p. 116.

28.   *Le Thomisme*, 6th ed., 73 n. 22 and 94; "Prolégomènes," 59-64.

29.   Ibid., 90-96; Pegis, "Four Medieval Ways to God," *The Monist* 54 (1970): 317-58, at 345.

30.   *Le Thomisme*, 6th ed., 95. "Prolégomènes," 66-68, solves the problem differently; cf. Pegis, "Coherence," 113-16.

31.   "Four Ways," 346; cf. 340-42.

32.   "Henry," 228.

33.   See Cajetan's view that the first way concludes to God not per se but merely per accidens by concluding to the predicate 'prime mover,' which is an attribute of God (in Aquinas, *Summa theologiae cum commentariis Thomae de Vio Caietani*, in *Sancti Thomae de Aquino Opera omnia: iussu impensaque, Leonis XIII. P.M. edita* [Rome, 1882-], vol. 4, 1.2.3, p. 32). Others hold that the proof from motion concludes merely to one attribute that all will admit belongs to God; whereas the existence of what Christians mean by 'God' is demonstrated only by the end of the subsequent questions on the divine nature. See William Lane Craig, *The Cosmological Argument from Plato to Leibniz* (London, 1980) 159, 170; Edward Sillem, *Ways of Thinking about God: Thomas Aquinas and*

Some Recent Problems (London, 1961), 72-78, 140, 172-75; R. P. Phillips, *Modern Thomistic Philosophy: An Explanation for Students*, vol. 2 (Westminster, Maryland, 1935), 283.

34. See Vincent Edward Smith, "The Prime Mover: Physical and Metaphysical Considerations," *Proceedings of the American Catholic Philosophical Association* 28 (1954): 78-94, at 86-89. Smith's position is based on the view that to begin with the nominal definition of God is proper to a theological procedure, as is found in the two *summae*, and not to a philosophical one. See Antonin Finili, "Is There a Philosophical Approach to God?" *Dominican Studies* 4 (1951): 80-91; Thomas O'Brien, "Reflexion on the Question of God's Existence in Contemporary Thomistic Metaphysics," *The Thomist* 23 (1960): 1-89, 211-85, 315-447, at 426.

35. See, for example, article 36, that the celestial mover is part of a self-mover; article 61, that an intelligence, full of forms, impresses these forms on things below; and, article 73, that the celestial bodies are moved by an intrinsic principle, the soul, through an appetitive power like an animal's (Roland Hissette, *Enquête sur les 219 articles condamnés à Paris le 7 mars 1277* [Louvain, 1977]).

36. See *Quaestiones de anima*, ed. J. Robb (Toronto, 1968), 8 ad 3; *De spiritualibus creaturis* 6, in *S. Thomae Aquinatis Quaestiones disputatae*, vol. 2, ed. P. Bazzi et al. (Turin-Rome, 1953); ST 1.70.3.

37. *In octo libros Physicorum Aristotelis expositio*, ed. P. M. Maggiòlo (Turin-Rome, 1950), 8.5 l. 9, n. 13 (1049); cf. l. 10, n. 6 (1055); 8.6 l. 13, n. 3 (1079); 8.7 l. 14, n. 11 [1096]). See also CG I.13, n. 101, 106 (*Liber de veritate catholicae fidei contra errores infidelium, seu Summa contra gentiles*, vol. 2, ed. C. Pera et al. [Turin-Rome, 1961]); *Scriptum super libros Sententiarum*, ed. P. Mandonnet and M. Moos (Paris, 1929-1947), I d.8.3.1 ad 3.

38. For a first self-mover, see *In Phys.* 8.5 l. 11, n. 5-6 (1066-1067); 8.6 l. 12, n. 8 (1076); 8.5 l. 9, n. 13 (1049). For the dual conclusion, see *De substantiis separatis* c. 2, ll. 19-26, in *Opera omnia*, vol. 40.

39. *In Phys.* 8.6 l. 13, n. 6 (1082); 8.4 l. 7, n. 7 (1027).

40. Ibid., 8.10 l. 21, n. 12 (1152); cf. n. 9 (1149); n. 13 (1153). See my "Averroes on the Prime Mover Proved in the *Physics*," *Viator* 26 (1995): 107-34, at 118-20.

41. See especially *De spir. creat.* 6 ad 9; ST 1.70.3c, ad 5.

42. See, for example, *In Phys.* 8.5 l. 9, n. 4 (1040); 8.6, l. 12, n. 1 (1069). CG I.13 n. 112 (Unde etiam); n. 101 (Quia vero); and CG II.70 n. 1472 (Probat enim); cf. III.23 n. 2036 (Adhuc. Omne). See also *De potentia* 6.6 ad 11, in *Quaestiones disputatae*, vol. 2; *In duodecim libros Metaphysicorum Aristotelis expositio*, ed. M.-R. Cathala and R. Spiazzi (Turin-Rome, 1950), 12.7 l. 7, n. 2517.

43. *In Phys.* 8.10 (267b26) l. 23, n. 9 (1172), quoting Romans 9:5.

44. "Sed quia Deus non est pars alicuius moventis seipsum, ulterius Aristoteles, in sua *Metaphysica*, investigat ex hoc motore qui est pars moventis seipsum, alium motorem separatum omnino, qui est Deus" (CG I.13 n. 108). In "Coherence," 80-82, 111-12, Pegis dismisses this passage as forming no part of

the *Physics'* proof but as merely a retrospective schema.

45. *In Met.* 12.7 (1072a25-26) l. 6, n. 2517-2518.

46. This reading, traceable to Báñez and others, was once widely held but has not received much attention of late. For some who have held this view, see Réginald Garrigou-Lagrange, *God: His Existence and His Nature*, trans. Bede Rose from 5th French eds. vol. 1 (St. Louis, 1934-1936), 245-47, 261-67; Robert Leet Patterson, "The Argument from Motion in Aristotle and Aquinas," *The New Scholasticism* 10 (1936): 245-54; *The Conception of God in the Philosophy of Aquinas*, (London, 1933), 57-59, 64-70; O'Brien, 407-08; George P. Klubertanz and Maurice R. Holloway, *Being and God* (New York, 1963), 230-37; R. Masi, "De prima via s. Thomae," *Doctor Communis* 18 (1965): 3-37, at 25; William A. Wallace, "The Cosmological Argument: A Reappraisal," *Proceedings of the American Catholic Philosophical Association* 46 (1972): 43-57, at 44; James A. Weisheipl, "Thomas' Evaluation of Plato and Aristotle," *The New Scholasticism* 48 (1974): 100-24, at 122-23; cf. also "The Principle *Omne quod movetur ab alio movetur* in Medieval Physics," in Weisheipl's *Nature and Motion in the Middle Ages* (Washington, D.C., 1985), 95, an article reprinted from *Isis* 56 (1965): 26-45.

For scholastics who maintain this 'metaphysical' interpretation see Ferrara, *In Contra gentiles* I.13, p. 38b, in Aquinas, *Opera omnia* (Leon.), vol. 13; John of St. Thomas, *Cursus theologicus* (Paris, 1932), 2, d. 3, a. 2 (419).

47. *Analytica priora et posteriora*, ed. W.D. Ross (Oxford, 1964), 2.8 93a15-29.

48. *Expositio libri Posteriorum* (in *Opera omnia*, vol. 1˙) 2.10 l. 8, ll. 91-130 (see Aristotle, *Post. an.* 2.8 93a31-b18); *Super Boetium De Trinitate* 6.3, ll. 114-129, in *Opera omnia*, vol. 50.

49. *In Post. an.* 2.10 l. 8, ll. 91-130.

50. For another example of this demonstration of existence, see *In Met.* 6.1 l. 1, n. 1151.

51. See ST 1.3.4 ad 2; CG I.12 n. 78 (Nec hoc).

52. See ST 1.12.4c, 12c.

53. ST 1.2.2c.

54. *In De Trin.* 1.2 ll. 89-92: "et sic se habet cognitio effectus ut principium ad cognoscendum de causa an est, sicut se habet quiditas ipsius causae cum per suam formam cognoscitur." See also ibid., ad 5; CG I.12 n. 79 (In rationibus).

55. ST 1.2.2 ad 2.

56. See the texts cited above, n. 51.

57. Henry, *Summa* 22.5, f. 134C, E; Scotus, *Ordinatio* I.3.1, 6.

58. Ibid., 7.

59. See, for example, Owens, *Aquinas on God*, 142-43, 156, 163.

60. Cf. Van Steenberghen, *Dieu dans s. Thomas*, 290, 292-93.

61. ST 1.13.8.

62. ST 1.13.8 ad 2: Sed ex effectibus divinis divinam naturam non possumus cognoscere secundum quod in se est, ut sciamus de ea quid est; sed per modum eminentiae et causalitatis et negationis, ut supra dictum est. Et sic hoc nomen Deus significat naturam divinam. Impositum est enim nomen hoc ad aliquid significandum supra omnia existens, quod est principium omnium, et

remotum ab omnibus. Hoc enim intendunt significare nominantes Deum.

63.   SN I d.3, div. 1.

64.   This is Finili's argument, at 82.

65.   *In De Trin.* 1.2c.

66.   See Jules Baisnée, "St. Thomas Aquinas's Proofs of the Existence of God Presented in Their Chronological Order," in *Philosophical Studies in Honor of the Very Reverend Ignatius Smith, O. P.*, ed. John K. Ryan (Westminster, Maryland 1952), 29-64.

67.   *In ST* 1.2.3, n. III.

68.   I take it that the proof of God's existence for Aquinas need not expressly draw the conclusion that a name of God is proper, that it belongs only to God; but it needs to know that name in the way that it turns out to be proper to God. Perhaps Cajetan has the first point in mind, but he fails to preserve the second, as we shall see Báñez arguing.

69.   *Disputationes metaphysicae* 29.2, n. 5.

70.   Ibid., 29.3, n. 2; cf. 29.1, n.

71.   *In ST* 1.2.3, n. III.

72.   Domingo Báñez, *Scholastica commentaria in Primam partem Summae theologiae s. Thomae Aquinatis*, ed. L. Urbano (Madrid, 1934), 1.2.3, p. 115a.

73.   *In Met.*, prol. (*Secundo ex*); *In De Trin.* 5.1c, ad 6; CG 2.54, n. 1296 (Sic igitur); ST 1.85.1 ad 2.

74.   For 'motion' in such cases, see especially ST 1-2.9.1, 6; 10.1; 109.1c. For Aquinas, even such spiritual motions involve potency and act, and succession. See ST 1.9.1: any motion or mutation (a) involves potency, (b) composition, being partly the same, partly different, and (c) the acquisition of something that before it lacked. Nevertheless, such a change not only needs no body but also need not be measured by time, such as in the case of angel's acts of knowing (cf. ST 1.9.2; 10.6; 53.1, 3; SN I d.37.4.1, 3).

A perfect act or an operation *as such* is not a motion, since what belongs to all motion in the proper sense is to be the act of what exists in potency (ST 1.18.1c; SN I d.8.3.2). Thus, divine operations are referred to as 'motions' only by way of likeness, insofar as motion is the act of a subject (ST 1.18.3 ad 1; 9.1 ad 1-2). Nevertheless, perfect operations of creatures can be truly called motions to the extent that they are acts of a passive potency. For a discussion of the properties of physical motion as distinct from spiritual, cf. *Sentencia libri De anima* (in *Opera omnia*, vol. 45) 2.5, c. 10, 1.37, to c. 11, l. 151 (Marietti, n. 352-367); 3.7, to c. 6, ll. 1-17 (n. 765). Aristotle also speaks of the perfect activities of the soul in terms of motion: regarding sensation, *De anima* 2.5 417a6-20; 2.11 424a8-11; 3.4 429a13-17 (in *Aristotle: On the Soul, Parva Naturalia, On Breath*, trans. W. S. Hett [Cambridge, Massachusetts, 1935]); regarding imagination, 3.3 429a1-2; regarding intellection, 3.4 429a13-22; 3.7 431b2-5; *De memoria* c. 2 542a27-b13; *Aristotelis Metaphysica*, ed. W. Jaeger (Oxford, 1957) 12.7 1072a26-27, 30; and, regarding desire, *De anima* 3.10 433b17-18.

75.   Certum est enim et sensu constat aliqua moveri in hoc mundo. Omne autem quod movetur, ab alio movetur. Nihil enim movetur, nisi secundum quod est in potentia ad illud ad quod movetur; movet autem aliquid secundum quod est actu. Movere enim nihil aliud est quam educere aliquid de potentia in actum;

de potentia autem non potest aliquid reduci in actum, nisi per aliquod ens in actu; sicut calidum in actu, ut ignis, facit lignum, quod est calidum in potentia, esse actu calidum, et per hoc movet et alterat ipsum. Non autem est possibile ut idem sit simul in actu et potentia secundum idem, sed solum secundum diversa; quod enim est calidum in actu, non potest simul esse calidum in potentia, sed est simul frigidum in potentia. Impossibile est ergo quod secundum idem et eodem modo aliquid sit movens et motum, vel quod moveat seipsum. Omne ergo quod movetur, oportet ab alio moveri (ST 1.2.3c).

76. See Jean Paulus, "Le Caractère métaphysique des preuves thomistes de l'existence de Dieu," *Archives d'histoire doctrinale et littéraire du moyen âge* 9 (1934): 143-53, at 145-48.

77. Aristotle's original (257b6-12) is based on two propositions: (1) what is mobile is potentially being moved; (2) what moves is already in act. But a specific potency and act are envisioned by Aristotle: the form or property already actually possessed by what moves, and, correspondingly, the form only potentially possessed by what is mobile. For, Aristotle's rejection of anything strictly self-moved turns on his denial that anything both possesses and lacks the same form. In fact, Aristotle actually draws his reasoning out simply through an example: since what causes heat is hot, he concludes, nothing strictly heats itself; otherwise, it will simultaneously, in the same respect be both hot and not hot. Furthermore, because the argument turns on the possession of form, Aristotle admits that it works in the same way only for 'univocal' causes, that is, for as many causes as possess a property univocally the same as their effect.

78. See above n. 44.

79. "Aquinas's Parasitic Cosmological Argument," *Medieval Philosophy and Theology* (1991): 119-55, at 146-52.

80. See the appendices to the Leonine edition, pp. 6*-7*.

81. See Viannet Dècarie, "Les Rédactions successives de la *secunda via* de 'Contra gentiles' I,13," in *De Deo in philosophia s. Thomae et in hodierna philosophia*, vol. 1 (Rome, 1965), 138-44, at 142.

82. Albert himself affirms subordinate intelligences that are 'pure acts' (*De causis et processu universitatis a prima causa* 2.2.4, p. 97.41-43; cf. 2.1.8, p. 69.83, in *Opera omnia*, vol. 17.2, Cologne ed. (Münster i. Westf., 1951-), that are absolutely immobile, lacking all potency even in their knowing (*Metaphysica* 11.2.30, p.521.41-59; 11.2.28, p. 518.34-37; Cologne, vol. 16).

83. ST 1.54.1c.

84. Cf. ST 1.76.1c: nothing acts except insofar as it is in act; hence, a thing acts by that whereby it is in act.

85. ST 1.77.1c.

86. ST 1.54.3c; 59.2c; cf. 57.1: an angelic intellect is always actualized by species in potency to second act.

87. ST 1.105.3-4; 1-2.9.6; 109.1c.

88. For this conclusion, see SN I d.3, div. 1; d.8.3 ad 4; ST I.9.1-2.

89. I take CG I.16 n. 133 (Item. Videmus) to provide the ultimate basis for the first way's causal proposition. Thus, the latter is not an immediate first principle, *contra* MacDonald, 129-32. See my "Why Motion Requires a Cause: The Foundation for a Prime Mover in Aristotle and Aquinas," in *Philosophy and*

the God of Abraham, 235-54.

90.   The closest parallel that I have found is Maimonides's fourth way, *Guide for the Perplexed,* tran. S. Pines (Chicago, 1963) 2.1, 249.

91.   I wish to thank Kevin White and Lawrence Dewan for commenting on drafts of this paper. Dewan suggested to me the response to the objection at n. 83.

AMERICAN CATHOLIC PHILOSOPHICAL ASSOCIATION

FINANCIAL STATEMENTS

DECEMBER 31, 1995 and 1994

Prepared by

David J. Rada, C.P.A.

AMERICAN CATHOLIC PHILOSOPHICAL ASSOCIATION
BALANCE SHEET
December 31, 1995 and 1994

ASSETS

	1995	1994
Current Assets:		
Cash	$38,364	$18,383
Money Market	37,704	35,842
Accounts receivable	0	611
Inventory and supplies	1,212	1,250
Prepaid assets	0	2,542
Total Current Assets	77,280	58,628
Other Assets:		
Investments	46,030	30,057
Office equipment, net of accumulated depreciation of $4,076 in 1995 and $2,718 in 1994	0	1,358
TOTAL ASSETS	$123,310	$90,043

LIABILITIES AND FUND BALANCE

	1995	1994
Current Liabilities:		
Accounts payable	$ 121	$ 197
Deferred income-dues and subscriptions	43,176	40,866
Total Current Liabilities	43,297	41,063
Fund balance	80,013	48,980
TOTAL LIABILITIES AND FUND BALANCE	$123,310	$90,043

The accompanying notes are an integral
part of the financial statements.

AMERICAN CATHOLIC PHILOSOPHICAL ASSOCIATION
STATEMENTS OF REVENUE, EXPENSES AND CHANGES IN FUND BALANCE
for the years ending December 31, 1995 and 1994

	1995	1994
Support and Revenues:		
Membership Dues:		
Constituent	$22,460	$23,331
Library and Institutional	4,775	4,950
Student	1,860	2,060
Emeritus	1,905	1,740
Associate	2,199	2,400
Life	500	0
Total Membership Dues	33,699	34,481
American Catholic Philosophical Quarterly:		
Subscriptions	17,280	16,535
Sales	2,019	1,045
Sales	850	525
Total American Catholic Phil. Qtly.	20,149	18,105
Proceedings:		
Subscriptions	2,558	1,416
Sales	1,794	2,646
Total Proceedings	4,352	4,062
Annual Meeting	9,983	10,065
Interest and dividends	4,219	2,578
Postage and mailing	1,033	1,777
Donations	1,798	2,543
Other	1,659	3,087
Increase (decrease) in fair market value		
of investments	14,272	(1,651)
Total support and revenue	91,164	75,047

	1995	1994
Expenses:		
Publications:		
American Catholic Phil. Qtly.	14,938	17,558
Proceedings	5,602	4,975
Total Publications	20,540	22,533
Salaries	16,214	10,897
Annual meeting	7,740	7,310
Postage	6,546	5,688
Printing and duplicating	318	394
Accounting services	450	350
Office supplies and expenses	5,472	1,483
Telephone	644	746
Dues	475	393
Depreciation	1,358	1,359
Other	373	1,079
Total expenses	60,131	52,368
Excess (deficiency) of revenue over (under) expenses	31,033	22,679
Beginning fund balance	48,980	26,301
Ending fund balance	$80,013	$48,980

The accompanying notes are an integral
part of the financial statements.

AMERICAN CATHOLIC PHILOSOPHICAL ASSOCIATION
SCHEDULE OF REVENUE AND EXPENSES
OF ANNUAL MEETING
for the years ending December 31, 1995 and 1994

	1995	1994
Revenue:		
Registration and banquet	$ 7,715	$ 6,418
Book exchange, exhibits and advertising	1,632	1,496
Donations	636	2,151
Total Revenues	$ 9,983	$10,065
Expenses:		
Program, travel and other	2,836	1,980
Banquet expenses	4,654	5,080
Matchette awards	250	250
Total Expenses	7,740	7,310
Excess of Revenues over Expenses	$ 2,243	$ 2,755

The accompanying notes are an integral
part of the financial statements.

AMERICAN CATHOLIC PHILOSOPHICAL ASSOCIATION
STATEMENT OF CASH FLOWS
for the years ending December 31, 1995 and 1994

	1995	1994
Cash Flows from Operating Activities:		
Excess (deficiency) of revenue over (under) expenses	$31,033	$22,679
Adjustments to reconcile net excess of expenses over support and revenues:		
Depreciation	1,358	1,359
Decrease (Increase) in fair market value of marketable securities	(14,273)	1,651
(Increase) Decrease in accounts receiable	611	(11)
Decrease in inventory and supplies	38	38
(Increase) Decrease in prepaid assets	2,542	(2,542)
(Decrease) in accounts payable	(76)	56
Increase in deferred income-dues and subscriptions	2,310	4,852
Cash Provided by (Used in) Operating Activities	23,543	27,482
Cash Flows From Investing Activities:		
Purchase of investments	(1,700)	(7,000)
Net Increase in Cash	21,843	20,482
Cash at Beginning of Year	54,225	33,743
Cash at End of Year	$76,068	$54,225

The accompanying notes are an integral
part of the financial statements.

1. Accounting Policy

   The Association's financial statements have been prepared on the accrual basis of accounting. Accrual accounting reports income as it is earned rather than when it is received; and reports expenses as incurred, rather than when they are paid.

2. Investments

   Investments are carried at fair market value.

3. Income Taxes

   The Association is exempt from Federal income taxes under Internal Revenue Code Section 501 (c) (3).

AMERICAN CATHOLIC PHILOSOPHICAL ASSOCIATION

Seventieth Annual Meeting
MINUTES OF THE 1996 EXECUTIVE COUNCIL MEETING
Rodondo Beach, CA, Crowne Plaza,
Redondo Beach & Marina Hotel
Friday, March 22, 1996

The meeting of the Executive Council opened at 9:00 a.m. with a prayer. In attendance were: Drs. Thomas C. Anderson, President, Linda T. Zagzebski, Vice-President, Thérèse-Anne Druart, Sr. Marian Brady, S.P., Sr. Prudence Allen, R.S.M., Dominic J. Balestra, Robert G. Kennedy, Mary C. Sommers, Kevin Staley, Patrick L. Bourgeois, Michael D. Barber, Ronald K. Tacelli, S.J., Maria T. Carl, Joseph J. Godfrey, S.J., Elizabeth Morelli, Eileen C. Sweeney, and Robert E. Wood. Also attending (as incoming non-voting observers) were Drs. Daniel Dambrowski, John Deely, John Drummond, Jorge Gracia, Sandra Rosenthal, and Kevin White.

The minutes of the 1995 Executive Council meeting were approved as well as the reports of the Secretary and of the Treasurer. The Treasurer, Sr. Marian Brady, S.P., was commended for her great work. The Executive Council also approved the report of the *ACPQ* and *Proceedings* editor, Dr. Robert E. Wood. As Dr. William Frank, Associate Editor of the *ACPQ* and of the *Proceedings* has resigned, Dr. Lance Simmons was appointed to replace him. A motion passed that the Executive Committee decide in consultation with the Editor whether the size of the regular issues of the *ACPQ* should be increased.

The Executive Council discussed the report of the ACPA Committee on the Future of the Association (Drs. Robert G. Kennedy, chair, Theresa Sandok, and Kevin Staley). A motion that "the ACPA shall change the customary date of its annual meeting to early November beginning in 1999" passed but on the condition that there would some consultation of the membership. This new date would allow the Placement Bureau to be more effective and would lessen conflicts with other meetings such as The Metaphysical Society regularly attended by our membership. The Executive Committee also adopted the following motions: 1. "The ACPA shall provide placement opportunities for graduate students and professors as well as to colleges and universities, in so far as a Committee appointed by the President may be able to implement this" [subsequently the President appointed Drs. Robert G. Kennedy and Kevin Staley to constitute such a Committee]; 2. "In order to increase active participation at the annual meeting: (a) the ACPA may expand the number of presentations at its annual meeting and (b) at its discretion, the Executive Council may choose to limit the number of papers from the meeting to be published in the *Proceedings*"; and 3. "The ACPA is committed to emphasizing the importance of its Catholic character, both in theory and in practice. More concretely, the following means should be adopted:

(a) The President shall energetically seeks ways to increase in the Association the membership and active participation of Catholic philosophers, as well as non-Catholic philosophers interested in and sympathetic to the Catholic tradition.

(b) The President and the Program Committee for the annual meeting shall encourage papers for the meeting that address topics of current interest to Catholics, or controversial topics where Catholic philosophy has a special perspective or a particular contribution to make.

(c) With due regard for the diverse views and concerns of the membership, the Executive Council shall seek appropriate ways for the Association to bring the voice of Catholic philosophy to bear on current issues."

Finally, the following motion passed: "In order to enhance the activities of the Association and to serve the interest of the membership, the President shall seek out and promote collaboration with other societies, such as the American Maritain Association, the Fellowship of Catholic Scholars, Society of Christian Philosophers, etc."

Drs. Patrick L. Bourgeois and Michael D. Barber were elected to the Executive Committee.

Dr. Louis Dupré was elected Aquinas Medalist for 1997.

The Executive Council rejected a request from Fr. Ronald D. Lawler, O.F.M. Cap, that the ACPA Committee on Philosophy and Priestly Formation be exempted from the necessity to take a turn for a Friday meeting slot instead of a Saturday one. The 1995 Report of the Committee on Groups Using the ACPA Name was slightly amended (final version enclosed).

The Executive council passed a resolution that the *Bylaws* be amended from "the Nominating Committee shall present a slate of three (3) names for the office of Vice-President (President-Elect)" to "the Nominating Committee shall present a slate of at least two (2) names for the office of Vice-President (President-Elect)."

The Council approved Dr. Linda J. Zagzebski's proposal that the theme for the next meeting (Buffalo, NY, March 21-23, 1997) be *Virtues and Virtue Theories*.

A motion that a hard copy of the membership list be published every five years was approved.

The Council approved Pittsburgh, PA, as the site for the 1998 Annual Meeting, which will be hosted by the Franciscan University of Steubenville. Dr. John F. Crosby will head the Local Committee. Minneapolis/St. Paul, MN, was approved as the site for the 1999 Annual Meeting, which will be hosted by the University of Saint Thomas. Dr. Michael Degnan will chair the Local Committee.

The Council passed a motion that the *Proceedings of the ACPA* have the right of first refusal for the ACPA selected papers as well as for the papers of the plenary speakers.

The JCCLSS (Joint Committee of Catholic Learned Societies and Scholars) and FISP (Fédération Internationale des Sociétés de Philosophie) reports were adopted.

The meeting closed at 2:00 p.m.

Respectfully submitted,

Thérèse-Anne Druart
ACPA National Secretary

SECRETARY'S REPORT (1995)

Generally 1995 was a good year for the ACPA: the membership has increased, subscriptions to and sale of isolated copies of the *ACPQ* and the *Proceedings* have risen, the annual meeting in Washington brought a record number of registrations, moving the National Office from the Administration Building to Leahy Hall allowed us to make better use of modern technology, and all of these have resulted in a fairly sound financial situation. Best of all, though, is the participation and help provided by so many members of this Association who worked for it in many ways, accepted to run for elections, and provided facilities, grants or donations. The Secretary is very grateful to all of them.

I. Membership

For 1995 the Association counts 1077 active members [active members are those who paid at least *some* dues in the last two years; for the number of people who actually paid dues each year see the Treasurer's report] (992 in 1997; 944 in 1993): 207 Professors (199, 194), 151 Associate Professors (143, 135), 141 Assistant Professors (144 in 1994, 158 in 1993), 109 Emeriti and Emeritae (94, 77), 25 Lecturers (17 in 1994, 4 in 1993), 31 Instructors (18 in 1994, 6 in 1993), 117 Students (116, 107), 99 Life Members (101, 103), 129 Associates (88, 89), 14 Institutions (13, 15), and 55 Libraries (58, 56). **So our membership nearly increased by 9% this year and by 14% in two years.** This is most encouraging but does not necessarily lead to an increase in dues. In fact we collected less dues in 1995 than in 1994.

The membership shows an increased number of Emeriti and Emeritate, who rightly pay lower dues. This highlights their loyalty but should entice us to increase our efforts to recruit more younger members. The category that shows the most significant progressive decrease is Assistant Professors. As there is an increase among Instructors and Lecturers this may simply reflect general recent academic hiring trends, i.e., replacing full-time faculty by part-time faculty.

The number of Associates, who pay low dues, is increasing significantly and shows that the ACPA appeals to people outside the official academic setting; however, for administrative purposes this category also includes some 20 seminaries in Eastern Europe to which we are offering one year of the *ACPQ* and *Proceedings* thanks to a grant received by the Editor, Dr. Robert E. Wood.

N.B. Publications are mailed only to those who have paid dues and to Life

Members but the two yearly mailings concerning the annual meeting are sent to all active members.

II. Publication

a. ACPQ: in 1995 four issues of the *American Catholic Philosophical Quarterly* (*ACPQ*) were edited by Dr. Robert E. Wood assisted by Dr. William A. Frank and typeset by Mrs. Edna Garcia. They include one special issue on Heidegger, edited and partially subsidized by John D. Caputo, which is in high demand. A team under the director of Dr. Robert E. Wood typesets the text in Texas and Capital City Press in Vermont prints and mails it. As usual, Dr. Wood and his team worked very hard.

The 1993-1995 distribution of the *ACPQ* is as follows:

	1995	1994	1993
Members of the ACPA	1077	992	944
Subscribers	523	530	534
Exchanges	33	32	29
Totals	1633	1554	1507

N.B. In 1995 we sold 114 (103 in 1994) isolated copies of the *ACPQ*.

b. *Proceedings*: Dr. Thérèse-Anne Druart edited volume 69 of the *Proceedings of the ACPA*, "The Recovery of Form." Again, Dr. Robert E. Wood and his team ensured the timely production of this volume.
The 1993-1995 distribution of the *Proceedings of the ACPA* is follows:

	1995	1994	1993
Members of the ACPA	1077	942	944
Subscribers	88	68	63
Exchanges	33	32	29
Totals	1198	1092	1036

N.B. In 1995 we sold 128 (48 in 1994) isolated copies of the *Proceedings of the ACPA*.

The Secretary acknowledges with deep appreciation the free and generous facilities granted by the University of Dallas for the Editor and Associate Editor of the *American Catholic Philosophical Quarterly* and its supplement, the *Proceedings of the ACPA*.

The National Office is studying the possibility, financial cost and marketing techniques to add a CD-Rom version of our publications to their current paper form.

III. Annual Meeting

The sixty-eighth Annual Meeting of the ACPA took place March 24-26, 1995 at the Radisson Barceló Hotel in Washington, D.C. Dr. Robert E. Wood, President, had chosen *The Recovery of Form* as the theme for the meeting. Papers read at the meeting and reports of the official business of the Association are now published as volume 69 of the *Proceedings of the ACPA*.

Since our 1992 meeting in San Diego, registrations for the meeting had oscillated between 180 and 200. In Washington, D.C. we reached 216 registrations, but for the first time we barely made our room contract with the hotel (probably many people were local or could stay with friends and did not need an hotel room). If for some reason attendance would happen to be much lower than usual we would have to pay significant fees for the meeting rooms.

Budgeting for the annual meeting is very difficult. Prices vary from one city to another (Los Angeles is very expensive but Buffalo is not) and the ACPA bill is much affected by how much money the local committee provides for properly ACPA expenses and whether we can be tax exempt for the meeting (this varies from state to state). Eventual travel and hotel expenses for Aquinas medalists vary widely according to location and circumstances.

Making the convention at least pay for itself has been one of our aims, since not so long ago the ACPA floundered financially because of huge expenses for plenary speakers. We made a profit on the Washington meeting because we are tax exempt in D.C. and because there were no travel expenses for the Secretary and Treasurer and very few expenses for the National Office. The cost of the meeting does not, however, include the many hours necessary to typeset the program, prepare the material for the Executive Council, etc.

Since at least our 1991 Boston meeting, the number of satellite societies or groups meeting with us has increased (13 in 1996, 14 in 1995, 14 in 1994, 10 in 1993), and therefore, we contract for more breakout rooms than we need for ACPA sessions. The proliferation of smaller societies seems to be a general trend since the APA faces it too.

The Book Exhibit had become smaller and in 1995 Mr. Mark Rasevic, our Office Manager, made great efforts to increase it (we vastly increased the list of publishers we contact, etc.). This bore some fruit at our Washington, D.C. meeting.

The sixty-ninth Annual Meeting on *Philosophy of Technology* attracted 57 submitted papers (63 in 1995 and 50 in 1994). Many young scholars submitted papers; twelve—a record number—of them asked to considered for our annual prize, now called the *ACPA Young Scholars Award* and kindly underwritten by the American Maritain Association. Dr. Philip Buckley (McGill University) won it. The Secretary is most grateful to the Program Committee chaired by Dr. Gregory R. Beabout and including Dr. John D. Jones and Dr. Elizabeth Morelli who are most organized and efficient.

Dr. Mary Catherine Sommers obtained for us a grant of $200 from Delta Epsilon Sigma to help us with the expenses of one of the plenary speakers.

The Secretary wishes to thank the Local Committee, chaired by Linda T. Zagzebski and including Dr. May Elizabeth Ingham, C.S.J., Dr. Elizabeth Morelli, and Dr. Scott Cameron, as well as Loyola Marymount University for its generous financial support.

IV. National Office

a. Organization. In 1993 under the impetus of Dr. Michael Baur, Dr. Marian Brady, S.P., and Mr. Mark Rasevic the office was completely reorganized. Since that time we have been operating much more efficiently, professionally, and at lower cost.

In 1997 we had to leave the Administration Building which was scheduled for demolition, and we moved to Leahy Hall which had just been renovated. We have two small offices but lost a closet which contained boxes of back issues. We have requested storage space since we are overwhelmed if not overpowered by boxes but have not yet received an answer. The new offices are more modern and allowed us to get a dedicated fax line (202-319-6408) and a second computer. We also had the opportunity to acquire at low cost a small, second-hand copier. As it turned out that the University was paying for new custom-made shelving, we did not have to use the money originally budgeted for shelving.

b. Membership drive and promotion of the *ACPQ*. Copies of recent special issues of the *ACPQ* along with a letter of the President inviting them to join the ACPA were mailed to all heads of philosophy department in Catholic Colleges and Universities. Faculty at such departments who were not yet members of the ACPA were also contacted. This brought some positive results.

c. Membership list. The *Bylaws*, III, B, 8 stipulates that one of the duties of Secretary is "to maintain, update, and publish the membership of the Association." We do maintain and update the membership list but we have not published it since 1992 (see *Proceedings*, vol. 66, 1992, after p. 291). Publishing it in this manner is impractical: the additional pages push the *Proceedings* to another postage bracket and increase the weight to the point of tearing the plastic bags used to mail it. We have contacted Capital City Press and intend to publish it separately as a booklet type-set at the National Office but printed and mailed by this company. We wonder whether the list should include not only the address but also the e-mail address, the work and fax phone numbers and whether it should be published every two or three years. We have already budgeted the money to publish it in 1996.

d. **Liaison with national and international societies.** The secretary represented the ACPA at the FISP (Fédération Inter-nationale des Sociétés de Philosophie) meeting to prepare for the 1998 World Congress of Philosophy to be held in Boston. Sr. Marian Brady, S.P., the Treasurer, participated actively in the JCCLSS activities.

e. **The office of the Treasurer.** After two terms of four years each, Sr. Marian Brady, S.P., our Treasurer, will at the end of our Redondo Beach meeting turn her books and files over to Dr. Kevin White. Sr. Brady accomplished marvels. Like Plato's Demiurge she has made order out of chaos. She completely revamped our financial recording system, spent countless hours tracking expenses, establishing budgets, preparing studies of cost effectiveness and through it all remained calm and poised. Her advice, good sense, humor and wisdom have been invaluable to the Secretary. She has been initiating Dr. White to this complex and time consuming job.

1995 again brought a high surplus of income over expenditures—in fact much higher than projected—but it significantly lower than the previous one. We have also to replete our reserves which were depleted when we lost some $50,000 in a few years.

The Secretary wishes to thank Sr. Marian Brady, S.P., Dr. Kevin White, Mr. Mark Rasevic our Office Manager, Mr. Martin Neomianu and Ms. Mary Troxel his helpers for all their work, their suggestions, and initiatives. She also acknowledges with deep appreciation the free facilities and generous services of the School of Philosophy at The Catholic University of America for the Secretary and Treasurer/ Business Manager of the *American Catholic Philosophical Association*.

V. Elections

The results of this year's election are as follows:

Vice-President - President Elect:
    Dr. Jorge Gracia (SUNY - Buffalo)

New Executive Council Members:
    Dr. Daniel Dambrowski (Seattle University)

Dr. John Deely (Loras College)
Dr. John Drummond (St. Mary's College)
Dr. Thomas Hibbs (Boston College)
Dr. Sandra Rosenthal (Loyola University of New Orleans)

These individuals are to be congratulated, and all those who permitted themselves to be nominated for the offices in the Associate deserve special thanks. We are a voluntary association kept alive by the enthusiasm and support of such individuals.

This year 249 ballots were cast out of 832 voting members (Students and Associates are not entitled to vote; 243 out of 787 in 1994, and 281 out of 748 in 1993).

The Secretary has not words enough to thanks the Treasurer, Sr. Marian Brady, S.P. She wishes to express her gratitude to the President, Dr. Thomas C. Anderson for his enormous work and patience, to Dr. Robert E. Wood, Editor of the *ACPQ* and Past President, for all his labor, to Dr. Linda T. Zagzebski, the Vice President for her work as chair of the Local Committee, to Mr. Mark Rasevic, the Office Manager for his business acumen and organization, and to his helpers for their work at the National Office. It is a pleasure to serve the ACPA but "it ain't easy" and without such help the Secretary could not have managed.

Respectfully submitted,

Thérèse-Anne Druart
Associate Professor
ACPA National Secretary

REPORT OF THE COMMITTEE
ON GROUPS USING THE ACPA NAME

The types of groups associated with the ACPA which were the subject of this report:

1) Regional Conferences of the American Catholic Philosophical Association, e.g., the Western New York Regional Conference;
2) Societies which meet in conjunction with the Annual Meeting of the American Catholic Philosophical Association, e.g., the American Maritain Association, Philosophers in Jesuit Education, etc.;
3) Standing Committees appointed by the President, e.g., ACPA Committee on Philosophy and Priestly Formation;
4) Working Groups formed by members of the ACPA, e.g., the Working Group on Business Ethics.

The following concerns were raised at the 1994 ACPA Executive Council Meeting in discussion of Agenda item 9: "use of the ACPA's name by various groups and association":

1) That the ACPA has no involvement in the operations, officers of these groups or in the content of their programs;
2) That their activities, publications, etc. might be taken as "official";
3) That their association with the ACPA might involve the organization in any financial or other responsibilities;
4) That there should be fairness in treatment of the different types of groups with respect to scheduling and other matters involved with the planning of the annual meeting;
5) That any guidelines which are laid down for the use of the ACPA name should not inhibit the growth, relevance or impact of the ACPA.

The Committee appointed by the President (Thomas R. Flynn) to suggest policies on the use of the ACPA name: Mary C. Sommers, chair, Dominic Balestra, K. R. Staley.

Proposals for consideration by the Executive Council with respect to these concerns:

1) Appropriate Designations.

a) Only Regional Conferences of the ACPA may designate themselves ACPA or of the ACPA, e.g., "Western New York Regional Conference of the ACPA."

b) Standing Committees appointed by the President and Working Groups started by members may use the designation (**ACPA**) after their name, e.g. "Women's Working Group (ACPA)." This indicates that they are constituted by members of the ACPA.

c) Any society or association given a place on the program of the Annual Meeting by the Secretary of the Association may use some equivalent of **'meeting in conjunction with the ACPA'** in their bulletins, newsletters, etc.

2) Conditions for using the ACPA designation:

a) All groups who use the **ACPA** designation in any way must have a description and/or constitution and by-laws on file with the national office. These materials should be sent to the Secretary of the Association for approval by the Executive Committee;

b) Names of officers or chairs or designated liaisons should be on file as well, and the ACPA should be apprised of changes in governance;

c) A copy of any program, newsletter or press release which uses the ACPA designation should be forwarded to the Secretary of the Association;

d) Standing Committees and Regional Conferences "should send an annual report to the National Office listing their officers and activities." (as *per* Executive Council motions, 25 March 1995). This report will be taken as fulfilling b)-c);

e) The ACPA national organization, its officers, personnel, etc. will assume *no* financial or other obligation to assist the activities of groups using the ACPA designation. Such obligations include, but are not limited to: paying for the use of meeting rooms, including meeting announcements in ACPA mailings, publishing papers presented at meetings;

f) The ACPA Executive Council reserves the right to withdraw approval for use of the ACPA designation from groups whose activities are contrary to the objective of this organization, which "is to promote the advancement of philosophy as an intellectual discipline consonant with Catholic tradition."

3) Procedures for groups wishing to use the ACPA designation:

a) Societies or associations wishing to meeting in conjunction with the ACPA and to send out announcements to that effect must follow the procedures established by the Secretary of the Association for this purpose. These procedures for inclusion in the program should be revised to meet conditions 2) a-c;

b) Standing Committees appointed by the President may use the designation (ACPA). Their membership, charge and subsequent report as recorded in the minutes of the Executive Council meeting will be taken as meeting conditions 2) a-c for the first year;

c) Working Groups, which are substantially constituted by members of the ACPA, may be established for various purposes consistent with the objective of the Association. Notice of their formation meeting conditions 2) a-c should be sent to the Secretary of the Association. Subsequent use of the designation (ACPA) is taken as agreement to fulfill these conditions;

d) New Regional Conferences may be formed by submitting a constitution and by-laws to the Secretary of the Association;

e) No group may make public statements or publish documents using ACPA designation without prior authorization from the ACPA Executive Council.

4) The ACPA welcomes the participation of such groups in the Annual Meeting and gives them fair consideration in establishing the program. Inclusion in the program is to be understood as a courtesy extended by the Association.

a) All requests for inclusion of such groups in the program must come from members in good standing of the ACPA;

b) The inclusion of these groups in the program is dependent upon the number of rooms available at each time slot;

c) The primary consideration in scheduling should be rotations of the popular Saturday slots. All groups, without exception, are subject to rotation;

d) Other considerations based on the commitments or convenience of the participants will be considered secondary and may be used at the discretion of the Secretary of the Association.

. as revised on 3/22/96.

Available Back Issues

Proceedings of the
American Catholic Philosophical Association

STUDIES

No.	Year	
I	N.D.	*Philosophy of Law of James Wilson*
II	1946	*Physics and Philosophy*
III	1952	*Conventional Logic and Modern Logic*

VOLUMES

No.	Year	
1	1926	*The First Annual Meeting*
2	1926	*Second Annual Meeting*
3	1927	*Third Annual Meeting*
4	1928	*Fourth Annual Meeting*
5	1929	*Fifth Annual Meeting*
6	1930	*Sixth Annual Meeting*
7	1931	*Seventh Annual Meeting*
8	1932	*Eighth Annual Meeting*
9	1933	*Ninth Annual Meeting*
10	1934	*Tenth Annual Meeting*
11	1935	*Philosophy of the Sciences*
12	1936	*Christian Philosophy and the Social Sciences*
13	1937	*Philosophy of Education*
14	1938	*Causality in Current Philosophy*
15	1939	*Philosophy of the State*
16	1940	*Problem of Liberty*
17	1941	*Philosophy and Order*
18	1942	*Truth in the Contemporary Crisis*
19	1943	*Philosophy in Post War Reconstruction*
	1944	*None Printed*
20	1945	*The Philosophy of Democracy*
21	1946	*The Philosophy of Being*
22	1947	*The Absolute and the Relative*
	1948	*None Printed*
23	1949	*Philosophy and Finality*
24	1950	*The Natural Law and International Relations*
25	1951	*The Nature of Man*
26	1952	*Philosophy and the Experimental Sciences*
27	1953	*Philosophy and Unity*
28	1954	*The Existence and Nature of God*
29	1955	*Knowledge and Expression*
30	1956	*The Role of Philosophy in the Catholic Liberal Arts College*

31	1957	*Ethics and Other Knowledge*
32	1958	*The Role of the Christian Philosopher*
33	1959	*Contemporary American Philosophy*
34	1960	*Analytic Philosophy*
35	1961	*Philosophy and Psychiatry*
36	1962	*Justice*
37	1963	*Philosophy in a Pluralistic Society*
38	1964	*History and Philosophy of Science*
39	1965	*Philosophy of the Arts*
40	1966	*Scholasticism in the Modern World*
41	1967	*The Nature of Philosophical Inquiry*
42	1968	*Philosophy and the Future of Man*
43	1969	*Truth and the Historicity of Man*
44	1970	*Philosophy and Christian Theology*
45	1971	*Myth and Philosophy*
46	1972	*The Existence of God*
47	1973	*The Philosopher as a Teacher*
48	1974	*Thomas and Bonaventure*
49	1975	*Philosophy and Civil Law*
50	1976	*Freedom*
51	1977	*Ethical Wisdom East and/or West*
52	1978	*Immateriality*
53	1979	*The Human Person*
54	1980	*Philosophical Knowledge*
55	1981	*Infinity*
56	1982	*The Role and Responsibility of the Moral Philosopher*
57	1983	*The ACPA in Today's Intellectual World*
58	1984	*Practical Reasoning*
59	1985	*Realism*
60	1986	*Existential Personalism*
61	1987	*The Metaphysics of Substance*
62	1988	*Hermeneutics and the Tradition*
63	1989	*The Ethics of Having Children*
64	1990	*Ways to World Meaning*
65	1991	*Religions and the Virtue of Religion*
66	1992	*Relations: From Having to Being*
67	1993	*The Importance of Truth*
68	1994	*Reason in History*
69	1995	*The Recovery of Form*
70	1996	*Philosophy of Technology*